Electricity Acts

by Leonard S. Hyman

© 2017 by Leonard S. Hyman

Leonard S. Hyman, author.

This publication is designed to provide accurate and authoritative information in regard to the subject matter covered. It is sold with the understanding that the publisher is not engaged in rendering legal, accounting, or other professional service. If legal advice or other expert assistance is required, the services of a competent professional person should be sought. (From a *Declaration of Principles* jointly adopted by a *Committee of the American Bar Association and a Committee of Publishers.*)

Electricity Acts / Leonard S. Hyman

    pages cm

ISBN 978-0-910325-38-7

1. Electric utilities--Great Britain--History. 2. Electric power--Great Britain--History. I. Hyman, Leonard. II. Title.

HD9685.G72 H96 2017

333.793/20941

2017023869

First Printing, July 2017

Printed in the United States of America

Front cover images: © Can Stock Photo/darrin and stillfx

Electricity Acts

A cautionary tale and case study of how British electricians pioneered the technology, the government regularly interfered, privatization produced big profits and electricity consumers usually ended up as losers.

by Leonard S. Hyman

… while the Author had shown that it was possible to manufacture for something like twopence a unit, Mr. Preece thought they ought to know something of the reason why the consumer was called upon in London to pay 7 1/4 [pence], and in a great many other places 8 [pence] per unit.

– "Discussion of Electrical Energy,"
Proceedings of the Institute of Civil Engineers
(The INST.CE, VOL. CVL, 1891, p. 35)

Dedicated to the memory of Shimon Awerbuch, who plied the trade of energy economics on both sides of the Atlantic, who planned to edit this book before his untimely death in an air crash.

Table of Contents

Part Four

Part Five

Tables

Note: FY: Fiscal Year

Figures

Author's Preface

I first thought about writing this book after working in a Merrill Lynch investment banking team whose presumptuous goal was to steal the British electricity privatization assignment from Goldman Sachs. We did not succeed, but in the process of trying, I met all the principal players in the industry, the think tank experts who furnished ideas to the government, the regulator, members of Parliament, government officials, union members and I even offered advice at Number 10 Downing Street, advice unaccepted. As one ever condescending civil servant put it, we Americans were there to "flog" (sell) the stocks, not offer advice.

How did the electricity industry in the United Kingdom reach a point at which it required such major surgery, and why did the Tories think they had the formula to rejuvenate it? As both an early proponent of electricity deregulation and part of Merrill Lynch's British Telecom team, I thought that I had the answer to the latter question, but not the former. I began to dig into the industry's background and concluded that what happened followed from what previously happened, which sounds like the joke whose punch line is, "It's elephants all the way down." Those questions mattered because the UK's electricity restructuring set the standard, influenced California's disastrous deregulation, and prompted imitation in the rest of the world. Yet, why did 25 years of restructuring effort, hoopla and imitation produce such feeble gains for consumers, the supposed potential beneficiaries of all the sound and fury?

So I decided to tackle the questions from start-up to present, doing it to create a template for examining the electric industry overall, emphasizing how the industry organized, evolved and ran as a business, how and why the UK government meddled so often, and who gained from those interventions and where, in the end, that left consumers. But, did the government really meddle or is that just a conceit cooked up by free market types? Well, more than 20 major Electricity Acts, between 1882 and 1989 – one every five years – probably qualifies as meddling. In the United States, during the same period, the Federal government enacted five or six major statutes that affected electric industry structure. People in Washington did not concern themselves about dealings between crooked politicians and public utilities in Illinois or who would build the electrified subway in New York. They let the states experiment and choose appropriate solutions. To a significant degree, they still do. Maybe that wonderful confidence of the central government in London that it had the answer characterized the UK's electricity market more than anything else, making this more the story of parochial politicians and plodding bureaucrats than visionary business leaders and technological wizards.

Studying the UK's electricity industry has definite advantages. It depended largely on domestic fuels for most of its existence, which removed international complications from the story. The central government made all the policy decisions, usually with a clear ideological bias. Consumers had the money to buy the product, local manufacturers could produce the required equipment and the enormous capital market in London could furnish all the money needed for investment. Thus, outside factors (other than government actions) probably had relatively little impact on the industry's development.

Moreover, as outsiders looking in, we can more clearly analyze and observe developments, having no vested interest in outcomes, no reputation to protect, no spurned arguments to rehash. We can also compare the UK and American experiences. After all, both countries are major industrial nations, burn the same fuels, have big capital markets, high levels of income, access to the same technologies and investors, and even employed some of the same entrepreneurs. How did they differ and did that difference affect the outcome? Looking forward, the British government and its electric industry have fearlessly pioneered new industry structures and continue to do so, the latest being a monumental attempt to deal with energy security and global warming, and the sticker shock that comes with the chosen solutions. Most of the world faces the same energy problems. Can we learn from the British or do we take the attitude that if it is not invented here it is not worth thinking about?

Full disclosure: I was probably the first Wall Streeter (in a 1983 book) to lay out the outlines of a competitive electricity market and I was a member of one of the first state panels that contemplated deregulation (Pennsylvania). I fervently believed that sharp competitive managers, as opposed to DOUGs (dumb old utility guys) would shape up the electricity sector. However, I also held the quaint notion that competitive markets should benefit consumers. Isn't that what competition is all about, introducing new products, reducing costs to stay competitive, lowering prices in order to keep the customer? During electricity restructuring (no longer called "deregulation") policy makers overlooked key aspects of electricity markets and industry organization when they did the deed, and benefits to consumers have fallen short of expectations. Consumers may have ended up paying more for less.

The book follows chronological order, dividing the story into four periods, with a summary a for each comparing developments in the United Kingdom to those in the United States. A comprehensive analysis ends the book. Look at it as a biography of an industry, the interactions of technologists, politicians, investors and customers, but not a history dominated by the big man or the big woman, the Napoleon or Elizabeth of electricity. The electric industry produced only a handful of commanding figures. The long succession of energy secretaries rarely stayed in office long enough to see their projects through to fruition. Prime ministers Attlee and Thatcher made big decisions about ownership, but the product did not change.

As for the big issues in the book – the war on coal, electricity versus pristine environment, the nuclear solution, building too much or too little, market or regulation, individually generated or centralized power, too little or too much regulation, political interference and laws that benefit insiders, protecting consumers or picking their pockets, and choosing the right technology – they seem startlingly contemporary. Today's problem solvers often come up with yesterday's solutions. Do they solve the problems better?

To ensure clarity, I explain technical and financial terms either in text, footnotes or appendices. Readers with only a glancing knowledge of the topic might want to first read the appendix "Technics and Metrics." I present tables of data and graphs to back up my conclusions but readers allergic to numbers can simply follow the text. I have documented all sources and explained adjustments to data. Numbers have to play a role in any discussion of business and markets. They show what happened and say more than the slick, spin-filled presentations so common in the UK nowadays.

Finally, I began this book, in earnest, with the help of the late Shimon Awerbuch, who plied the regulatory, environmental and capital market trades,on both sides of the Atlantic, in government, consulting and academia. Shimon died in an air crash with his family, so he never saw the final product. I wish to thank Mark Erdman (Hermelee Law), Alex Henney (EEE Ltd), Ellen Lapson, CFA (Lapson Advisory), Stephen C. Peck (SCP Analytics) and William I. Tilles for encouragement, advice, vital information and detailed criticisms, Jen House, Kailey Krystyniak, Steve Mitnick and Phil Cross at Public Utilities Reports for encouraging and facilitating the publication, my son Andrew for solving authorial problems and my wife, Judith, for putting up with the nonsense for so long.

Leonard S. Hyman, CFA
Sleepy Hollow, NY
2017

Introduction

The results, which I had by this time obtained with magnets led me to believe that the battery current through one wire did, in reality, induce a similar current through the other wire.[1]

— Michael Faraday

The British invented the electricity business. Humphry Davy demonstrated the first electric light in 1808. Michael Faraday developed the first electric generator in 1831. James Clerk Maxwell explained electromagnetic fields in 1861. Joseph Swan invented the incandescent lamp almost simultaneously with Edison in 1878, and Charles Parsons the modern steam turbine, in 1884. The British had a ready-made regulatory and organizational set-up for electricity based on the town gas model.[2] Yet, by the early 1900s, the United Kingdom had fallen behind other industrial nations, electrically. And British electricity cost more, too. Why the failure to keep the lead? I.C.R. Byatt, historian and utility regulator, wryly explained that "British electrical engineering was strong on the theoretical side"[3], implying that the problem lay in a peculiarly British inability to deploy technology. Yet the British industry employed technological pioneers and nothing prevented outsiders from moving in to provide what the locals could not, technologically or managerially. More likely, the industry fell behind because the country was wedded to coal and because of a confusing array of government policies that thwarted efficient technology rather than because of commercial diffidence.

After World War I, the government imposed the one reform that produced a striking technological improvement. It created a structure later utilized when the industry returned to private ownership in 1990. The reformers after World War I, however, never tackled

1. Michael Faraday, November 24, 1831 lecture to the Royal Society, quoted in Sir William Cecil Dampier, *A History of Science* (Cambridge: Cambridge University Press, 1948), p. 222.
2. Town gas was manufactured from coal, then distributed through pipes to consumers who burned it for light. Gas utilities sold town gas until natural gas became available via pipelines in the 1950s and 1960s.
3. I.C.R. Byatt, *The British Electrical Industry, 1875-1914* (Oxford: Oxford University Press, 1979), p. 185.

the parochialism and special interests that threatened to fragment electric supply and appropriated much of the benefits of the restructuring. The reformers saw the problems but could not reconcile all the vested interests. After World War II, the Labor government solved everything by nationalizing the industry, a solution it had advocated for two decades, no matter the problem. Socialism was the wave of the future, back then. True believers in capitalism later derided what they saw as a bureaucratic, socialized electric industry whose managers had little incentive to operate efficiently. The problems, though, arose more from the direct intervention of politicians who pushed the industry into making uneconomic decisions than from the socialist structure. Four decades later, a Conservative government put electricity back in private hands, thereby curing whatever ills that government ownership brought to the electric business.

Privatization, selling the industry to investors, allowed the government to formulate an industry structure and regulatory mechanism radically different from that used elsewhere. Generators sold power into an unregulated market and the regulator of the rest of the business focused on price rather than profits. Governments from all over (including California) decided to emulate the British model. British electricity managers looked abroad to find ways to employ skills honed in a unique market. Yet, within a decade, the government threw out the most distinguishing feature of the new market, the central buying mechanism called the Pool, because it was prone to market manipulation. The British sold most of the industry to foreigners (what did they know that the buyers did not?) and within another decade, the regulator and government decided that the electricity market could not send signals to encourage the right kind of long-term investment and politicians chimed in that the public also paid too much for electricity. Everyone else had bought into the British model because they thought it sent the right kind of signals to attract investment and it lowered prices. Will they follow the new direction?

The UK restructured the electric industry roughly once every 25 years. Did the public or others benefit? Arguably, there was one reform that was an unequivocal success in favor of the public (but still flawed) and all other successes lie largely in the eye of the beholder. Whatever the answer, though, expect more restructuring. British politicians never run out of ideas. They experiment with abandon. Everyone can learn from those experiments free of charge.

Part One:

Inventors, Investors and Interfering Politicians: Getting the Business Going

Chapter 1

In the Beginning (1800 -- 1887)

Although Swan was strictly first, so great and so close was Edison's achievement that it would be pedantic to exclude him.[4]

– R.A.S. Hennessey

Britain's electricians – a band of experimenters, inventors, telegraphers and visionaries – pioneered electric technology. The British gas industry provided a ready-made organizational and regulatory model for them. But parochialism and restrictive laws created barriers to economies of scale and frightened away investors. The United Kingdom's electricity industry soon fell behind those of its two principal industrial rivals: Germany and the United States.

Arc Light and Dynamo

Electricity became more than an intellectual curiosity in 1800 when Italian physicist Alessandro Volta invented a primitive electric battery, aptly named the voltaic cell. Then, Humphry Davy, working at Britain's Royal Institution, the scientific research and development agency of the day, conducted battery-powered lighting experiments. Davy could produce light electrically either by running a current through a filament which caused the filament to heat up and glow (incandesce) or by making the electric current jump a gap in the electric circuit which created a continuing spark (an arc of light) that filled the gap. Incandescence was more difficult to pull off so Davy and his immediate successors concentrated on arc lighting. His arc lamp consisted of two sticks of carbon (charcoal) wired to the battery. Electricity from the battery jumped the gap between the two sticks, to produce the glaring arc. (See Figure 1-1.) Davy showed off his arc light spectacularly in an 1808 lecture, at which he connected 2,000 battery cells to the circuit to produce a bright arch of light. Modern electricity supply began on that day.

---

4.   R.A.S. Hennessey, *The Electricity Revolution* (London: The Scientific Book Club, 1972), p. 10.

Figure 1-1. Arc Lamp

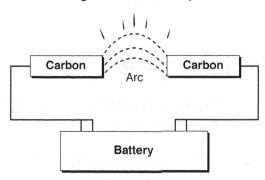

The voltaic battery, however, produced limited quantities of expensive electricity. In 1831, Michael Faraday, also of the Royal Institution and an attendee at Davy's 1808 demonstration, found an alternative when he co-discovered the principle of magnetic induction. (The American Joseph Henry discovered the principle first but Faraday beat him to publication.) Faraday found that moving a magnet within a coil of wires induced a flow of electric current in the wire.[5] A few months later he constructed the first electric generator, also called the magneto or dynamo, which used the principle of magnetic induction to produce electricity in a circuit.

A Parisian instrument maker, Hippolyte Pixii, advanced the idea by building his electric generator in 1832, a device with a hand-powered crank that turned a magnet mounted on a shaft between the wires of an electric circuit. The moving lines of magnetic force cut the wires as the magnet spun. That device, too, required improvement to advance beyond a laboratory toy, especially a better way to turn the shaft. Water power or a steam engine, perhaps? The coil from the electric circuit could go onto the shaft and spin between fixed magnets as an alternate configuration, cutting the lines of force as it revolved through them. (Moving the wires through a stationary magnetic field or moving the magnet between stationary wires accomplishes the same thing, getting the electric current into the wire.) An ingenious device on a ring maintained the connection between the revolving armature and the rest of the electric circuit. (See Figure 1-2.)

---

5.   Technically speaking, the motion of the wire cutting through the lines of force induces an electromagnetic force into the wire which causes a current to flow in the wire. Moving the wire faster through the lines of force or using a stronger magnet both induce a greater electromagnetic force.

Figure 1-2. Electric Generator

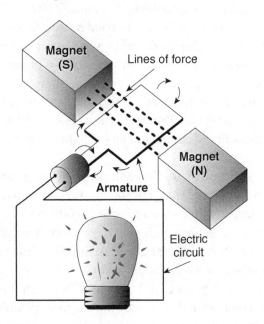

After four decades of efforts to improve the strength and speed of the electric generator, a Belgian, Zenobe Theophile Gramme, made the big breakthrough, doing "more than any other man to develop the generator and the motor commercially."[6] Gramme rearranged the dynamo's innards and powered it with a reciprocating steam engine. He installed his first machine in 1873. At an exhibition in Vienna in that year, according to legend, workers mistakenly wired an operating Gramme generator to an adjacent generator not yet set up to produce electricity and that second generator began to run backwards. Gramme had not only perfected the generator but he had discovered the principle of the electric motor, which was, according to James Clerk Maxwell who had written the equations that described electromagnetism, the greatest discovery of the time. Gramme made electricity work rather than just provide illumination. And his invention replaced the battery as the key source of power. In 1876, Gramme offered an improved generator – higher speed, lighter, greater output. Robert Routledge, that contemporary popularizer of Victorian science, wrote that Gramme's machines:

> ... solve the problem of cheap production of steady and powerful electric currents, so that electricity will soon be applied in processes of manufacture where the cost of electrical power has hitherto placed it out of the question.[7]

---

6. Richard Shelton Kirby, Sidney Witherington, Arthur Burr Darling and Frederick Gridley Kilgour, *Engineering in History* (NY: Dover, 1990), p. 155.
7. Robert Routledge, *Discoveries and Inventions of the Nineteenth Century* (NY: Crescent Books, 1989), p. 417.

Even with an improved dynamo, the arc lamp had its limitations. Used for lighthouses, street lamps and lighting in cavernous Victorian railway stations, it burned too brightly for use in homes, offices or stores so Victorians tried to "subdivide" the light. The early arc light had still other drawbacks. The spark had to jump between carbon rods placed head to head. As the rods burned up, the space between them widened, eventually becoming too wide for the spark to jump the gap. The arc lamp, therefore, required a regulator, a mechanism to push the rods together as they burned up, in order to maintain the correct distance between them. In addition, the arc lamps were strung in series along the circuit, so if one lamp failed, that cut the circuit and all the lights went out.

Paul Jablochkoff, an ex-officer of the Russian army living in France who worked for Gramme, solved the circuit and regulator problems. He connected six candles (arc lights) in a subsidiary circuit and attached each of those circuits to a main circuit powered by a Gramme alternating current generator. If a light failed, only the lights in the same subsidiary circuit would go dark, not all the lights. He eliminated the regulator by placing the carbon rods side by side, separated by kaolin, a clay. (See Figure 1-3.) As both rods burned down (direction of arrows in figure) so did the kaolin between them, allowing the arc to descend as the rods burned down. No moving parts required! Jablochkoff first displayed his wares at the Paris International Exposition of 1878. Soon thereafter, his "candles" lit London's Victoria Embankment. He also tried to produce low-powered candles to compete with gaslight but they cost twice as much. I.C.R. Byatt wrote off Jablochkoff's arc light as a failure because it did not make use of " the comparative advantage of arc lighting – the ability to produce a high-powered light cheaply". [8]

Arc lamps installed in 1878 lit London's Gaiety Theatre, Holborn Viaduct, Billingsgate Fish Market and Waterloo Bridge. At night soccer games at Sheffield players complained that the arc lamps blinded them.

Figure 1-3. Jablochkoff Candle

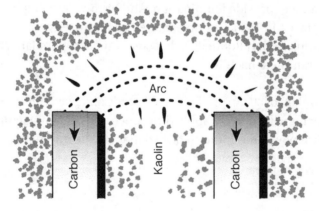

---

8.   Byatt, *op. cit.*, p. 13.

In 1879, inventor, engineer and entrepreneur R.E.B. Crompton started Britain's first electrical manufacturer, producing generators and arc lamps. In the same year, the local government turned off Holborn Viaduct's arc lamps in order to restore gas lighting. Producing the same product as gas – lighting – and then selling it for a higher price was not a promising business model. Nor was the alternative, selling it at a competitive price and losing money on every sale.

The Battle of the Bulbs: Edison vs. Swan

Victorian inventors targeted the incandescent lamp as the product for the subdivided, low glare market. They filed at least 24 patents on that lamp between 1838 and 1879. They attempted to place a filament (usually of carbon or platinum) in a glass bulb, pump the air out of the bulb to prevent oxidation of the filament, and then heat the filament electrically until it glowed (incandesced). Thomas A. Edison of the United States, and Joseph W. Swan, of Newcastle, England, produced the first workable lamps, almost simultaneously.

Swan, a chemist and businessman, from 1848 to 1850 carbonized strips of paper and tried to "render ... the strips incandescent under a vacuum". [9] He failed, but tried again, in 1860, using smaller strips of carbon, an improved vacuum pump and a larger battery. He later recalled that "I did succeed in making a flat carbonized paper arch bright red hot ... I did not go any further at that date". [10] Encouraged by improved dynamos and development of the Sprengel air pump, which could produce the necessary vacuum, Swan returned to the quest in 1877. By October, he and co-worker, Charles Henry Stearn, had incandesced carbonized strip filaments in a vacuum within a glass bulb. In December 1878, Swan managed to run a lamp for a half-hour before the filament broke. He then displayed the bulb at a lecture to the Newcastle Chemical Society.

Across the Atlantic, Thomas A. Edison observed arc lighting and new dynamos in a Connecticut workshop in August 1878. Encouraged by the same developments that spurred Swan to action, he aimed at producing both a practical incandescent lamp and an entire lighting system. (Edison claimed, in a patent lawsuit, to have worked on the problem earlier. He produced no evidence for the claim despite his habit of keeping meticulous notes of all experiments. Swan, who did not keep meticulous notes, also claimed precedence in his patent appearances. Much of the British claim that Swan invented the incandescent light before Edison is based on Swan's ability to recall unrecorded events.) On October 5, 1878, Edison filed a patent for a filament light bulb. By the end of that month, he announced his plans for an entire electrical system. The shares of gas light companies fell. The press applauded. Edison telegraphed his London agent:

---

9.   C.N. Brown, *JW Swan and the Invention of the Incandescent Electric Lamp* (London: Science Museum, 1978), p. 3.
10.   Brown, *op. cit.*, p. 11.

> I have just solved the problem of the subdivision of the electric light ... When the brilliancy and cheapness of the lights are made known to the public – which will be in a few weeks... – illuminating by ... gas will be discarded. [11]

In fact, Edison had no operating system, no light, and he was working with a platinum filament.

Swan must have known of Edison's claims. (By late 1879 or early 1880, Edison's correspondence and notebooks showed that he knew that Swan was working on a carbon filament. Edison's court testimony claims that by that time he, too, was using carbon.) Early in 1879, Swan delivered a trio of lectures during which he turned on a lamp that burned until the lecture ended. By October of that year, though, Edison's team had produced a bulb with carbonized thread filament that burned for 14 hours. Edison then filed a patent for a carbon filament lamp, although he still did not have a practical lamp. On New Year's Eve, 1879, though, Edison put on a spectacular public demonstration of electric lights and a complete power system at his Menlo Park, New Jersey laboratory.

In January 1880, Swan filed his first patent, for removing gas incorporated in the carbon filament while pumping gases out of the bulb, not a patent for a light bulb. In the same month, he patented a method to prevent leakage in the lamp. Then a disastrous fire at his business diverted him from his research. By October, though, he held a lecture at the Newcastle Literary and Philosophical Society, where he turned off 70 gas jets, relit the room with 20 incandescent lamps, said that he had a better filament than Edison, and declared that "his process would be less costly than gas" and " ... could be measured as easily..."[12] Swan did not disclose the filament's composition. In November, he filed a patent on a parchmentized thread filament. By December, in contrast, Edison was ready to produce an entire commercial electric lighting system.

Swan established the Swan Electric Light Co. in February 1881, began manufacturing lamps in April, set up a joint venture with Crompton, and opened a factory in France. At the Paris Electrical Exhibition of 1881, Swan won a gold medal. Edison, who exhibited an electrical system, from generator through light bulbs, won a higher award, the diploma of honor.

Moving faster than Swan, Edison, in December 1878, made a deal with J.P. Morgan to obtain and manage patents in Great Britain and parts of the British Empire. Edison filed the patents and then displayed his generating and lighting equipment at the International Electric Exhibition of 1882 at London's Crystal Palace. Morgan then established English Electric Light Co., Ltd. to take over Edison's British patents. In April, the company

---

11. Robert Friedel and Paul Israel with Bernard Finn, *Edison's Electric Light: Biography of an Invention* (New Brunswick, NJ: Rutgers University Press, 1986), p. 13.
12. Brown, *op. cit.*, p. 34.

opened the Holborn Viaduct generating station, a forerunner of New York's Pearl Street Station. The Holborn operation lit 1,000 lamps (400 in the General Post Office) at first, providing street and interior lighting along Holborn Viaduct. English Electric Light set its prices close to those of gas lighting. It lost money on electricity sales. English Electric Light's managers hoped to make money by selling licenses and equipment to other electric companies.

Meanwhile, in May 1882, Swan regrouped his operations into the Swan United Electric Light Company, Ltd. Edison's business immediately charged Swan with patent infringement, prompting one trade journal to ask, "Is this the beginning of a sham battle preparatory to a combination of the two companies?"[13] In 1882, however, Parliament passed a law that limited the life of utility franchises, making them less attractive than entrepreneurs had hoped, which put a damper on prospects for both Swan and Edison's lighting businesses. By late 1883, Swan had made large profits selling lamps and installations, not from running electricity supply operations. Edison's company had lower sales and a money-losing venture at Holborn Viaduct. In October 1883, the two groups merged into The Edison and Swan United Electric Light Company, Limited. As General Electric's historian, George Wise, put it, with patents in hand,

> they sued their competitors and successfully cornered the light-bulb market, ensuring a profit for their company. They had only to compete with the gas-lighting companies...[14]

Although Swan had the larger company, and the British bought the Swan lamp, Edison took top billing. Initially, Edison opposed any billing for Swan.

Gas Model

The gas utility industry provided the business and regulatory models for electric supply. In the mid eighteenth century, the British began to "carbonize" coal on a large scale. They heated it to a high temperature in absence of oxygen, creating two products, coke for use in blast furnaces and coal gas (mainly methane) which when lit produced a flame suitable for illumination. William Murdock, trained by James Watt and in charge of steam engines at the mines in Redruth, Cornwall, lit his workshop with coal gas in 1798, the beginning of gas lighting. Early nineteenth century entrepreneurs built carbonization facilities to produce gas for sale. Some attempted to deliver it in containers. They soon settled on a delivery system similar to water utilities. They manufactured the gas in a central facility, transporting it to consumers through a network of underground pipes. The

---

13.   Brown, *op. cit.*, p. 44.
14.   George Wise, "Swan's way: a study in style," *IEEE Spectrum*, April 1982, p. 70.

industry made rapid progress. In 1804, Philippe Lebon patented a gas manufacturing process. By 1807, gas lights illuminated London's streets. Samuel Clegg, in 1815, invented the gas meter. With it, the gas company could charge for actual consumption, rather than by the number of burners on the premises. Then, in 1825, Thomas Drummond created the incandescent gas light. He placed a stick of lime in the flame. The lime glowed white hot, hence the word "limelight."

Parliament authorized London's first public supplier of gas in 1810 – possibly the first public utility to deliver energy anywhere – in the Gas Light and Coke Company Act. (British law required parliamentary authorization for each new company, a problem that would dog the public utilities for decades to come.) Parliament issued 200 more charters by 1830. The early charters neither limited profits nor required the company to serve all those who applied for service. Competition would protect customers from abuses such as high prices, Parliament reasoned. To assure competition, Parliament overlapped the territories of gas suppliers so that two served each neighborhood. Competition from candles and whale oil, would, no doubt, help to control gas prices as well. After a while, just in case competition did not work as planned, Parliament inserted limits on both the price of gas and the dividend that the company could pay.

Competition did not protect consumers.[15] W. A. Robinson, writing about the economics of competition and regulation over a century ago, commented:

> Experience has... shown that competition... is never long-lived. Either one of the competitors is ruined, and the survivor makes the customer pay for the cost of the contest, or the two coalesce, and levy rates to pay a dividend upon the double capital.[16]

Unlike competitive businesses, gas suppliers could not seek new customers elsewhere or move assets to another place when the original market failed to produce adequate profits. If gas suppliers built smaller systems that do not overlap with those of other suppliers they would not face the problem of having too much investment per customer but they would not compete, either. They would have a monopoly. They could charge whatever

---

15. The arithmetic did not work for the competitive model. For instance, two gas companies decide to serve an area with 100 customers. Each puts in plant to serve all the customers at a cost of £100 per customer, or £10,000 for the entire business. Each has to collect £1,000 per year (£20 per customer) to cover expenses. (A single company serving all 100 customers could break even charging £10 per customer.) Each of the two competing companies each snare 50 of the potential customers. If they charge the price that a single company would charge, £10 per customer, each would collect only £500 per year, and each would have twice as much investment as needed to serve its customers and they would operate at a loss until one goes out of business, after which the survivor could raise its prices. The companies would have to double prices (to £20 per customer) in order to break even.. Thus the consumer would be better off served by one firm.

16. Philip Chantler, *The British Gas Industry: An Economic Study* (Manchester: Manchester University Press, 1938), p. 66.

they wanted, up to a price that would attract the entry of a competitor or cause consumers to revert to candles.

If competition could not protect consumers from unjustly high prices, and the companies could not survive in a competitive environment, then the corporate charter had to include provisions to prevent the monopolistic supplier from overcharging, something politicians figured out even while still espousing the competitive model. Initially, the government limited the distribution of profits over a prescribed level. The Nottingham and Oxford Gas Acts of 1818 prohibited the company from paying a dividend in excess of 10% on capital and required it to contribute all profits over 10% toward public lighting or civic improvements. In effect Parliament capped return on investment at 10%. The company's owners could not extract any more out of the company even if it earned in excess of 10%. The Bristol Gas Act of 1819 fixed a maximum price for public lighting. Every charter from 1840 onwards included dividend limitations. The Gasworks Clauses Act of 1847 set a 10% lid on the dividend other than for payments to make up for previous shortfalls below the 10% level, but did allow companies to accumulate reserves of up to 10% of capital. Charters thereafter also fixed a maximum price for gas sold.

No semi-intelligent twenty-first century investor would accept the terms set by the gas acts: fixed price for gas and fixed return on capital invested. Investors took the risk that operating costs might increase but not the price of gas sold and the risk that cost of capital would rise, leaving investors with inadequate returns.[17] Nineteenth century investors, however, lived in a different world. While the economy grew, cost of coal (the principal ingredient in the manufacture of the gas) was in decline for most of the century, as was the cost of living, (so no inflationary pressure on wages). The cost of capital, as evidenced by yield on the risk free bench mark, UK government bonds, declined through the century, too. If investors could collect a fixed price for their product (whose cost of production was falling) and a fixed return on investment (while other returns dropped), they had a good deal. This formula for return influenced electricity regulation. For decades in the twentieth century, American utility regulators granted 6% returns almost reflexively, no matter what. The gas acts left a legacy. (See Table 1-1.)

---

17. The company would require an act of Parliament to raise the dividend to reflect higher capital costs. Without a higher dividend, the share prices would drop to provide the new market return to investors. If price fell below par value, the utility might face legal difficulties when raising new capital and that could discourage capital expansion.

Table 1-1. Prices, Yields and Economic Indicators, in the UK. (1801-1900) (%)

Period	Coal price (average annual change)	Cost of living (average annual change)	Government bond yield (average)	Real GDP (average annual change)	Real GDP per capita (average annual change)
	[1]	[2]	[3]	[4]	[5]
1801-1810	-1.7	0.6	4.8	2.6	1.4
1811-1820	-1.7	2.0	4.6	-0.3	-1.7
1821-1831	-2.3	-1.7	3.6	2.2	0.9
1831-1840	-2.0	1.2	3.4	2.1	0.9
1841-1850	-3.7	-3.2	3.2	1.6	1.1
1851-1860	1.7	3.1	3.2	2.4	1.4
1861-1870	-1.2	-0.6	3.3	2.4	1.7
1871-1880	-1.5	-0.6	3.3	1.8	1.6
1881-1890	2.5	-1.9	3.0	1.7	0.9
1891-1900	2.1	0.5	2.7	2.1	1.1
1801-1900	-0.7	-0.5	3.5	1.9	0.9

Notes (by columns):

1. Average price of best coal at ships' side, London, 1800-1885. Wallsend, Hatton in London, 1885-1900.
2. Composite cost of consumables.
3. Yield on British Government Consols.
4 and 5. 1801-1820 estimated by author based on data in Mitchell.

Sources (by columns):

1. B. R. Mitchell (with Phyllis Deane), *Abstract of British Historical Statistics* (Cambridge University Press, 1962), pp. 482-483.
2. B. R. Mitchell, *British Historical Statistics* (Cambridge: Cambridge University Press, 1988), p. 168.
3. _____, *op. cit.*, p. 678.
4 and 5. Samuel H. Williamson, "What Was the U.K. GDP Then?", *Measuring Worth* 2015.

By modern accounting standards, those utilities earned less than a 10% return because they did not systematically deduct depreciation expense (wear and tear on assets) from earnings. The return actually allowed, calculated by subtracting an allowance for depreciation expense, came to around 7-8%. Government bonds (a virtually risk free investment) then yielded about 3%, so gas utility investors earned a 3-5% risk premium for owning a utility stock versus a government bond that was in line with historic averages. The maximum return, if earned, exceeded cost of capital, as indicated by the rise in share price over issue price (par value) in anticipation of that high profit.

The gas companies figured out how to use high stock prices to get around profit limitations. To raise funds, they offered new shares at par value to existing shareholders, who could then sell the newly acquired shares at the higher market price. Profit on sale of new shares was like an additional dividend. Furthermore, the government did not control the size of gas company capitalizations, so companies could sell more stock than needed to fund their expansions and could charge more for gas in order to earn the dividends on the unnecessary capital. (American regulated utilities adopted similar tactics years later. Critics called it padding the rate base.) Before the government ended the stock sale racket in 1876, some gas shares with 10% dividends sold at twice par value, to yield only 5%. Existing shareholders had the right to buy new shares at half the market price. Under the 1876 rules, the gas companies had to auction off new shares to the public, and use any surplus over par value for the benefit of consumers.

But that is getting ahead of the story, because by the 1850s, the gas industry needed a new regulatory framework. Price and dividend limits had not worked as expected. The gas companies had an obligation to serve the street lighting load but, beyond that, they could choose their customers, so they served only part of the public. Competition had worked in that it prevented companies from earning anticipated returns. Unpaid dividends accumulated in arrears. By the mid 1850s, London's gas suppliers no longer competed for customers. The government needed a formula to control prices and still encourage the companies to seek out efficiencies that would benefit customers. Dr. John Chalice, representing the South London Gas Consumers Mutual Protective Society, presented his idea for a "sliding scale" pricing plan at an 1858-1859 investigation of the industry: gas companies could raise dividends if they could lower the prices they charged for gas. The government, eventually, adopted that price framework.

The Metropolitan Gas Act of 1860 ended competition between gas suppliers, granted a monopoly franchise to each supplier, fixed the price of gas (with an adjustment allowed for changes in production costs), required the gas utility to serve all who demanded and could pay for service, and limited to six years the period for payment of dividends in arrears. Consumer benefits were slim because the consumers had to pay the dividend arrearages and the government did not look into why some gas companies required more capital than others to serve customers.

The monopoly granted, incidentally, referred to a monopoly of the right to dig up the streets in order to put in gas mains. In the words of Corbett Woodall, engineer in chief of the Phoenix Gas Company in London in 1879:

... that monopoly however does not interfere with any individual consumer or ratepayer; he may erect his own gasworks if he choose, running the risk of being indicted for a nuisance.[18]

The Sheffield Act of 1866 endorsed the sliding scale:

... experience has shown that the increase of the rate of the dividend on the company's capital in proportion to the decrease in the general charge for gas has worked beneficially for the public... [19]

The City of London Gas Act of 1868 consolidated the industry into fewer companies, set a maximum dividend of 10%, and a maximum price for gas, with an escape clause that allowed price modification during periods of extraordinary conditions. (Four years later, during a coal shortage, gas companies raised prices to reflect a sharp increase in coal costs.) In addition, the Act standardized the quality of gas at 16 candlepower (cp).[20] Without a standard product, price controls had no meaning. The industry closed down small gas works, and in 1870, Gas Light and Coke opened the massive Beckton Gas Works, the largest in the world to serve customers from a central location. That facility operated until 1969. Victorian engineers built for the ages.

From 1877 onwards, all charters standardized both price for gas and dividend rate, with the provision that the gas company could raise its dividend by a prescribed amount if it could reduce prices by a prescribed amount. If the company raised prices over a certain level, it would have to lower the dividend. Parliamentarians figured that they had aligned the interests of investors and consumers, no mean feat.

Technological developments opened up new markets and strengthened the industry's competitive position. In 1850, Robert Wilhelm Bunsen invented his eponymous burner, creating the means to use gas for heating, cooking, water heating and industrial lighting. In 1885, Austrian chemist Carl von Welsbach strengthened the competitive position of gas lighting with his invention of the Welsbach mantle, a fabric structure coated with oxides of thorium and cerium that became incandescent when heated by a Bunsen burner.

---

18. House of Commons, *Report from the Select Committee on Lighting by Electricity together with the Proceedings of the Committee, Minutes of Evidence and Appendix*, Ordered by the House of Commons to be Printed 13 June 1879, p. 110.

19. Irvin Bussing, *Public Utility Regulation and the So-Called Sliding Scale* (NY: Columbia University Press, 1936), p. 22.

20. Candlepower (cp) is the measure of light produced by "the standard sperm candle burning 120 grains of sperm per hour." A standard gas burner consumed five cubic feet (cf) per hour. Thus, a gas burner taking five cf in one hour, with a rating of 16 cp, produced the same level of light as 16 sperm candles burning a total of 1,920 grains (16 x 120) of sperm oil. The industry needed that standardization. Quality of gas varied depending on manufacturing process and type of coal burned. Routledge claimed that testing showed wide variation: 12.1 cp in London, 22.0 cp in Liverpool and 28.0 cp in Glasgow. (See Routledge, *op. cit.*, p. 642.)

The gas industry continued to grow, thanks to those non-lighting loads and the Welsbach mantle, despite inroads from electricity and the disadvantage of operating with gas production standards based on lighting requirements that discouraged growth of gas heating. Fortunately, the industry managed to improve gas manufacturing processes in the 1880s. In 1890 gasworks introduced carbureted water gas (a volatile carbon substance sprayed over steam introduced into the manufacturing process raised the caloric value of the gas), which lowered manufacturing costs, maintained illuminating power, and created a fuel useful in industrial processes.

After 1890, those lower costs led to calls to revamp the pricing formulas. The gas companies could charge less and still earn acceptable returns. The Board of Trade, which supervised the industry, resisted a change in the formula. It admitted that " a bad bargain may have been made" on behalf of consumers but argued that any change could disturb the "faith placed... by investors in the course taken by Parliament." [21] Not cricket to change the rules, in mid game.

This "bad bargain" dilemma never goes away. Regulators who lower prices to reflect conditions unanticipated when the regulatory arrangement was made confer immediate benefit on consumers. They also send a signal to investors who committed funds based on an understanding with regulators. The investors now know that the regulators will break the agreement with them when opportune to do so. That uncertainty about the commitment of the regulator increases risk, and therefore, cost of capital for new investments. Regulators have to weigh the benefits of the immediate price reduction against higher cost of capital in the future. Regulatory risk becomes part of the calculation of cost of capital.

On the other hand, costs and conditions might change enough to lower profits to a dangerous level that leaves the utility with insufficient funds to pay suppliers of raw materials or unable to raise capital needed to serve its customers. Under those circumstances, regulators have a choice: raise prices above those previously agreed to, or leave the utility in a position where it cannot take new customers or might have to cut off service. Should the regulator cite the old bargain and let the chips fall where they may?

The simple, multi-year fixed formula, historically favored in the United Kingdom, then, requires modification whenever conditions change significantly. The pricing formula, thus, remains in effect only under "normal" circumstances. In 1899, a House committee proposed a revised sliding scale with lower prices that took effect in London in 1901. Future regulators made similar decisions. At some point, practical politics trumps legalisms and everyone involved agrees to the new arrangement.

---

21.   Chantler, *op. cit.*, p. 93.

As for the growing competition faced by the gas industry, even in the 1890s some experts had trouble seeing the picture. Robert Routledge, that doyen of Victorian technology, wrote:

> Without any rivalry from the electric-light, gas as a domestic luminant has now met with a competitor on the grounds of cheapness in the mineral oil...[22]

He meant not electricity but paraffin oil, made from petroleum or lignite, sold as "solar oil" or "photogen." As soon as its manufacturers learned how to remove the odor and build a safe lamp, paraffin oil would eclipse gas. The futurist at work.

Electric Supply Circumscribed

Parliament investigated electric supply in 1879. Dr. Lyon Playfair chaired the Select Committee. He had studied chemistry with the great Liebig in Germany, was a chemistry professor, as a government official advocated using poison gas during the Crimean War, and served as Postmaster General and president of the prestigious British Association for the Advancement of Science. Lord Lindsay, another key member, was one of those improbable Victorian polymaths: Scottish peer, Tory politician, astronomer, Fellow of the Royal Society, electrical experimenter, investigator of spiritualism and explorer. The investigation was extraordinarily thorough, although its reputation was not burnished by the claim of one witness that the telephone would not flourish in Britain because "Here we have a superabundance of messengers, errand boys, and things of that kind."[23]

Committee members did their homework. They grilled scientists, engineers, gas industry executives, navy officers and business people who had actually used electric lights, all in great detail. They examined cost of electric vs. gas lighting (electricity cost about four times as much), characteristics of electric lighting (colors looked natural), whether to permit gas utilities to offer electric service (no), the impact of electric lighting in public places upon fashionable women's attire (they had to wear clean dresses every time or be found out) and why the Board of Trade forbade the use of electric lights on British ships (mariners on other ships would mistake them for lighthouses). The committee saw electricity as a business that offered an expensive product largely manufactured on the consumers' premises without connection to a public network. The gas companies thought they might sell gas to electric generators, but that was the only potential connection between the isolated electric generators and a regulated network at the time.

---

22. Routledge, *op. cit.*, p. 643.
23. House of Commons, *op. cit.*, p. 69.

16

The committee explored the duties of the franchised monopolist, including the need to serve consumers under onerous conditions. A gas company executive said in that understated English fashion, "You are aware that we do occasionally have fogs in London, which do not give us a long warning."[24] On a winter day, when the gas company was selling 40-50% of its production, fog could raise demand to 250% of production, "so that we are obliged to have a reserve in store to meet such a contingency."[25] What could an electricity supplier do under such circumstances? Would customers want to wait the two hours it would take to get the additional generators operating? The gas company executive neatly summed up one of biggest problems that would dog the electricity business for its entire existence, the inability to store the output in an economical manner to serve customers during peak or emergency periods.

The committee concluded that:

> The general nature of the electric light has been well explained... It is an evolution of scientific discovery which has been in active progress during the whole of the century... A remarkable feature of the electric light is, that it produces a transformation of energy in a singularly complete manner. Thus the energy of one-horse power may be converted into gaslight and yields a luminosity equal to 12-candle power. But the amount of energy transformed into electric light produces 1,600-candle power. It is not therefore surprising that while many practical witnesses see serious difficulties in the speedy adaptation of electric light... the scientific witnesses see in this economy of force the means of great industrial development... that... is destined to take a leading part in public and private illumination...

> Scientific witnesses also considered that in the future the electric current might be extensively used to transmit power... to considerable distances, as that power applied to mechanical purposes during the day might be made available for light during the night...

> Compared with gas, the economy for equal illumination does not appear to be conclusively established... Unquestionably the electric light has not made that progress... to enter into general competition... for... domestic supply. In large establishments, the motors to produce the electric light may be readily provided, but, so far... no system of central origin and distribution... has... been established...

> In considering how far the Legislature should intervene in the present condition of electric lighting, your Committee would observe... that in a system which is

---

24. _____, *op. cit.*, p. 110.
25. *Ibid.*

developing with remarkable rapidity, it would be lamentable if there were any legislative restrictions calculated to interfere with that development.

If... local authorities have not power... to take up streets and lay wires... your Committee thinks that ample power should be given them...

Gas Companies...have no special claims to be considered as the future distributors of electric light...

Your Committee... do not consider that the time has yet arrived to give general powers to private electric companies to break up the streets.... But... the Legislature should show its willingness, when the demand arises, to give all reasonable powers for the full development of electricity as a source of power and light. [26]

The Playfair committee had a clear view of electricity's potential but saw no immediate need for the government to act because neither the existing technology (on-the-premises generation for large scale lighting) nor the economics of the product (substantially more expensive than gas) called for action – which meant not yet granting electric companies a right to dig up streets. The committee favored giving the municipal authorities first crack at electricity distribution, when the time came, but expressed reservations about their willingness to promote a new technology. As for Edison, neither the committee nor the expert witnesses ignored the Wizard of Menlo Park. They knew what he was working on (subdivision of light and network distribution) but had no way to sort out claims from reality. Edison filed his most important patents a few months after the report came out. [27] For the next two years, thanks to the committee's hands-off attitude, the electric supply business grew, unfettered by legislation.

By the end of 1879, arc lamps lit Liverpool and London's Embankment. Late in 1880, Swan installed incandescent lighting in Cragside, the Northumbria mansion of Sir William Armstrong, the first home served by hydroelectricity, probably the first residential installation outside of those in Swan's and Stearn's homes lit by incandescent lamps. The owner was a leading Newcastle industrialist, inventor of the hydraulic engine and the breech loading gun, the father of modern artillery and some claim the model for Andrew Undershaft, the munitions king in Shaw's *Major Barbara*. He was the perfect early adopter of a new product. If he wanted it, others would follow. In the same year, electric railways went into service in Brighton and at Ireland's Giants Causeway. In 1881, the City of London tested arc lighting, and the council in Godalming, Surrey established the first

---

26. House of Commons, *op. cit.*, pp. i-iv.
27. Arthur C. Clarke, British science fiction writer and technologist, claimed that one expert of the time declared that "[Edison's ideas are] good enough for our transatlantic friends... but unworthy of the attention of practical and scientific men." The quote took on a new life when included in Christopher Cerf and Victor Navasky, *The Experts Speak* (NY: Pantheon Books, 1984), p. 203. (Author was unable to locate it in the Select Committee transcripts.)

public electric supply. The owner of a local water mill provided power in return for free lighting. Historian Brian Bowers wrote:

> The main streets were lit by Siemens arc lamps and the side streets by Swan filament lamps. The installation created much public interest, but the River Way was neither adequate no reliable, and a steam engine was brought in. [28]

In February 1882, Robert Hammond established an electric company in Brighton that holds the world's record for longest, continuous electricity supply. In April, Edison opened the Holborn Viaduct station. Before that, isolated and miniature projects prevailed, Richard D'Oyly Carte lit up the Savoy Theatre for Gilbert and Sullivan's *Patience* in 1881. The audience read the libretto during performances of *Iolanthe* (1882) because the theater could not dim the house lights. On stage, the fairies wore battery-powered Swan electric lights wired into their hair. Yet, despite the introduction of the new light bulb, the cities of Birmingham and Leeds installed arc lights. With all that activity, Parliament decided that the time had come to pay attention to the new business.

The Board of Trade regulated British industry. Joseph Chamberlain, a zealous reformer and imperialist, headed it. He promoted the Electric Lighting Act of 1882. The bill drew on the Metropolitan Gas Company Act of 1860, the Tramways Act of 1870 and the Public Health Act of 1875. Chamberlain favored local government ownership of utilities. The bill had that bias. [29]

The Act provided that:

- Electricity undertakers could open streets in order to lay lines.
- Local governments could raise money for the purpose of supplying electricity.
- Electricity suppliers could obtain licenses by Act of Parliament, if they so chose.
- The Board of Trade could grant a supply license for seven years, with consent of local authorities.
- The Board of Trade could grant a license through a provisional order (it required later ratification by Parliament) with or without the consent of local authorities.
- The concession had no time limit, but the local authority could buy out the con-

---

28. Brian Bowers, *Electricity in Britain* (London: The Electricity Council, 1986), p. 8.
29. The British referred to the entity providing electricity service as the "undertaker" or "undertaking." Municipal government, called the "local authority" owned some undertakings, while private investors owned others called "companies." The nomenclature became more complicated later on, when various government agencies called "boards" took over parts of the business and "power companies" attempted to concentrate on the generation side of the business. In this book, I refer to all agencies owned by the local government as "municipal" and those owned by investors as "companies," and all entities providing electricity service to the public on a regulated basis as "utilities".

cession at the end of 21 years, and every seven years thereafter, at a value set at the time of purchase.

- The Board of Trade made clear that "value" meant scrap – not going concern – value.

Critics blamed the Act's last provision for throttling the industry by causing investors to fear that they would not recover their investment by the end of the concession. The Act, however, had still other provisions that adversely affected the electricity market's growth:

- Electric utilities could not prescribe the type of equipment used by their customers, which prevented them from offering a standardized service.

- Customers could demand connections even when uneconomical for the utility, a discouraging burden for as firm just starting up.

- The Board of Trade would determine the service territory, price of service (the Act set a maximum price) and maximum voltages (a safety measure).

Municipal governments, often owners of profitable gas utilities, shied away from investing in the new field. Only the rich would take electricity, anyway, because electric lighting cost at least twice as much as gas. The municipalities could, however, hinder others from getting an electric franchise by opposing them at the Board of Trade.

Thirty three years after the passage of the Act, Charles P. Sparks, aptly named electrical engineering pioneer, still harbored sour thoughts about that legislation. Speaking at his inauguration as president of the Institution of Electrical Engineers, he noted that Parliament had ignored the lessons learned from the consolidation of the gas industry and "Parliament... decided that our industry should be a parochial one."[30] As for the 1882 Act:

> The effects of legislation in... stifling the new industry... were far reaching. Our American and Continental competitors seized the opportunity to forge ahead... In the meantime manufacturers in Great Britain, unable to proceed with the equipment of undertakings operating under statutory powers, had either to turn their attention to the equipment of separate works... or to operate without statutory powers by using private wayleaves of uncertain tenure. By this method of development the electric supply industry was started on unsound lines... and the effects of this false start are still felt.[31]

Potential electricity operators rushed into the new business – getting in on the ground floor of the next big thing, they must have thought. Then they sobered up. The bubble burst. Parliament issued 69 provisional orders in its 1883 session but only four in 1884, and, even worse, only seven of those 73 still remained in force in 1895. Of the five licenses

---

30. Charles P. Sparks, "Inaugural Address," *The Journal of The Institution of Electrical Engineers*, Vol. 54, No. 252, 1 December 1915, p. 2.
31. *Ibid.*

granted between 1883 and 1887, only one remained in force by 1895. After the initial enthusiasm, investors quickly realized the defects of the law. They had staked claims, so to speak, but did not put further money into the business. (See Table 1-2.)

Table 1-2. Provisional Orders and Licenses. (1883-1887)

Year	Provisional orders applied for	Provisional orders approved by Board of Trade	Provisional orders still in force (1895)	Licenses applied for	Licenses approved by Board of Trade	Licenses still in force (1895)
	[1]	[2]	[3]	[4]	[5]	[6]
1883	106	69	7	10	1	0
1884	4	4	0	0	0	0
1885	1	0	0	2	2	0
1886	2	1	0	1	1	1
1887	0	0	0	1	1	0

Sources:
Garcke Manual 1896, p. 394, *Garcke Manual 1906*, p. 38.

The 1882 Act discouraged formation of public electricity suppliers. Individual entrepreneurs found ways to fill the breach without provisional orders and licenses. (The Holborn Viaduct operation, for example, continued into 1886 despite the Act.) They could not achieve economies of scale or standardize their services and equipment, but they could supply electricity to selected customers. They sold electricity to the wealthy, and installed private generators for their customers. Sir Coutts Lindsay, who owned the Grosvenor Art Gallery on New Bond Street in London installed a generator to light his gallery. The neighbors wanted electricity, so he strung wires to them, over the rooftops, thereby escaping the onerous rules of the 1882 Act because he did not have to open the streets in order to lay cable. Lindsay hired the dashing, young Sebastian Ziani de Ferranti, as his chief engineer. Ferranti expanded Lindsay's network. Then, in 1887, the Lindsay family backed Ferranti in the formation of the London Electric Supply Co. – an ambitious leap from neighborhood to metropolitan service on a vast scale.

Developing Technology

Lighting customers and companies had to choose from a confusing array of competing technologies: alternating or direct current, arc or incandescent lighting, and even different voltages. Lucien Gaulard and John D. Gibbs, French and American partners, in 1883, developed a transformer to step down the high voltage alternating current economical for long distance transmission to a lower voltage suitable for distribution to local customers. They solved the problem of how to economically wed transmission and distribution, and their machine could handle the different voltages that individual customers could demand. Gaulard and Gibbs devised a complicated answer to a problem made unnecessarily complicated by Parliament's insistence on freedom of voltage choice for electric consumers. By 1885, however, Ganz & Company, a Hungarian electrical equipment manufacturer, developed a simplified transformer. The Hungarians took the same approach as Edison, designing a transformer as an integral part of a system, while Gaulard and Gibbs designed something that connected to an assortment of devices out there. But, as the historian Thomas P. Hughes put it:

> ... another major explanation for the ingenious and complicated Gaulard and Gibbs design and the ingenious and relatively simple Ganz & Company design is that the Electric Lighting Act of 1882 did not apply in the Austro-Hungarian Empire. [32]

Finally, Ferranti challenged the Gaulard and Gibbs patent, which led to its revocation in 1890.

Early electric companies served customers with overhead lines in order to avoid parliamentary restrictions, but to do so they had to provide AC electricity at high frequencies (80 to 100 cycles per second) in order to reduce costs and raise operating efficiencies. As Charles P. Sparks noted, "Although not recognized at the time, the choice of these high frequencies made the development of single-phase alternating-current motors almost impossible." [33] Swiss and Hungarian engineering companies figured out the virtues of low frequency (40-50 cycles) AC long before the British, and even in 1915, Sparks complained that "the effect of developing alternating-current systems on a wrong basis has not yet been eradicated".[34]

Charles Parsons, a Tyneside engineer and son of the Earl of Rosse, revolutionized electric generation in 1884 by inventing the steam turbine, still the mainstay of the business. The turbine was simpler, more compact, weighed less, used less fuel, cost less to build, and ran at higher speeds than the reciprocating engine. In 1890, he installed the first steam turbine for power generation at a Newcastle electric company he had founded, a decade before German or American utilities put turbines into service. Displacing the reciprocating engine with the turbine was the nineteenth century equivalent of replacing the vacuum tube with a transistor.

---

32. Thomas P. Hughes, *Networks of Power: Electrification in Western Society, 1880-1930* (Baltimore and London: The Johns Hopkins University Press, 1983), p. 96.
33. Sparks, *op. cit.*, p. 7.
34. *Ibid.*

Figure 1-4. Reciprocating Engine Electric Generator

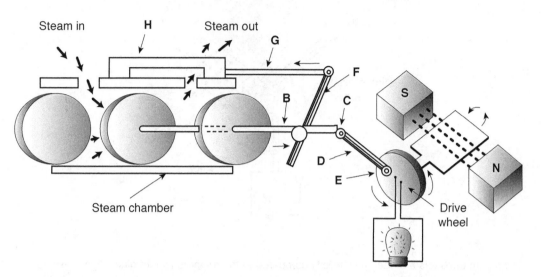

Steam in   H   Steam out

Steam chamber

Drive wheel

The reciprocating engine steam generator combined a steam engine and generator in a complicated combination that looked like a Rube Goldberg invention. Steam enters into a cylindrical steam chamber. The pressure of the steam pushes a piston (A) to the right, which pushes the piston rod (B) to the right as well. The movement of the piston rod pushes the wrist pin (C) which sets in motion the crank shaft (D), which is attached at the crank pin (E) to the drive wheel. The push of the crank shaft causes the drive wheel to turn, which causes the armature (with the electric circuit wire) to revolve between the magnets, thereby causing the electric current in the circuit. Meanwhile, as the piston rod moves to the right, an attached connecting rod (F) on a pivot moves to the left, which pushes the valve rod (G) to the left, which causes the valve (H) to switch the steam intake into an exhaust mode and the steam exhaust into a steam intake. Steam then pushes into the right side of the cylinder, forcing the piston to the left, which causes the piston rod to move in that direction, pulling the crank shaft and thereby continuing to turn the wheel. The engine gets its name from the back and forth (reciprocating) motion. So many moving parts had to decrease reliability and efficiency.

Figure 1-5. Steam Turbine Electric Generator

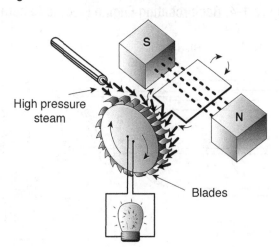

The steam turbine is a model of simplicity compared to the reciprocating engine. High pressure steam pushes the blades attached to a wheel (akin to water pushing down the on the buckets attached to a waterwheel) and the rotation of the shaft of the wheel causes the armature to revolve between the magnets The wheel's rotation turns its shaft, onto which is attached the electric circuit. The turbine gets its name from the Latin word for spinning.

Robert Davidson pioneered electric rail transportation when, in 1840, he operated a battery-powered vehicle on the Edinburgh & Glasgow Railway at four miles per hour. Thomas A. Edison built a well-publicized but small electric railway at Menlo Park in 1880. Charles Joseph Van Depoele, a Belgian working in the United States, and Frank Julian Sprague, a former US Navy officer who had worked for Edison turned the electric train into something practical. Between 1879 and 1884, their companies developed the overhead trolley and electric traction motor, and built electric railway systems. In 1888, Thomson-Houston, a predecessor of General Electric (of the USA) bought out almost all important street railway patents.

The British took advantage of German technology at first, using the third rail conduction and cars shown by Siemens and Halske at the Berlin Trade Fair of 1879. In 1883, Magnus Volk built a line in Brighton, and followed that up with an electric railway car that not only rode seven meters high on stilts above the surf but also – thanks to UK regulations – required a sea captain on board. William Siemens, the British agent of the Siemens clan, built his line at Giants Causeway in Ireland, in the same year. Americans preferred the overhead trolley to the third rail, built systems at home, improved the engines, and exported their products. The British manufacturers had less enthusiasm for the product, possible because their home market customers showed even less enthusiasm. The United Kingdom continued to franchise horse-drawn street railways after the United States and Germany had turned to electrification.

The Battle of the Systems, alternating current (AC) versus direct current (DC), distracted the industry more than choice of transformers or the virtues of putting horses out of business. [35] Edison favored DC, claimed that AC was dangerous, and underwrote the development of the AC electric chair to prove it. DC had significant advantages for use in arc lighting, traction and motors. It had significant disadvantages, too: the DC dynamo could not produce high voltages with the same ease as the AC generator, and long distance DC transmission and the step-down of voltages both required complicated systems.

In the early 1880s, Siemens produced an AC motor. Nikola Tesla, a Serbian immigrant to the United States who formerly worked for Edison, saw no reason to produce AC in the generator, convert it to DC, transmit it as DC, then convert it back to AC in a motor. In 1888, Tesla patented a complete polyphase AC system, involving generation, transmission and electric motor. (The polyphase electric generator contains several – usually three – coils of wire into which the revolving magnet induces electricity. Industrial customers take power from all three of those lines. Residential customers usually take from one line.) By the early 1890s, American industry had developed Tesla's system into the Niagara or universal power plan, which standardized the service offering and allowed American utilities to benefit from load diversity and economies of scale – the same benefits predicted by the experts during the Playfair investigation.

Conclusion

British scientists developed concepts that transformed electricity from scientific novelty to useful tool of society. They understood the direction the industry had to take. A British inventor patented a light bulb almost simultaneously with Edison, and Edison set up an electricity supply operation in London before New York. A British firm installed the first steam turbine. British enterprises, whether municipal or private, would have had little difficulty raising money for electrical investment in London, the financial capital of the world. The government, however, adhered to an ideological policy that hindered the realization of product standardization, load diversity and economies of scale. The British electricity supply industry fell behind that of other countries. The urban transportation industry clung to the horse.

From an organizational and regulatory standpoint, the UK's policy makers and business people had all the advantages of the gas industry experience. They knew about over-

---

35. Direct current moves in one direction and remains unchanging in characteristic. Alternating current switches direction and varies in voltage over the back and forth cycle.

capitalization, metering, setting up a network business, gaming the regulatory system, incentive regulation, the difficulty of making competition work, rigid rules that could disadvantage consumers and the advantages of a franchised monopoly. Yet, with all that background, Parliament passed a law that stifled the establishment of the public electric supplier. It stacked the deck in favor of municipal suppliers despite the fact that many of them had little interest in promoting an electric service that would compete with their gas operations. It created uncertain conditions for investor-owned companies because a hostile Board of Trade oversaw the industry and nobody could know in advance the value of the asset on termination of the franchise.

The electricians and editorialists of the day plainly expressed their frustrations: [36]

- *The Statist* – The electric infant was strangled at birth.
- *Electrician* – ... the fanatical dread of monopoly has resulted in there being no business to either monopolize or compete for.
- *The Times* – This bill will deprive the public of the benefits of electric lighting. The local authorities will not dare to embark on comparatively unknown undertaking, and the private companies will refrain from doing so...

Those protesters got it wrong. Parliament almost strangled the electric supply industry in the six years between its first and second attempts to regulate the business. But it did not succeed. The electrical entrepreneurs did not give up that easily.

---

36.   Hennessey, *op. cit.*, p. 32.

Chapter 2

Growth and Scale (1888-1909)

We hold that electrical enterprises should have their limits … set by economical considerations…, and that arbitrary boundaries, mostly of medieval ecclesiastical origin, should not limit … electrical systems. [37]

– James Swinburne

After almost throttling the electric industry at its birth, Parliament proceeded to revise the law, but not enough. According to Hennessey, the new law "encouraged great enterprise, but because it still saw electricity as a local matter it … cramped … the industry. Many of its ill effects lasted until 1948". [38] Yet several entrepreneurs took daring steps to enlarge the market and reorganize the industry. Perhaps they could make up for lost time.

The 1888 Act and Electricity Regulation

In 1884, managers of industry giant Edison & Swan told the Board of Trade that they would not implement any generating projects due to the provisions of the 1882 Act. In February 1888, the President of the Board of Trade complained that of the 64 franchises handed out to private companies and 17 to municipalities under the 1882 Act, none were in operation. To rectify previous policy errors, Parliament passed the Electric Lighting Act of 1888.

The new law still required the electricity undertaker (the "utility" in modern parlance) to inform local authorities of its plans, but it also stretched the period of operation before compulsory purchase to 42 years, with a re-opener of the purchase clause every 10 years thereafter. The Board of Trade, though, retained its religious belief in competition as the best means of control. Major F. A. Marindin – Royal Engineer in the Crimean War, member of the Board of Trade Railway Inspectorate and one of the great football referees of the age – chaired a special committee that advised the Board of Trade on how to implement the Act. Its report, issued in 1889, recommended that the Board of Trade give local government authorities preference to operate the local electric utility and license of competing AC and DC systems in the same territory. The committee viewed AC as unproven and unreliable. It could not understand the value of high voltage AC transmission. (West-

---

37. Presentation by Institution of Electrical Engineers to the Board of Trade in 1902. Quoted in Hughes, *op. cit.*, pp. 233-234.
38. Hennessey *op. cit.*, pp. 32-33.

inghouse in the United States and Brown in Germany built their first long distance AC projects in 1891.) It did not anticipate the value of load diversity or economies of scale.

Marindin's committee split the London market into many small territories designed for service by small undertakings that operated low voltage DC systems that could not transmit economically more than a half mile from the power station. Charles P. Sparks argued that:

> The main technical reason for this decision was the unsuitability of the alternating-current system for motive power, although it was obvious from the character of the central area that the demand for motive power could never attain important dimensions in that part of the Metropolis. [39]

Sparks said that the report focused on details instead of "the more important question of supplying London as a whole" and overlooked "the question of system."[40] Marindin had declared that:

> It would... be wisest... to give fair scope to all the proposed systems... and it may be predicted... that whatever system proves itself to be the best under all circumstances will eventually be adopted by all.[41]

To that point, Sparks commented, "The difficulty and expense of change of system was not foreseen." [42] Nor was the loss of business that the undertakings suffered because traction companies whose lines cut through many jurisdictions decided to put in their own power stations rather than deal with a multitude of small electric utilities.

As for regulation of the undertakings, although the select committee taking evidence for the 1888 Act heard testimony in favor of sliding scale regulation similar to that imposed on the gas industry, the Act did not include sliding scale provisions. In 1893, however, a private company buying a municipally owned plant in Chiswick agreed to a sliding scale: it would apply half of any earnings that exceeded a 10% dividend to reducing the price of electricity, a profit sharing arrangement that American regulators adopted over a century later.

After 1898, several firms sought to incorporate as "power companies" to serve large areas. The Special Acts that established these companies provided for sliding scale regulation which allowed a 0.25% dividend increase for every 1.25% that the price fell below a

---

39. Sparks, *op. cit.*, p. 3.
40. *Ibid.*
41. *Ibid.*
42. *Ibid.*

maximum or standard price. Early in the twentieth century, similar sliding scale tariffs became common for local distribution companies.

The Colossus of London

Even before the 1888 Act, the Lindsay family decided to back S.Z. de Ferranti in a venture that would quickly test the Board of Trade's tolerance. They established the London Electric Supply Company in 1887. Its business model was the Gas Light and Coke Co., which supplied northern and eastern London from the gigantic Beckton central gas manufacturing plant.

Ferranti proposed a mammoth undertaking on the banks of the Thames at Deptford. He planned to install 10,000 horsepower (hp) generating units of an untested design operating at 10,000 volts (versus 700 hp units at Grosvenor Gallery operating at 2,500 volts). Ultimately, Ferranti expected total plant capacity to reach 120,000 hp – equal to 89,520 kilowatts (kW). Deptford could power one million lamps, more than all those in London. The backers of the project moved ahead, although an undertaking of this scale and scope did not fit into the narrow confines favored by the government. About a dozen companies applied for franchises to serve parts of London after passage of the 1888 Act. London Electric Supply Co. petitioned to serve a large part of the market. Local authorities that wanted to serve their own areas shut the company out of those markets. The Marindin Committee cut the company's territory, allowed DC competition in it, fretted that AC transmission would interfere with telephone and telegraph circuits and warned of the danger of depending on one large AC station as opposed to many small DC generators. The Marindin Committee had struck a blow for parochialism, for small, inefficient local electricity supply enterprises providing all manner of non-standardized and competing services and for all that made the United Kingdom so electrically backward.

The cautious bureaucrats, though, were right about Deptford. The company took on too many customers before completing the plant, causing overloads at Grosvenor Gallery "culminating in a fire in 1890," according to historian John E. Smart, "which completely destroyed the station. Deptford was still not ready and as a result all supplies to consumers were cut off for three months, inevitably the number of consumers dropped by three-quarters." [43] With a smaller service territory and fewer customers, the directors scrapped Ferranti's plans for 10,000 hp generators. Ferranti resigned. The company had to cope with inadequate water supplies, equipment breakdowns, short weighting of coal from suppliers and faults in the transmission mains. The company never completed Deptford

---

43.   John E. Smart, "The Deptford Letter-books: An Insight on S.Z. de Ferranti's Deptford Power Station," Science Museum Paper, Science Museum, London, 1976, p. 2.

as planned. It did not pay a dividend until 1905. London Electric Supply eventually, however, did become the capitol's major electric supplier.

When Ferranti announced Deptford, the press hailed him as the Edison of England, a Michelangelo of industry. One cartoon depicted him as a Colossus astride the Thames, in classical robes with a light bulb in one hand and a generator in the other. He had grasped the value of economies of scale before others. He realized the value of location for easy access to fuel supply, of choosing the best power plant location and then transmitting the electricity rather than trying to find a generating plant site at the load center. Yet, he chose to furnish single-phase service using reciprocating engines. Parsons, though, had invented the steam turbine in 1884, and Tesla patented a polyphase system in 1888. If Ferranti had completed Deptford according to plan, he might have locked London into obsolete service on a huge scale, for years to come. Edison constructed an entire system. Ferranti attempted a huge insertion into an existing system. Ferranti was less the British Edison and more the electricity industry's equivalent of Isambard Kingdom Brunel, the great Victorian engineer some of whose massive projects failed commercially in part because they did not fit into the market.

With no market integrator or large supplier in sight, London ended up with a hodgepodge of private and governmental systems, offering single phase AC and DC, at numerous frequencies and voltages. Admittedly, DC systems gained few economies from scale, so smallness in itself caused no disadvantages. But, by insisting on competition, customer choice for all aspects of service and priority for local government electricity suppliers, the Board of Trade denied Londoners the savings derived from large-scale, standardized operation.

Outside London

Outside the capitol, municipal governments set up electric undertakings, took over existing facilities and applied for orders to serve (thereby shutting private suppliers out of the market) without providing service afterwards. Bradford, in 1889, built the first municipally owned power station. In 1891, Brighton began to compete with the private supplier, which it bought out in 1894. Municipalities elsewhere followed suit. From the early 1900s through World War I, municipal corporations served two-thirds of the market. In 1894, Joseph Chamberlain argued:

> As it is difficult ... to reconcile the rights and interests of the public with the claims of ... a company seeking ... the largest attainable private gain, it is most

desirable that … the municipality should control the supply, in order that the general interest of the whole population may be the only object pursued. [44]

The big cities owned electricity undertakings that ranked among the largest and most efficient, but they stayed within city limits, leaving large parts of the country either unserved or served by tiny, inefficient suppliers. That gap in service opened up a market for the power company, which would build large scale generating stations and sell its output to customers outside the municipal boundaries, to municipal electricity suppliers seeking a cheaper source than their own small scale generation, and directly to large customers within the boundaries of other electricity undertakings. Manufacturers in Chesterfield established the first such enterprise, the General Power Distribution Co. They planned not only to serve a large area (much of which had no electric supply) but also to compete with established electricity undertakings. The latter, naturally, objected to the plans. General Power sought authorization from Parliament in 1898. The bill failed. The fact that fewer than 2,000 people out of a population of one million had electricity service in the proposed service territory meant less to Parliament than the opposition of municipal governments fearing competition from the new entrant.

In the same year, Parliament set up still another committee to examine the electric industry, this time chaired by Viscount Cross, a veteran Conservative politician. The committee differentiated local distribution from bulk power supply for major customers. It recommended that "where sufficient public advantage is shown", Parliament should approve the formation of power companies that serve wide areas that encompass "numerous local authorities", especially in those instances "involving plant of exceptional dimensions and high voltage".[45] The committee also argued against letting local authorities purchase power company assets after 42 years, because the power companies would operate in a territory that covered many municipalities, so purchase would fragment the service territory. That same issue tilted the government toward nationalization of the entire industry almost four decades later.

In 1900, the House of Commons dealt with five power company bills. Local authorities objected to all of them. Sir James Kitson, an industrialist who chaired the committee examining the bills, proposed a clause for all such bills. The companies could sell power in bulk to existing electric undertakings and directly to consumers in unserved areas, but not directly to consumers in the service territory of an existing undertaking without consent of the undertaking, which could not withhold consent unreasonably. If it did, the Board of Trade could authorize the power company to serve the disputed customers at a lower price than that offered by the local electricity undertaking. The Kitson clause did not placate the municipalities. They pressured local members of Parliament to gut

44. Leslie Hannah, *Electricity before Nationalization* (Baltimore and London: The Johns Hopkins University Press, 1979), p. 23. [Cited as "Hannah" in future references.]

45. _____, *op. cit.*, p. 25.

the four power company bills enacted in 1900 by restricting the activities of the companies and extracting from their territories the large municipalities. Emile Garcke, engineer, member of the same debating club as Fabian socialist Sidney Webb, chronicler of the industry and managing director of British Electric Traction, complained that:

> the character of the undertakings ... has been determined, not by engineering conditions or the interests of the public, but by the parochial spirit and ambition of local authorities ... [46]

Whatever the spirit of the law, local authorities resisted buying from power companies and they withheld permission from the power companies to serve large customers within municipal limits.

After the passage of the 1888 Act, the market for utility-furnished electricity boomed, and while all utilities sold more, the municipal power entities overtook the large private power industry within short order. Local authorities obtained two thirds of the Provisional Orders issued from the passage of the Act through the end of 1895 although they accounted for only one third of sales at that date. With Parliament so hostile to investor-owned power companies, the municipal suppliers took two-thirds of the market by the early 1900s. (See Table 2-1.)

Table 2-1. Electricity Sales by Ownership (1896-1909)

Year	Electricity sold by companies (million kWh)	Electricity sold by municipalities (million kWh)	Total electricity sold (million kWh)	% sold by companies	% sold by municipalities
	[1]	[2]	[3]	[4]	[5]
1896	20.6	9.7	30.2	65	32
1900-1901	56.9	68.8	125.1	45	55
1905	137.1	311.1	448.1	31	69
1909	339.6	615.4	955.0	36	64

Notes:
May not add due to rounding.
Board of Trade Units (BTUs) are kilowatt hours.

Source:
Frederick C. Garrett, ed., *Manual of Electrical Undertakings and Directory of Officials, 1925-1926* (London: Electrical Press Limited, n. d.)
(Garcke's Manual.)

---

46. Hannah, *op. cit.*, p. 27.

Northeastern Enterprise

In the industrialized northeast, on the Tyneside, the home of Swan and Parsons, the business took a different course. Newcastle, unlike most large cities, had no municipal electricity undertaking.

Parsons installed the first utility turbine in 1890, at his own enterprise, the Newcastle and District Electric Lighting Co. In the same year, J. Theodore Merz – local businessman well connected to the important families in the area, director of British Thomson-Houston (an electrical equipment manufacturer), and historian as well (author of *The History of European Thought in the Nineteenth Century*) – established the Newcastle-upon-Tyne Electric Supply Co. (NESCO). The third utility in the area, Walker and Wallsend Union Gas Co., also supplied electricity. The shipbuilding Richardson family (to which Merz' wife belonged) controlled Walker and Wallsend. By 1898, Merz viewed NESCO – with its small single-phase AC generator and minimal service territory – as an obsolete enterprise. He wanted NESCO to operate a large generating station located away from congested areas and on the river for easy fuel access. Merz, with his family connections, also saw the business opportunities in serving industrial load.

In 1899, Walker & Wallsend Gas hired J. Theodore Merz's son, Charles (an engineer who had supervised the electrification of the Irish city of Cork) to design and construct a power station that would serve industrial load. To do so, he formed a partnership with William McLellan that became one of the industry's leading engineering firms. The great physicist Lord Kelvin opened Merz and McLellan's Neptune Bank station – the country's first three phase AC generating facility – in 1901. After that, NESCO took advantage of the power company bills, applied to serve the Tyneside region, beat out competing applicants, and bought the Neptune Bank station.

Charles Merz systematically developed technology to create the regional system that his father had envisioned. First, he had to select the generation frequency for Neptune Bank. The most common frequencies then used were 25 hertz (Hz), excellent for motors but it caused lights to flicker, or 100 Hz, good for lighting but hard on the motors of the day.[47] Merz decided to standardize all operations at 40 Hz. He had to garner customers for his system. The North Eastern Railway (Sir James Kitson being one of its directors) was one of his big wins. By 1904, NESCO completed another generating station, , a startlingly innovative design by Merz and McLellan. The steel-framed building allowed for future expansion. The four turbines ran independently of each other. Two units had 1,500 kW capacity each, and the other two 3,500 kW each. The two big units were twice the size of any installed to date. Merz and McLellan placed a control room for the entire NESCO system

---

47.   Hertz means cycles per second, that is: how many times the alternate current reverses per second, in plain English.

in Carville, another first for Britain. Carville became the lowest cost power generator in the country.

Merz and McClellan not only designed power stations but also rounded up customers for NESCO, convinced them to switch from steam to electricity, showed them how to convert to electricity, and designed and installed the electrical equipment that the customer needed. NESCO was part of Newcastle's industrial establishment. It could close the deals.

NESCO expanded by purchasing neighboring power companies and connecting isolated power sources. In 1905, NESCO put in service 22 kilovolt (kV) transmission lines, the highest voltage in the United Kingdom. Beginning in 1907, NESCO took electricity generated from the waste heat of coke ovens and blast furnaces. Charles Merz had created the first – and until nationalization the only – efficient, modern, regional electric system in the United Kingdom.

The Market for Electricity

First electricity suppliers sold lighting, an expensive product. Industrial and transportation firms generated their own electricity, to the extent that they used it at all. After the turn of the century, industrial and transportation customers began buy from the utility, as power perhaps because the utilities began to see the wisdom of charging a price that attracted those customers that took the electricity at times other than the peak lighting period. By 1909, industrial (power) and transportation (traction) accounted for a combined total of 62% of sales. (See Table 2-2.)

Table 2-2. Utility Sales Breakdown by Customer Class (1895-1909)

Year	Electric utility sales (million kWh)	Lighting % of sales	Power % of sales	Traction % of sales
	[1]	[2]	[3]	[4]
1895	37.8	100	0	0
1896	52.3	100	0	0
1900	180.4	84	7	9
1905	646.6	55	18	27
1907	956.0	44	33	23
1909	1,123.4	38	42	20

Notes and sources (columns):
1. Leslie Hannah, *Electricity before Nationalization* (Baltimore: Johns Hopkins University Press, 1979), pp. 427-428, after Byatt.
2-4. I.C.R. Byatt, *The British Electrical Industry 1875-1914* (Oxford: Clarendon Press, 1979), p. 98.

In the early 1880s, lighting accounted for most electricity sales although electric light was several times more expensive than gas light. The industry had no other product to sell and it did not sell that product on the basis of price. Development of the incandescent gas mantle later in the decade maintained gas light's competitive edge even as the price of electricity declined. Electricity in 1904 probably captured less than 20% of the lighting market, becoming price competitive with gas only after the invention of the tungsten filament lamp in 1911.

Electrification of transportation started slowly. By 1896, only 80 miles of railway and tramway (trolley) track had been electrified (vs 12,000 miles in the United States). In that year Emile Garcke established British Electric Traction (BET) to purchase and electrify tramways. By the early 1900s BET controlled over half the market not operated by municipal transit agencies. Private operators, though, generally avoided electrification, for fear that they could not recover their investment before the end of the 21 year franchise period, after which the local government could buy them out. For that reason, municipal authorities accounted for the bulk of the doubling of electrified mileage (from 1,200 miles to 2,400 miles) between 1900 and 1907. Ridership rose dramatically, too, and the transit industry's purchases of electricity sextupled.

The London Underground system began operation in 1863, powered, improbably, by steam engines. Parliament, in 1884, rejected the formation of an electric powered line. Later, the electrified City and South London line earned poor returns for its backers. Early in the twentieth century, the Americans moved in. General Electric built the Central London Railway. Charles Tyson Yerkes, the Chicago street car magnate who ran into trouble for buying votes of Illinois legislators, formed the Metropolitan District Electric Traction Company. J.P. Morgan almost established another line. The Americans over expanded the system, lost passengers to the newly developed motorbus, and earned poor returns on their investments too.

Suburban and main lines electrified in the early 1900s too, but competition from electric tramways cut into their business. When a suburban building boom ended, the urge to electrify rail transportation expired.

The electricity undertakings had a difficult time marketing to industry. Their sales people had to make the industrialist understand how electricity differed from older sources of energy. Before electricity, a central source of energy (a mill or steam engine) turned a shaft. Belts attached to it transferred circular motion to individual machines. The machinery operator could not start work until the factory's engineers raised steam to turn the shaft. Given that all power came off that central shaft, the factory had to energize the entire central power system, even to perform small tasks. When the shaft stopped turning, all machinery stopped.

Initially, factory owners hoped to save money by replacing the steam engine turning the central shaft with an electric motor. That idea, though, produced small savings. Electricity's advantage lay in its ability to power many or few machines, separately, without need for a central shaft, at variable speeds, or at a constant speed with a changing load. Using electricity led to improved factory layout, measurement and control of production. Engineering, transportation equipment and shipbuilding industries moved first to electrify and purchase from the local electricity supplier. Not coincidentally, many of those firms operated in the northeast. After a slow start, industrial sales rose fivefold from 1900 to 1909, by which year NESCO accounted for 12% of total and 23% of industrial electricity sales in the UK.

Within a decade, electricity suppliers metamorphosed from lighting companies with incidental industrial and transportation loads to suppliers to industry that also sold to lighting and transportation customers.

Pricing the Product

Electricity suppliers had to price their product competitively. In early years, consumers wanted lighting. Electric lighting cost twice as much as gas lighting in the late 1880s, only reaching parity in the early 1900s. As long as the industry marketed electricity as a luxury good, price did not matter. To achieve economies of scale, though, electricity suppliers had to reduce price in order to penetrate the mass market. (See Table 2-3.)

Table 2-3. Gas vs Electric Lighting Costs (1888-1909) (d/hour)

Year	Gas lighting price [1]	Electric lighting price [2]
1888	0.23	0.45
1895	0.23	0.33
1900	0.23	0.28
1905	0.21	0.22
1909	0.23	0.22

Note:

The British monetary system divided the pound (£) into 20 shillings (s). The shilling, in turn, was divided into 12 pence (d). Approximate cost of 16 candlepower (the equivalent of a 60 Watt bulb) per hour.

Sources:

Byatt, *op. cit.*, p. 24.

Hannah, *op. cit.*, p. 429.

B. R. Mitchell, (with Phyllis Deane), *op. cit.*, , p. 370.

Crompton, Swan's partner, acted as consulting engineer to electric utilities and ran his own electricity undertaking in London. In 1891, he developed the concept of the load factor, the average load on the system as a percentage of the maximum load. Crompton calculated the load factor of the system at 10%, meaning that for all practical purposes most of the equipment stood idle for most of the day, so any additional sales made during slack periods would add to profits because the company would not have to make any additional investments to meet that demand. Lighting already accounted for most of the demand. Crompton decided to promote other uses of electricity, such as heating and cooking. His manufacturing firm introduced an electric oven in 1891. To peddle the culinary innovation, he helped to establish Miss Fairclough's School of Electric Cookery in London in 1894.

Crompton, in lectures, took a less aggressive stance on load factor than his American counterparts who "were happy in having load-factors much higher than had hitherto obtained in England".[48] He implied that they might have to deal with higher maintenance and repair costs because they ran their machinery so much. He also explained to his fellow engineers why the steady tram and industrial loads were cheaper to serve (a fact more apparent in US than UK pricing). When discussing the savings achieved by operating at a higher load factor, he omitted capital costs because "they did not much interest engineers." [49] Yet those savings were what mattered, and if the engineers did not understand them, did they understand the economics of the electricity industry?

The City of London Electric Lighting Co. sold or rented ovens to consumers, installed a separate oven circuit for them, and priced electricity for ovens at half the lighting price. That company operated an AC system. It could not store electricity generated during off peak periods for use at peak periods. Consequently, it had idle plant available at off peak periods and would benefit if it could find customers to take electricity from that plant at any price that would cover production (variable) costs plus contribute to overhead. Crompton preferred DC, because he could install fewer generators, run them all the time at full output, and store excess production in storage batteries until needed at peak periods.

The electricity suppliers had a complicated pricing problem. First, the tariff had to apportion the cost of providing service between fixed costs (determined by the need to have enough plant in place to meet to meet peak loads) and variable costs (determined by output). Second, the utility had to determine the revenue it required (price per unit times the volumes of units sold) that would to cover the total cost of providing the electricity.

John Hopkinson, one of the leading electrical engineering academics of the time, devised a solution that revolutionized the business. He put it simply in his address, "The Cost of Electricity Supply" to the Junior Engineering Society in 1892:

48. Institute of Civil Engineers, "Discussion of Electrical Energy", *Proceedings*, Vol. CVL, p. 33.
49. _____, *op. cit.*, p. 35.

> The ideal method of charge … is a fixed charge per quarter proportioned to the greatest rate of supply the consumer will ever take and a charge by meter for the actual consumption. Such a method I urged in 1883 and obtained the introduction into certain Provisional Orders of a clause sanctioning "a charge which is calculated partly by the quantity of energy contained in the supply and partly by a yearly or other rental depending upon the maximum strength of the current required to be supplied."[50]

Hopkinson proposed a two-part tariff. One part, to pay "standing" (or fixed) costs, covered those costs required to install and maintain the utility's plant, costs which did not vary with level of output (in other words, cost of capital, maintenance and labor). The second part covered "running" (or variable) costs that rose and fell with level of output (primarily fuel).

Arthur Wright, the Brighton electricity utility's engineer, after reading Hopkinson's paper, developed a Hopkinson tariff for Brighton. The utility would measure maximum demand by meter. Within five years, one quarter of the utilities used Wright's tariff concept. Samuel Insull, the Chicago utility tycoon, when on vacation in his native England, spotted the Wright tariff, which he brought back to the United States.

In order to determine the total revenue truly required to cover all costs, the utilities needed an accounting system that counted all costs, and they did not have one. As a starter, the companies often earned returns below the maximum set so they did not cover cost of capital. Possibly more importantly, the utilities in the United Kingdom (in common with those in the United States) had a cavalier attitude toward accounting for depreciation, the loss of value of plant and equipment over time due to wear and tear and obsolescence. Not only did companies face loss of value from wear and tear, they also might have to sell their plant and equipment at the end of their franchises to the local municipalities at a price determined at time of sale. Would the buyers pay the original cost of plant and equipment, or more likely a lower price that reflected age and condition of assets? The income statement should have shown an expense that reflected the annual decline in value of the company's assets, rather than maintaining a pretense that value remained unchanged until that last moment when the company scrapped the old equipment or sold it at fire sale prices to the municipality. It often did not.

The Joint Stock Companies Registration and Regulation Act of 1844 required companies to "declare dividends out of profits," the Companies Consolidation Act of 1845 prohibited payment of dividends that would reduce capital stock, the Joint Stock Companies Act of 1856 said companies should pay dividends out of profits, and the Companies Act of 1862 provided model articles of association that prohibited dividends payments except

---

50. James Greig, *John Hopkinson, Electrical Engineer* (London: Her Majesty's Stationery Office, 1970), p. 17.

from profits, and authorized directors to set aside a reserve for "repairing and maintaining the works…"[51] Accounting historian Maxwell Aiken described the "doctrine of capital maintenance" [52] as a means to set aside funds to protect creditors, that is to prevent the company from paying out dividends that would leave the company without sufficient funds to pay creditors later on.

In February 1889, however, Lord Justice Lindley of the English Court of Appeal handed down a decision, *Lee v. Neuchatel Asphalte Company,* which allowed the quarry in the case "to omit depletion … when calculating profits available for dividend distribution…"[53] (Depletion of a natural asset is akin to depreciation of plant and equipment.) That precedent ended the capital maintenance principle. Directors could ignore the depreciation of assets when calculating the income available to pay dividends. They could kick the can down the road, so to speak, pay out all the cash earnings as dividends without regard to whether the company could pay its debts or fund the purchase replacement assets in the future.

In 1896-1906, on average, consumers paid about 3.9 d per kWh and utilities had 29.0 d of capital invested per kWh sold. After paying for fuel and other direct expenses, utilities had 1.5 d per kWh left to pay those who provided capital (interest to creditors and dividends to shareholders), a 5.2% return on capital. Considering that risk free government bonds at the time yielded about 3%, investors collected a barely reasonable premium over risk free returns for putting money into electric utilities.

Most utilities of the day paid out roughly 5% on capital but did not really earn 5% because they calculated profit without deducting depreciation. Assume that plant and equipment had a lifetime of 40 years, an optimistic assumption but close to the maximum duration of a franchise. That asset life produces a depreciation rate of 2.5% per year, which would subtract about 0.7 d per kWh from the operating income figure which is used to provide a return to the owners and creditors of the utility (assuming that the utility in question made no provisions whatsoever for depreciation expense). The utility, then really earned a true return of 2.8%, less than what investors could earn on government bonds. In the case of a utility that had borrowed part of its capital and had preferred stock outstanding, there would have been nothing left for the common stockholder if realistic accounting procedures had been used. (See Table 2-4.)

51.　Maxwell Aiken, "An accounting history of capital maintenance: legal precedents for managerial autonomy in United Kingdom", *Accounting Historians Journal*, June 1, 2005, p. 4.

52.　_____, *op. cit.*, p. 1.

53.　_____, *op. cit.*, p. 2.

Table 2-4. Average Revenue, Cost and Investment per kWh (1896-1906)

Revenues and Expenses (d/kWh)

	Revenues collected from customers	3.9
less	Fuel costs	0.6
less	Other direct operating costs	1.8
equals	Operating income used to pay dividends to shareholders and interest to creditors	1.5
less	Estimated depreciation expense	0.7
equals	True operating income	0.8

Analysis of Operating Income (d/kWh)

	Reported operating income	1.5
less	Interest expense	0.9
equals	Net income	0.6
less	Preferred dividends	0.1
equals	Net income for common stock	0.5
less	Estimated depreciation expense	0.7
equals	True net income for common stock	-0.2

Capital and Returns

Capital invested (d/kWh)	29.0
Operating income return on capital invested (%)	5.2
True operating income return on capital invested (%)	2.8

Note:
Total electric industry. Estimates by author.

Sources:
Garcke's Manuals
Byatt, *op. cit.*, p. 129.

By modern accounting standards, early electric utilities did not earn cost of capital. Byatt noted that "By 1910 there was a general feeling that electric power tariffs were below cost..."[54] And they certainly looked that way, assuming that investors should earn returns commensurate with risk taken (more than government bonds) and that companies should pay dividends out of income not capital. Did the utilities charge too little, or did someone syphon off the profits, perhaps the electrical equipment, engineering or banking firms that sold their services and equipment to the utilities, which they often controlled or had arrangements with those who did? They could have milked profits from utilities by overcharging for services or equipment, leaving the other shareholders with low profits. Thus, unnecessarily high expenses rather than unduly low prices may account for the low profits, but either way ordinary utility shareholders lost out.

---

54. Byatt, *op. cit.*, p. 135.

The admittedly inadequate numbers show an industry that earned low returns and paid out more in dividends than it really earned. It was a capital intensive industry that had to raise more and more capital not only to buy new equipment but also to pay returns to existing shareholders, sort of an industrial Ponzi scheme. Yet regulated utilities can run that way for decades (American utilities for over a century with few failures) as long as customers have no choice but to buy from them and regulators can assure them a steady flow of income. The problems come when consumers no longer want the product or when the capital markets no longer have enough faith in the system to finance sale of new debt to pay off old debt.

Hopkinson and Wright's ideas about accounting for costs and allocating them to particular customers encouraged the electricity suppliers to view their markets in a different light. The industry began to price more aggressively in order to attract non-peak (meaning non-lighting) customers. By 1900, tram enterprises paid roughly one-third the lighting price. Daytime industrial customers paid low prices too. Pricing policy encouraged enough off peak demand to raise the industry's load factor to 23% in 1907. (NESCO, with its enormous industrial and transportation loads had a load factor double the national average.)

The engineers that ran the electric companies thought that they had finally figured out how to allocate the total costs between customer groups. Unfortunately, the accountants who kept the books did not correctly calculate the total costs thanks in part to their omission of depreciation from the cost calculation, and the engineers might not have gotten it right, either, to the extent that they did not understand capital costs. Whether right or wrong, however, the electric companies started to realize that price was a powerful marketing tool.

London Revisited

At the turn of the century, England's electrical engineers deplored London's electricity system as a backward embarrassment with its dozens of suppliers, a multitude of voltages and frequencies, AC and DC systems, old equipment, high costs and politically determined boundaries. True, London lacked big industrial loads, and the transit companies generated much of their own electricity, which reduced diversity of load and economies of scale for the utilities. Yet, utilities that wanted that additional load might have attracted it with the right inducements.

The London County Council (LCC) and the local councils agreed that the LCC could supply electricity in a particular area if requested by the local council, but could not compete with local authorities. The policies of the LCC, though, changed with its political majority. The Conservatives disliked municipal enterprise and favored private power companies. The Liberals, as staunch advocates of municipal trading, opposed private monopolies. In

1902-1903, the LCC tried, but failed to get legislation out of Parliament that would enable the LCC and the local authorities to work together in the purchase of electricity suppliers.

When Charles Merz decided to enter the London market in 1904, he gathered backers from the Tyneside industrial establishment to establish a company that would apply the principles developed at NESCO to London's market. Merz and his investors established the Administrative County of London and District Electric Power Company. (Merz, when naming the company, had noticed that Parliament took up power bills in alphabetical order.) Merz and McLellan searched for the factories in London that might take electricity from the new enterprise, surveyed for potential customers, determined savings possible from consolidation of electricity undertakings, proposed the configuration of a new system and then showed how badly London compared to other cities: annual per capita usage there was 42 kWh, versus 343 kWh in Boston, 282 kWh in New York and 198 kWh in Chicago.

At that time, 39 undertakings with a combined capacity of 202,000 kW served London. Merz proposed to build three riverside generating stations with a combined capacity of 201,000 kW, employing larger turbines than ever previously installed, with capital costs per kW half those of the normal power station and running costs one-third or less those of the typical power plant. Merz in no way intended to replicate the Deptford fiasco. NESCO had successfully installed large turbines, and it would install record-size units at its next power station. Merz, furthermore, had a business plan. With those expected low costs he could snare the traction load (then largely self generated) and build up the small manufacturing load, so far discouraged by high prices.

The three stations would aim to sell to existing utilities in bulk, to large transportation and industrial customers, and up to one-fifth of output to individual customers.

The new company proposed to charge 0.75 d/kWh on average and a maximum price of 1.5 d, a number probably close to the fully allocated cost of generation (including cost of capital) in London at the time. It would pay a sliding scale dividend depending on price of electricity, set at 8% for the baseline 0.75 d price. If the price of electricity exceeded 1 d, the dividend stopped altogether.

Merz made a clear distinction between lighting and the power business. He said the profitable, dividend-paying, monopolistic lighting company could not succeed in the power business:

> It is a special business; it will not be a monopoly … because there are … other forms of power. It … requires commercial push. The electric power has to be sold … [55]

---

55. Hannah, *op. cit.*, p. 46.

The LCC opposed Merz' bill because his new company would have thwarted its electrical ambitions. Municipal governments throughout the United Kingdom opposed the bill because it threatened municipal socialism, not to mention London's many government-owned electricity undertakings. Other power companies objected, too. Competition from a low-cost supplier, after all, threatened the value of an investment made as a result of legislation, and J.F. Remnant, speaking for the power companies, said that passage of Merz' bill:

> would enable it to be said that businessmen could no longer rely on Parliament to protect the interest, which it had … brought into being...[56]

Parliament had faced a similar problem in the days of canals. When it chartered a new canal it usually compensated the owners of older canals nearby by allowing them to levy a charge on traffic that passed through the junction between the new and old canal. These charges made the new canals less profitable and slowed the expansion of the canal network. Initially, newly chartered railroads had to compensate canal owners for the competition, but Parliament finally gave up on protecting obsolescence. [57]

Taking the estimates at face value, replacing the old power stations with the new ones could have reduced the average delivered price of electricity to all London electricity consumers by close to 25%, and more to the targeted customer group. But, again, the fairness issue arose. Owners of the old generators, when investing their money, agreed to accept a limited profit in return for protection from competition and confiscation during a specified period of time. The new generators envisioned by Merz, could have put them out of business, unless the London market doubled in size as a result of the lower prices and aggressive development of prospects. Given how fast the market was growing at that time (generation more than quadrupled between 1902 and 1907) and the fact that Merz aimed to sell his output to an under served market, he might have pulled off his expansion without exterminating the existing generators, but why would they want to take that sort of chance if they could get protection from the new competitor instead? At some point, though, perhaps the government must say to incumbent regulated monopolists, "You took a business risk because you assumed no change in technology and in economic environment. The franchise monopoly cannot protect you against such changes. You made a bet that conditions would remain unchanged, and you lost the bet. Sorry."

Yet, if the government abrogated franchises before the owners could recover their investments, then investors in future franchises would demand higher returns to compensate for the risk of early termination without recompense. The benefits of a 25% price reduction to customers surely outweighed an increase in cost of capital for future projects. Yet, in the early 1900s, cost of capital, calculated using twenty-first century methodology,

---

56. *Ibid.*
57. Charles Hadfield, *The Canal Age* (NY: Frederick A. Praeger, 1969), p. 152.

accounted for roughly 30% of the cost of electricity to consumers. Raising risk on capital investment could have boosted cost of capital enough to significantly affect the price of electricity, eating up a large part of savings that a new generator would have brought to the sector, and the new generator would have to take the risk that an even lower cost generator could knock it out of the picture in the future.

Conceivably, the new generating entity, with permission of regulators, could have paid off the old generators, using the savings produced by the new equipment to make the payments. (After all, that is what American regulators did, supposedly, almost 90 years later, when they forced American consumers to make whole the legacy utilities during the deregulation process.) So, why not have the new generator borrow the money to buy out the old generators and pay off the loan over 20 years? Assuming that sales remain unchanged, the answer is that the consumer might not see any benefit from the lower cost generation for close to 10 years, when the cost savings would, finally, exceed the payments to the legacy generators. It seems unlikely, moreover, that a commercial entity would want to sign on to a long term liability that could sink it if lower cost generators entered the market.

The real problem is that the Merz proposal would have changed the dynamics of the market place in an unknown way, perhaps bringing about a huge expansion of demand that would help all suppliers, perhaps threatening incumbents, certainly bringing short term benefits to consumers. When faced with such uncertainties, politicians and entrenched interests prefer predictable outcomes over opportunities.

Gas companies fought the Merz bill, too. Electric lighting prices had become competitive with gaslight prices. An electric company operating with the costs Merz had projected would have devastated the gas companies. The bill did not make it through Parliament before the Conservative government fell. The backers reintroduced it in 1906. The governing Liberals, though, had even less taste for it than the Tories. David Lloyd George, then President of the Board of Trade, dismissed Merz's proposal as "hardly within the region of practical politics." [58] Lloyd George had a different plan. He wanted private companies to fund the risky up-front investments. After the private investors had made a profit, the LCC would buy them out. While the Liberals worked out their ideas, the Municipal Reform Party won control of the LCC , and it opposed municipal trading, effectively burying Lloyd George's plan.

---

58.   Hannah, *op. cit.*, p. 47.

Conclusion

A quarter-century after inception, the British electric industry still offered high-priced, non-standardized services. Politicians valued parochial interests, the virtues of municipal ownership and the appearance of competition over technological improvements, economies of scale or low prices to consumers. They disliked business initiative. They thought in static terms unsuited for a rapidly evolving market. They feared disruption more than they appreciated opportunity. Electric industries in other industrialized states had moved forward, adopting new concepts of service. It is not as if the British had to invent anything to move forward.

Did the British electric industry suffer from a shortage of dynamic business leaders? Is this another case of letting others seize opportunity because trade is something gentlemen do not do? An industry may have few bigger than life leaders, but only one is required to force a breakthrough. There was only one Henry Ford. Merz and his Tyneside compatriots had plenty of big ideas and the wherewithal to execute, but British politicians stifled them.

Chapter 3
Searching for Structure (1909-1926)

The most conspicuous feature of this phase of our electrical history, consonant with the political and legislative environment, was the multiplicity of small generating stations, scattered and isolated throughout the country, supplying small and restricted areas, and each pursuing the uneconomic policy of independent development. [59]

— *Garcke's Manual* (1946-1947)

Although engineers, consumers and politicians knew – or should have known – that the electric industry operated inefficiently, to the public's loss, British policy makers had to hear even more embarrassing international comparisons and commission even more government studies to arrive at the obvious conclusion: that the industry had foregone beneficial economies of scale in order to maintain the dubious benefits of local control and multiple choice.

Before the War

Finally Parliament passed the Electricity Act of 1909 that made small changes in the law:

- Municipalities could jointly purchase power companies at the end of the 42 year franchise.
- Power companies could purchase land on a compulsory basis.
- Municipal suppliers could purchase electricity in bulk from other municipal suppliers without Parliamentary consent.

Municipal undertakings gained ground after the passage of the law. They sold to outlying areas. They enjoyed load diversity, thanks to bulk sales made to their sister agencies, the municipal tram lines. They had customers who consumed more electricity per connection. They achieved economies of scale reached by few power companies (with the notable exception of NESCO) and, consequently, they operated with lower costs and prices. The municipal utilities retained a cost advantage through the end of World War I, after which the companies managed to boost load factors and bring costs down to the municipal levels. (See Table 3-1.)

---

59. Frederick G. Garrett, ed., *Manual of Electrical Undertakings, 1946-1947 Edition*, (London: Electrical Press, 1947), p. 8.

Table 3-1. Municipal (Local Authority) vs Company (1909-1925)

	1909 [1]	1918 [2]	1925 [3]
Municipal % of total kWh sales	**64**	**70**	**64**
Average revenue per kWh (d)			
Municipal	2.02	1.35	1.59
Company	2.71	2.32	1.54
Average cost per kWh (d)			
Municipal	1.00	0.84	0.91
Company	1.45	1.56	0.84
Average sales per utility (MWh)			
Municipal	2,304	6,838	11,280
Company	1,806	3,033	8,149
Average connections per utility (thousands)			
Municipal	87	247	599
Company	109	201	356
Average maximum load per utility (kW)			
Municipal	1,890	4,269	5,852
Company	1,967	2,658	4,191
Average generating capacity per utility (kW)			
Municipal	3,142	6,797	9,278
Company	3,739	3,477	5,927
Average capacity per generating station (kW)			
Municipal	2,776	5,904	10,718
Company	2,701	3,513	6,785
Average load factor per utility (%)			
Municipal	14	18	22
Company	10	13	22

Notes:

1. Years — columns 2 and 3 are fiscal years 1918-1919 and 1925-26. 1909 chosen as year of significant electricity legislation, 1918-1919 as year when municipal utilities were at their peak, and 1925-26 as year of next major law.
2. Number of utilities varies by category. 1909 and 1918-1919 from sample collected by Garcke. 1925-1926 reflect more complete data collected by Electricity Commissioners, unless stated otherwise.
3. "Working expenses" – operating expenses only. Excludes depreciation, bad debt and other charges and interest.
4. "Connection" – 30 Watt equivalent connection. Total number of customers for all utility undertakings may have approximated 300,000 in 1909, 700,000 in 1918-1919 and 2,000,000 in 1925-26. Garcke sample data for all years.
5. "Load factor" defined as average load (total kWh sales divided by 8,760 hours in year) as a percentage of maximum load.

Source:

Garcke's Manual, 1925-1926.

In 1912, only 16 electricity suppliers had 10 megawatts (MW) or more of generating capacity, two years after NESCO had demonstrated the economic virtues and workability of the 10 MW generating unit. Only a handful (out of 478) of undertakings, then, could attain economies of scale. NESCO, with 91 MW, was the largest operator. Municipal operations held the next five slots in descending order of size. Power companies produced less than one third of the nation's output. With the exception of NESCO, they generally charged higher prices than the municipal systems. Investor-owned companies tended to expand by setting up holding companies that owned scattered power companies and undertakings. Manufacturers, bankers, engineering firms and foreign entrepreneurs often put together these operations in order to profit from sales of equipment and services to them and from sale of company securities to the public. Expanding the electricity market, lowering costs or promoting technological development were secondary motives for the organizers and promoters. The municipal power utilities had a simpler goal, to provide reliable and reasonably priced power to local citizenry, and despite lack of profit incentives and all the defects of socialism, they managed to do so.

Reformers pointed out that the British used less electricity per capita than inhabitants of other industrialized nations. (The French and Italians used even less, but that inconvenient fact would have blunted the argument.) British industrial electrification lagged and self-generation grew as rapidly as public supply, meaning that industrialists who saw the virtues of electricity often decided to make it themselves. In 1913, Samuel Insull of Chicago's electric utility, Commonwealth Edison, and George Klingenberg of Germany's giant manufacturing and utility combine, Allgemeine Elektricitäts-Gesellschaft (AEG) joined with NESCO's Charles Merz, to testify to the backwardness of the United Kingdom's industry. They produced a devastating comparison between London, Berlin and Chicago. London had too many small power stations, they cost too much to build and they burned coal inefficiently. The electric companies had failed to develop non-lighting loads and, as a result, they could not fully utilize the power stations. Due to those failings, the trio of industrialists concluded, Londoners paid too much for electricity. (See Table 3-2.)

Table 3-2. London, Berlin and Chicago (1910-1912)

	London [1]	Berlin [2]	Chicago [3]
Population (millions)	6.5	2.6	2.2
Number of power stations	64	6	6
Installed capacity (MW)	298	137	222
Average size of station (MW)	4.7	23.0	37.0
Peak load (MW)	186	95	199
Sales (GWh)	319	216	640
Load factor (%)	25	33	41
Sales by customer class (%)			
Lighting	61	24	19
Power	27	45	12
Traction	12	31	69
Total	100	100	100
Reserve margin (%)	60	44	12
Coal consumed per kWh sold (kg)	2.37	1.38	1.61
Price per kWh sold (s)	0.195	0.159	0.090
Operating expense per kWh sold (s)	0.087	0.081	0.045
Real value per kW installed (s)	662	356	477

Source:

Thomas P. Hughes, *Networks of Power* (Baltimore: Johns Hopkins University Press, 1983), derived from Figure ix. 9, p. 258.

Insull produced a table. The inset in it said, "As the output per capita increases, the load factor improves and the income (and cost) per k.w.h. decreases". [60] Of ten cities he examined, London had the lowest electrical output per capita, and higher prices than all cities except Boston and Brooklyn. Merz and McLellan inventoried London's electricity systems to find:

65 electricity undertakings
70 power plants
49 types of supply systems
32 transmission voltages
24 distribution voltages
70 pricing methods

---

60. Hughes, *op. cit.*, p. 259.

Thomas Hughes noted that:

> Londoners ... toasted bread in the morning with one kind, lit their offices with another, visited associates in nearby buildings using still another variety, and walked home along streets that were illuminated by yet another kind. [61]

Customers moving to a new home often found that their old appliances did not work in the new house because of different voltages or frequencies in the new location. Appliance manufacturers could not mass-produce their products. Instead they had to make small batches for each subdivision of the market. The utilities had to install multiple sets of wires with different circuits for different purposes (cooking, heating or lighting). Traction companies owned and operated power stations that ran when the utility's power stations had idle capacity that the utility could have sold to the tram operator and the tramway's generating station stood idle at times when the utility needed all the generation that it could find. Manufacturers of generating equipment did not produce large, efficient, standard sets because few electricity undertakings operated at a scale large enough to buy the big equipment. Furthermore, a legion of individualistic engineers advising those undertakings guaranteed that manufacturers could never have come up with standard equipment to satisfy everyone.

In addition, electrical undertakings eschewed interconnection – not that interconnection would have been easy, given the multiplicity of frequencies and voltages used, plus the AC vs DC problem, as well. Interconnections between utilities, though, would have served three purposes:

- Load diversity – Utilities that had their busiest (peak) periods at different times could sell electricity to each other when their own customers did not need it, thereby using generating plant that would have stood idle and avoiding construction of redundant facilities, and lowering costs to customers.

- Reliability – By tying their systems to other utilities, undertakings could keep less idle plant reserved for emergencies (reserves) because they could draw on their neighbors in time of emergencies.

- Economy – Interconnections would permit utilities to tap into cheaper power sources at any time, rather than to depend solely on their own stations.

Unfortunately, those benefits did not mesh with public policy – especially in London – where officials viewed size with suspicion, monopoly as bad, and private monopoly even worse. Cooperation might lead to monopoly. The utility had to tailor output to individual needs, rather than offer a low cost standard service that would serve more customers economically. Competition was good, but competition against an entrenched interest

---

61.   _____, *op. cit.*, p. 227.

(especially if municipally owned) was bad. To compound difficulties and further stifle innovation, every commercial venture required an Act of Parliament before it could go into business.

The industry's technology evolved to favor suppliers big enough to take advantage of economies of scale. In other countries large, monopolistic suppliers operated under government supervision. Politicians in the United Kingdom, however, abhorred monopolists and ignored the value of scale. At the behest of the LCC, Merz & McLellan developed still another reform proposal for London in 1914. The LCC again refused to back a coordination scheme for London's electric supply. The nation entered World War I with an inefficient, poorly interconnected electric utility industry. Industry and transport self-generated close to two-thirds of electricity needs, buying the balance from public suppliers. A small percentage of homes had electric service.

World War I

The Gun Ammunition Department of Lloyd George's Ministry of Munitions took charge of electricity policy. George McLellan (Merz's partner) acted as departmental advisor. McLellan foresaw shortages of fuel and generating capacity. He required industrial firms to buy power from the local utility rather than erect private generators, except where they could use waste heat from generation in industrial processes. The government encouraged the electrical undertakings to install new, large scale generating units, sometimes advancing financial aid and even providing construction plans. It encouraged interconnections and standardized AC service. During the war, electricity output doubled. Generating capacity rose by more than one half.

As President of the Board of Trade, Lloyd George helped bury plans to improve electric utility efficiency. As Prime Minister, late in the war, he asked his Reconstruction Committee to study the industry. The Coal Conservation subcommittee chaired by Lord Haldane took charge. Haldane, a Scottish Liberal member of Parliament who later joined Labor, was one of the most remarkable politicians of his day. He co-founded the London School of Economics, wrote on the theory of relativity, promoted aviation in its early days, and most importantly, reorganized the armed forces before World War I. In pre-war days, however, he spoke about Germany's "search for system" and "the capacity it has developed for organisation," and a British press campaign about his supposedly pro-German leanings temporarily shoved him aside.[62] Haldane exercised his organizational talent by entrusting the technical study to Charles Merz who, predictably, wanted to revamp the industry by forming sixteen generating entities that would operate large, efficient, interconnected

---

62. "Lord Haldane on the Germans", *Grey River Argus*, 3 Whiringa-ā-nuku, 1911, p.8.

power stations, run transmission and power sales functions, and leave to the hundreds of existing undertakings the low voltage distribution business.

The Merz report came out in April 1917. The Board of Trade launched its own investigations. Sir Charles Parsons, inventor of the steam turbine, chaired a committee to investigate the electric equipment manufacturing industry. It concluded that manufacturers, in order to successfully compete with the Germans after the war, had to achieve greater size and produce standardized products, but they could not as long as the UK electricity supply industry remained fragmented.

Then, another committee, chaired by Sir Archibald Williamson, another influential Scottish Liberal MP, (with Merz as a member), argued for the establishment of an Electricity Commission to supervise the industry and the formation of non-profit District Electricity Boards to generate all electricity and operate the regional transmission networks. The local undertakings would continue to handle low voltage distribution. That report came out in April 1918.

Next year, the Ministry of Reconstruction set up still another committee. Chaired by Sir Henry Birchenough, a silk manufacturer who had become active in African affairs, that panel proposed a "nationally owned supply network … a single unified system under state control."[63] It feared interference from local governments and pressed for a state corporation run on business lines.

Thus, at the end of the war, a multiplicity of committees had come to the obvious and already old conclusion that the electricity supply industry required reorganization and modernization. The means to the end, however, greater centralization through the creation of regional or national entities, raised the hackles of advocates of municipal trading and private enterprise.

Gas Loses its Edge

Before the war, gas industry costs fell while price to consumers held steady. During the war, costs rose. Companies stopped paying dividends. The government suspended sliding scale regulation for two years, beginning in 1918, and allowed the companies to raise prices. The Gas Regulation Act of 1920 restored the sliding scale and also allowed a reset of the standard price at any time. The South Metropolitan Gas Company Act of 1920 apportioned excess revenue (over the amount needed to cover costs and dividends) between customers, employees and shareholders. This profit sharing scheme was designed to encourage employees and owners to run the company more efficiently, and it guaran-

---

63. Hennessey, *op. cit.*, p. 44.

teed that consumers would share the benefits of the improved operations. Then, in 1926, the National Gas Council warned that sudden cost increases could imperil the credit standing of gas utilities.

Sliding scale regulation, designed to encourage efficient operations, showed its deficiencies during the 1890-1920 period. When costs fell unexpectedly during the 1890s, customers received minimum benefits. When costs rose sharply during World War I, the utilities had to raise prices to cover rising costs. The price increases forced them to reduce dividends, thanks to the sliding scale formula. The old, rigid formulas could not cope with sudden changes or extreme conditions unimagined when the formula was imposed. Thus, the Gas Regulation Act of 1920 provided for revision of standard price at any time, as opposed to once every five years. That law created a mechanism to deal promptly with inflation or deflation. The government gave up on the principle behind sliding scale regulation, while outwardly retaining the formula.

Sliding scale regulation had still another weakness. It did not address level of service. The low price set as a standard encouraged inadequate maintenance of plant. Chantler commented:

> This is a danger all the more relevant in the case of gas companies because they never have been allowed to accumulate depreciation funds, as such. [64]

The market changed. During the war, manufacturers stripped gas of chemicals needed in the war effort, which reduced its illuminating power. In 1916, authorities began to measure gas by calorific (heating) value, another indicator of the shift in use from lighting to heating. By 1919, lighting fell to 35% of sales. The Gas Regulation Act of 1920 fixed the heat content of gas. Yet lighting hung on, accounting for 20% of sales as late as 1938.

By the mid 1920s, the gas industry's rapid growth had ended. The lighting market faded slowly, and the industry had to deal with competition from an reorganized, more efficient electric sector.

After the War

At the end of the war, electricity undertakings still served only 6% of homes (compared to over 30% in the US). Technocrats in the cabinet urged the government to participate in the electric industry. Sir Eric Geddes, at North Eastern Railway during its electrification, now Minister of Transport, wanted to nationalize and electrify the railroads, put electric utilities under his ministry's jurisdiction, and create District Electricity Boards to super-

---

64.   Chantler, *op. cit.*, p. 95.

vise the industry while leaving the distribution functions in the hands of existing utilities. "The war against Germany is over" he declared "and the war against obsolete and inefficient industrial and social conditions is just commencing." [65]

The Electricity (Supply) Act of 1919 contained Geddes's proposals and gave District Electricity Boards the right to purchase power stations without consent of their owners. Municipal and private power advocates fiercely opposed the bill. George Balfour, Tory MP, engineer whose firm operated utilities and tramways, and staunch opponent of woman's suffrage, led the battle against state ownership. Final legislation did little more than create a body of Electricity Commissioners and permit utilities in a region to establish Joint Electricity Authorities (JEAs).

Sir John Snell, an engineer and civil servant, served as first chairman of the Electricity Commissioners. That body had to deal with post-war tariff increases, standardization, and the allotment of franchises for previously unserved areas. It encouraged the purchase of larger generating equipment and the interconnection of systems. The electric utilities, however, concentrated on catching up with what they had not done during the war, rather than on making optimal decisions. Coordination efforts and the formation of JEAs moved in fits and starts, thanks to the old antagonisms between company and municipal undertakings.

Electrical World, America's top electricity trade publication, collected comparative statistics for 1920 that painted a dismal picture. The UK, despite its wealth, ranked in the middle or bottom for electricity usage. The nation had assets in place and connected customers but customers used the product sparingly. In 1920, the UK's economy was roughly 75% of size of the combined economies of Germany, France and Italy, yet the UK's electric customers took only 36% the electricity of the continental trio. (See Table 3-3.)

---

65. Hannah, *op. cit.*, p. 69.

Table 3-3. International Comparisons (1920)

	UK [1]	USA [2]	France [3]	Germany [4]	Italy [5]
GDP per capita ($1990)	4,651	5,559	3,196	2,986	2,531
% of population in electrically lit dwellings	16.9	36.8	13.6	14.5	11.3
Central station customers as % of population	3.8	9.1	3.4	3.3	2.5
Electricity consumption per capita (kWh)	139	472	147	141	85
Electricity consumption per central station customer (kWh)	366	519	433	423	340
Electricity consumption per $1990 of GDP (kWh)	0.031	0.084	0.043	0.047	0.036
Generating capacity per capita (kW)	0.065	0.228	0.054	0.055	0.035

Notes:
1. 1990 Geary-Khamis $ used for deflated GDP.
2. GDP data used for kWh consumption per unit of GDP adjusted by author to reflect German boundaries of the time, assuming no change in GDP per capita from Maddison.

Sources:
1. GDP and GDP per capita data from Angus Maddison, *Monitoring the World Economy 1820-1992* (Paris: Organisation for Economic Cooperation and Development, 1995).
2. All other data from *Electrical World*, Jan. 6, 1921 (Vol. 81, no. 1), p. 30.

British electrical manufacturers desiring to expand globally lacked the large scale domestic base of business that would give them the economies of scale needed to sell competitively abroad. Were electrical goods manufacturers too diffident to peddle their products? Certainly the British public was not immune to sales pitches. Somebody in the past sold them on the virtues of tea drinking, eating fish on newspapers and consuming mysterious breakfast spreads.

Marshall E. Dimock, an American political scientist and New Deal official had a different take:

> Great Britain has found it difficult to make advances in electrification because, to a greater extent than in any other country, industry has been naturally attached to coal and hence to the steam and gas engines.[66]

The attachment went beyond industry to transportation, lighting and heating. The British had an advantage over other coal users. They had huge domestic supply and had built an extensive infrastructure around it. Why change from a technology that works, that they do well, to another that performs the same functions? Why remove functioning assets and replace them at great expense to accomplish what the old assets did? The British suffered the disadvantage of having a well developed legacy energy system in place.

Electricity penetration numbers measured the modernization, flexibility and efficiency that electric power brought to industry, and the spread of modern living's benefits to the general public. A large local market also supported a major industry trying to expand in a dynamic worldwide marketplace. The British government had good reason to worry. First rate countries had a vigorous electric sector.

By the middle of the decade, the two largest networks (NESCO and a consortium in the Manchester-Lancashire region) together produced less than the Bavarian, Ruhr or Commonwealth Edison (Chicago) systems. The rise of electric holding companies (the largest of which was Balfour Beatty's 1922 creation, Power Securities Corporation) did not lead to the formation of interconnected systems, either. The holding company operated not as a vehicle for the integration of component companies, but for the financial benefit of the founding parties.

In the early 1920s, the industry expanded its reach. Consumers bought electric heating and other household appliances. The Commissioners managed to close 101 old stations within seven years. More undertakings made bulk power purchases from other companies. But they still maintained high reserve margins because they lacked transmission that would allow coordination between utilities and because they still ran generating units at low utilization levels.

Electricity prices fell, operating costs fell even more and the utilities pocketed the difference. Profitability improved: the companies paid an average dividend of 4.0 % in the years 1909 to 1918-1919 and 6.9% in 1919-1920 to 1925-1926. For comparison, in those two periods, government bonds yielded 3.6% and 4.7% respectively. Finally, in the 1920s, investors began to earn real returns in excess of the cost of risk free capital, with dividends paid after setting aside a reserve for depreciation. The industry expanded dramatically, and reached real profitability. A combination of toothless regulation, moral suasion

---

66. Marshall E. Dimock, *British Public Utilities and National Development* (London: George Allen and Unwin, 1933), p. 196.

and patriotism produced improvements, although only part of the benefits filtered down to consumers. (See Table 3-4.)

Table 3-4. Market Expansion (1920-1925)

	1920 [1]	1925 [2]
Utility sales (GWh)	3,512	5,606
Maximum demand on utility industry (MW)	1,740	2,899
Utility generating capacity (MW)	2,546	4,221
Reserve margin (%)	46	53
Average price per kWh (d)	2.12	1.65
Real price of electricity (1920 d per kWh)	2.12	2.12
Operating cost per kWh (d)	1.47	0.88
Real operating cost (1920 d per kWh)	1.47	1.19
Sales per kW of capacity (hours)	1,379	1,268
Number of undertakings (utilities)	473	565
Average sales per utility (MWh)	7,425	9,922
Number of undertakings (utilities) relying solely on local supply	336	290
Number of undertakings relying solely on bulk supply purchases	60	142
Number of undertakings meeting increased with bulk supply purchases	77	125

Note:
Years – Fiscal years 1919-1920 and 1925-1926.

Source:
Garcke's Manual, 1946-1947, esp. p. 9. and *Garcke's Manual*, 1925-1926.

The Labor government of 1924, in which Lord Haldane served, wanted to strengthen the 1919 Act. Herbert Morrison, a Labor leader who had served on London's JEA, proposed to nationalize the industry in order to rationalize it. Labor also viewed standardization of frequencies (17 still in use) as a job-creating program. The government fell in 1924. The Conservatives came to power. They appointed Lord Weir, head of a Scottish engineering firm and former Secretary of State for the Royal Air Force, as chairman of a committee to study the reform of the electric industry. Weir selected two like-minded members to join him, eschewed public hearings, and within five months produced a devastating critique of the British electric industry, replete with exaggerated claims of the benefits of rationalization, and a table that showed that the denizens of the United Kingdom used less electricity per capita than residents of Shanghai or Tasmania, not to mention less exotic locations. (See Table 3-5.)

Table 3-5. Weir Committee Compares Electricity Usage (kWh)

Region	Usage per Capita
United Kingdom	200
Scandinavia	500
Northeastern USA	800
Canada	900
California	1,200

Source:
Garcke's Manual, 1946-1947, p. 11.

Parliament and a coal study committee published more figures (the more the merrier considering their provenance, consistency and accuracy) that demonstrated that the British used less electricity and paid far more for it. Could one follow from the other? (See Table 3-6.)

Table 3-6. Consumption per Capita (kWh) and Price (d per kWh)

Countries	Consumption per Capita	Price per kWh
Canada	900	0.72
Switzerland	700	0.60
USA	500	1.05
Sweden and Norway	500	—
Belgium	230	—
France and Germany	140	—
Great Britain	118	2.07
Italy	145	0.46

Notes:
Consumption per capita in kWh for 1926.
Prices in pence(d) per kWh for various years from 1922 to 1924.
Consumption for Italy estimated by author.

Sources:
Marshall E. Dimock, *British Public Utilities and National Development* (London: George Allen and Unwin, 1933), pp. 197, 198.

The Weir report went beyond platitudes about frequency standardization. Weir wanted a national "gridiron" (a high voltage network) owned by a state organization run along business lines, the Central Electricity Board (CEB), to connect the country electrically. The CEB would coordinate industry planning and buy power from designated, large scale, efficient generating stations for subsequent resale to all undertakings. The report warned that the industry would waste money on uneconomic projects without a quick implementation of its plan. Stanley Baldwin's Conservative government first buried the report, but almost a year later, in March 1926, introduced a bill based on the recommendations. Municipal power undertakings, companies that produced electricity of a nonstandard frequency, and the diehard private power advocates led by George Balfour predictably opposed the bill.

Regulation and Legislation

The Electricity (Supply) Act of 1926 nationalized no assets but put a state-owned entity in the center of the market. It delighted British industrialists who sought economic electric power supplies. The Labor Party backed the bill, seeing it as the first step toward nationalization. A young Clement Attlee, who as prime minister two decades later nationalized the industry, praised it. George Balfour, however, managed to add an amendment, Section 13, which prohibited the CEB from charging an undertaking more for electricity than it would have paid if there were no CEB, a rule that forced the CEB to compete against hypothetical generating costs.

The London Electricity Acts of 1925 made still another attempt to encourage utilities to reduce prices. It allowed them to pay a dividend that exceeded the allowed 7% rate as well as a bonus for employees if prices declined.[67] The formula, though, tended to reward utilities whose average price per kWh was falling because consumers bought more electricity and thereby moved to a lower price on the tariff schedule, which had less do with utility efficiency than with the size of the consumers' purses. Getting regulation right never was easy.

Conclusion

The United Kingdom's electric industry grew rapidly despite politically produced inefficiencies and impediments, but not to the extent possible given the UK's size and wealth.

---

67. The London and Home Counties Joint Electricity Authority set an electricity price for the utility at costs plus a 7% dividend on shareholder capital, called the "standard price". (If the utility did not like the price set, it could appeal to the Electricity Commissioners.) In the following year, the Authority multiplied the standard price by the volume of electricity sold in the year. If the actual revenue collected was lower than the revenue calculated using the standard price, the difference constituted the "consumers' benefit." If the company earned enough of a profit above 7%, it could pay an extra dividend of one sixth of the consumers' benefit and another one sixth to employees. Consumers of course, got their benefit by paying less than the standard price.

Neither unflattering comparisons with other countries nor prospects for cheaper electricity could move the government to remove the impediments until the need to gear up for World War I and the fear that Britain could not compete after the war focused the minds of the politicians. They then launched a plethora of inquiries, all of which recommended reforms. The authors of the 1926 Act did not take the obvious but politically dangerous step of throwing out the fragmented and inefficient industry structure. Rather, they superimposed a new state agency on the industry. Perhaps they hoped that old and inefficient entities would wither away over time, leaving a modern industry in its place. Whatever their hopes, they implemented an advanced system that took advantage of modern technology and produced benefits for long suffering consumers.

Chapter 4
Before the Grid: UK vs USA (1882-1926)

... we were very severely tested by ... an abnormally heavy load during the fog of Tuesday last ... [68]

– P. Walter D'Alton (manager of Deptford)

The United Kingdom's electricity industry got off to a slow start and stayed behind, not as disastrously behind as some said, but behind. A centralized government with definite ideas about industry organization diverted the industry's evolution from paths dictated by technology and business opportunities. A strong gas competitor slowed the growth of the electric industry, too. British industrialists and public transit providers took time to grasp the benefits of an efficient electric power system. Steps to fix the problems came late in the game.

Economy and Electricity Diverge

By 1882, the year that the electric industry began operations in the United Kingdom and the United States, the UK had already ceded its position as the biggest economy in the industrial world to the USA and was nearing the end of its reign as the world's manufacturing leader. But the UK was still a hegemonic power, with an empire circling the world, a scientific leader, and without doubt, the financial and commercial capital of the world. The UK was no longer the biggest, but it was still the richest. Runner ups France, Germany and Italy lagged far behind. Thus, when the British got going in the electricity business, which they had just about invented, they should have set up a world beating business, not an also-ran effort. People had a right to expect better. (Figures 4-1 and 4-2.)

---

68.   Smart, *op. cit*,. p. 11.

Figure 4-1. Gross Domestic Product by Country in 1882 (billions of 1990 dollars)

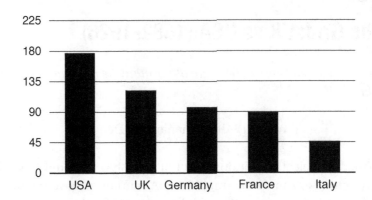

Figure 4-2. Gross Domestic Product per Capita in 1882 (1990 dollars)

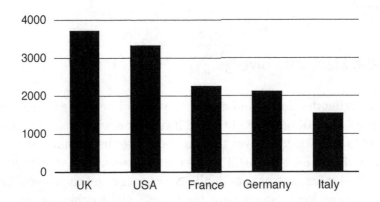

Notes and sources for Figures 4-1 and 4-2:
All data from Maddison, *op. cit.*, with Germany adjusted by author for population within boundaries of German Empire.

While Great Britain was the established power, the United States, as the up and coming rival, was more likely to install the newest technology because it did not have a legacy technology in place. Staying ahead, or close behind, required keeping up with the latest industrial improvements. The practical application of electricity was the most transformative improvement of the day and the British seemed slow to comprehend that. (See Figure 4-3.)

Figure 4-3. Electricity Generation (1882-1926) (billion kWh)

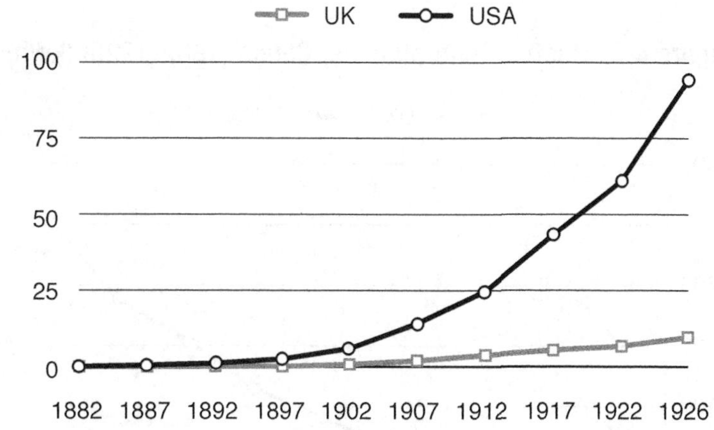

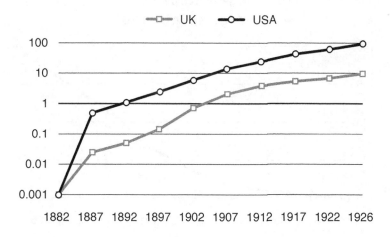

Source:
For data, see Appendix, Table B.

Using a semi-log scale (on which angle of line indicates rate of growth) shows more clearly how the UK electricity sector grew more slowly in the early years, when the government set up so many roadblocks, and it never caught up afterwards. (See Figure 4-4.)

Figure 4-4. Electricity Generation (1882-1926) (billion kWh) (Semilog Scale)

Source:
See Figure 4-3.

In 1882, the average Briton out produced the average American (11% more GDP per capita). Britons were richer and more productive, which made their relatively unenthusiastic embrace of electricity (an expensive product at the time) puzzling and perhaps symptomatic of the UK's relative decline. (See Figures 4-5 and 4-6.)

Figure 4-5. Electric Generation per Capita (1882-1926) (kWh)

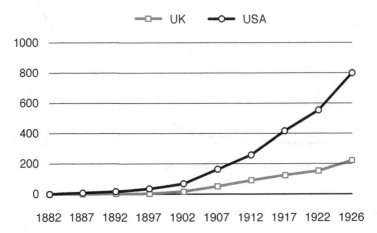

Source:
All data from Appendix, Table C.

Figure 4-6. Electric Generation per $ of real GDP (1882-1926) (million kWh per billion $ 1990)

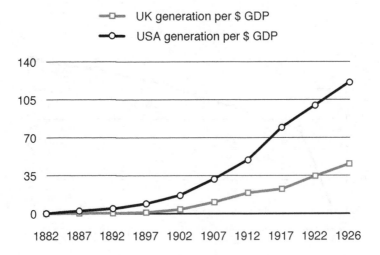

Source:
See Figure 4-5.

British and American electric companies shared ideas, entrepreneurs, technologists and financiers. They served population and industry concentrated in major cities at similar levels of economic development. But the political and regulatory environment for electricity differed. The British government exercised central control and took a more ideological stance. It favored municipal operation, which hindered the formation of large scale, regional utilities, and it set franchise terms that discouraged investment. Americans took a decentralized, economically oriented, pragmatic stance. Politicians in Washington, D.C. had little interest in electric matters. Electric companies did not need Congressional approval to set up shop. Developers, when necessary, paid off corrupt local officials to secure franchises. No need to go to Washington for help. Utilities drove down prices in order to secure customers. The municipal ownership movement petered out in America after the utilities embraced state regulation. Civic leaders wanted low prices more than ownership. Utility managers wanted protection from competition and from the depredations of venal local politicians and got those protections from state governments, and in return they accepted state regulation of prices and profits.

From a static point of view, the British system of incentive regulation should have made UK utilities more efficient than American counterparts operating under a cost-plus-allowed-profit regulatory framework. From a dynamic standpoint, though, the American system gave comfort to investors who poured money into innovations that lowered costs and prices and thereby encouraged more use of electricity. Neither electric industry had any trade secrets or financial advantages. Only government policy can explain the gap in performance between the two.

Pricing the Product

Pricing, as well, had to contribute to the slow uptake in the UK. British electricity suppliers charged more than the Americans, but not across the board. They charged industrial customers more and lighting and residential customers less. In the first two decades of the twentieth century British utilities charged industrial customers roughly 30% of the lighting rate, while American utilities charged them only 16%. Serving large customers — the manufacturing and traction firms — cost utilities less than serving small ones, and those firms were, undoubtedly more sensitive to price than residential consumers. Charging price-sensitive customers more than necessary and the balance of customers less makes no sense commercially. It drives away business, causing the price sensitive customer to self-generate or forsake electricity. Did the British utilities have different costs or did they have different commercial motivations?

Data from trade journals *Electrical World* and *Garcke's Manual* show that in the first two decades of the twentieth century, British utilities charged industrial and traction customers roughly 2.3¢ per kWh. Their average operating cost to produce and distribute all electricity ran to about 2.4¢ per kWh. In the United States, utilities sold electricity to large customers at about 1.3¢, while their average operating cost per kWh was about 1.7¢. American utilities, then, went farther than the British to shape the tariff to attract large customer loads and reap the benefits of the load diversity that those customers facilitated. American policy sped the electrification and modernization of industry. British policy favored the coal-and-horse based status quo. In the formative years, UK customers on average paid more, and big customers substantially more, than Americans (Figure 4-7). True, the British reduced prices at the same pace as the Americans, in real terms (as demonstrated on Figure 4-7's semilog scale), but they still started out and ended up paying more. Keeping pace did not mean catching up.

Figure 4-7. Price to Ultimate Users (1902-1926) (¢ per kWh)

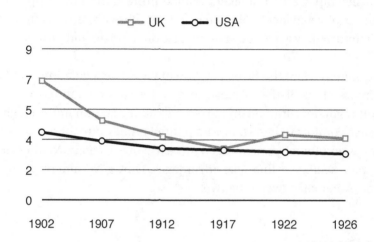

Notes and Sources:
All data from Appendix, Table G. Prices unavailable for all customers before 1902.

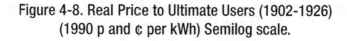

Figure 4-8. Real Price to Ultimate Users (1902-1926) (1990 p and ¢ per kWh) Semilog scale.

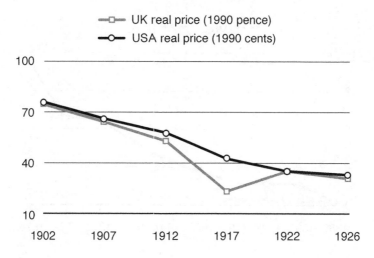

Source:
All data from Appendix, Table H.

Competing Against Gas

British electric companies competed against established gas utilities. American electric companies faced gas competitors that soon developed a new line of supply, natural gas, which had twice the thermal content of manufactured gas but sold at half the price per cubic foot of volume. Although in both countries gas consumption, usage per capita and usage per dollar of real GDP rose steadily, growth was more pronounced in the United States. American electric utilities had to compete with vigorous gas competitors that had a product to sell, perhaps forcing them to act more competitively than their UK counterparts. (See Figures 4-9, 4-10 and 4-11.)

Figure 4-9. Gas Consumption (Excluding Electric Generators) (1882-1926) (billion therms)

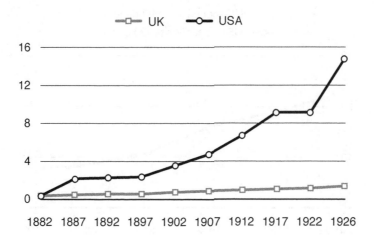

Source:
All data from Appendix, Table I.

Figure 4-10. Gas Consumption (Excluding Electric Generators) per Capita (1882-1926) (therms per capita)

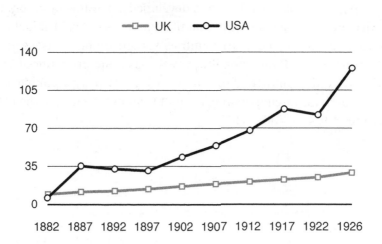

Source:
All data from Appendix, Table K.

Figure 4-11. Gas Consumption (Excluding Electric Generators) per $ of Real GDP (1882-1926) (therms per thousand $1990)

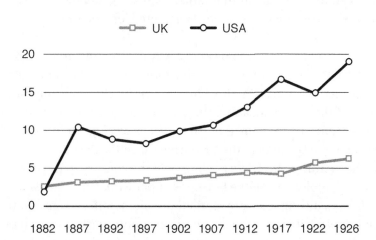

Source:
All data from Appendix, Table J.

In the UK and the USA, real gas prices held steady while real electricity prices fell. In the early 1880s, manufactured gas cost less in Britain than in the United States. Adding natural gas to the mix changed the picture in the USA, lowering the average price of gas. The British gas suppliers, then, had to depend more than the Americans on their ability to influence the laws that regulated the electric companies rather than on ability to sell a superior product. Having municipal owners that did not want competition and could prevent the entry of unfriendly (non-municipally owned) electric utilities helped.

Sizing Up the Customers

Electric lighting started as a novelty for the rich and a lure to bring customers into commercial establishments. Its popularity grew as light bulbs improved and became cost competitive with gas light, but consumers could easily live without electricity. For industrial and traction firms the new energy source had to provide clear economic benefits, such as lower cost, improved operating flexibility and ability to manufacture new products. They could produce their own electricity if the local utility sold power at unsatisfactory terms. Utilities, however, could offer good deals to industrial and traction firms because serving them cost less than serving small, lighting customers and because those firms often could be served by existing generators at times when sales to other customers were low.

Thus attracting big customers could add to utility revenues without a commensurate increase in costs and the utility could use the additional profit to reduce prices to all consumers. Serving the large customers, then, had a chicken and egg quality. Big users buying from the utility enabled it to lower average costs and prices but the utility had to offer prices that induced them to buy from it rather than self generate. That is, the electricity supplier had to sell its product, as Charles Merz noted long before.

The UK's electricity industry priced power in a way that put off large customers. Did that policy simply drive the large customers to generate their own power (Figure 4-12) or did they simply have less enthusiasm altogether for electricity than American firms? The proper measure of their electrification, then, would be the sum of their self-generation plus utility generation sold to them as a percentage of total generation and that percentage was far lower in the UK than in the US during the initial decades of electricity service (Figure 4-13). Furthermore, from the beginning of the century to the mid twenties, British manufacturers employed half the horsepower per worker as their American counterparts.[69] Thus, *in toto*, during this early period, British manufacturers and transportation firms were far less electrified than their American counterparts and used power from all sources less intensively. They finally caught up, but how much soon-to-be-obsolete plant did they install before they finally got the message?

Figure 4-12. Self Generation as Percentage of Total Generation (1887-1926) (%)

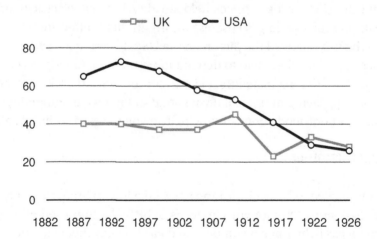

Source:
All data from Appendix, Table B.

---

69. Lászlo Róstas, *Comparative Productivity in British and American Industry* (London: Routledge, 1948), p. 68.

Figure 4-13. Self Generation Plus Utility Generation Attributable to Sales to Industrial and Traction Firms as Percentage of Total Generation (1887-1926) (%)

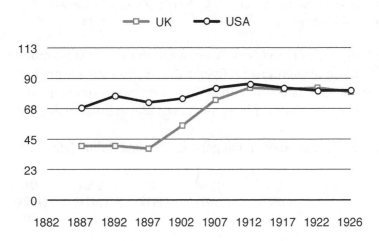

Source:
All data from Appendix, Tables B, E and F.

UK tramway (trolley or street car in American parlance) operators electrified less readily than American companies, too. They relied more on steam and other motive power sources. They finally caught up in terms of electrification by 1907, but even after the tram networks were essentially complete, in 1912, the UK still had only 62 miles of track per capita as opposed to 427 in the USA (Table 4-1).

Table 4-1. Tramway Electrification (1890-1926) (miles)

Year	UK Total (miles)	UK Electric (miles)	US Total (miles)	US Electric (miles)
	[1]	[2]	[3]	[4]
1890	949	8	8,123	1,261
1897	1,031	88	13,000	12,000
1902	1,845	1,484	22,577	21,901
1907	2,394	2,200	34,381	34,037
1912	2,637	2,500	41,065	40,808
1917	2,600	2,500	44,835	44,676
1922	2,579	2,500	45,931	43,789
1926	2,602	2,500	41,000	40,800

Note:
Author's estimates of missing data points.

Sources:
Columns 1 and 2 – *Garcke's Manuals*, various issues. Byatt, *op. cit.*, pp. 29, 32, 36. Central Statistical Office, *Annual Abstract of Statistics* (London: H.M. Printing Office, various years).
Columns 3 and 4 – *Statistical Abstract of the United States*, various issues.

American industry took to electricity faster, with electric power accounting for a larger percentage of industrial energy needs than in the UK. Industrial firms, like traction companies, initially self-generated more than they purchased. Even in the mid twenties, US manufacturers relied on electric power more than British firms. (Table 4-2.)

Table 4-2. Electrification in Industry and Transportation (1907-1924) (%)

	UK (1907) [1]	USA (1907) [2]	UK (1912) [3]	USA (1912) [4]	UK (1924) [5]	USA (1924) [6]
Industry						
% of energy requirements provided by electricity	10	19	23	38	49	66
% of electricity						
– purchased	41	33	42	33	58	65
– self-generated	59	67	58	67	42	35
Transportation						
% of electricity						
– purchased	28	0	29	33	72	38
– self-generated	72	100	71	67	28	62

Sources:
Columns 1,3,5 – Byatt, *op. cit.*, pp. 72-74, 95.
Hanna, *op. cit.*, pp. 427-428.
Central Statistical Office, *op. cit.*, various dates and tables.
Columns 2,4,6- – U.S. Department of Commerce, *Historical Statistics*, pp. 155-159. 179.
_____, *Statistical Abstract*, various dates.

In short, British industry did not see value in electricity, at first. Was that lack of value due to the price they would have to pay? If so, they would have generated the electricity themselves, rather than paying the utility an exorbitant price. British industrial and tram firms self-generated less of their needs than the Americans, who got more favorable prices from their utilities. British utilities did not, in general, offer attractive prices to the potentially large users in early years, but those users may not have looked like eager buyers, either. Eager sellers, however, know how to convince reluctant buyers, and the utilities may have been more at home in selling electric tea kettles than electric motors. Eventually, British utilities managed to achieve roughly the same sales mix as the Americans. They just took longer and they missed opportunities to sell into the transportation market (Table 4-3).

Table 4-3. kWh Sales by Customer Class (1882-1926) (%)

Year	UK Lighting [1]	UK Power [2]	UK Traction [3]	US Lighting [4]	US Power [5]	US Traction [6]
1882	100.0	0.0	0.0	100.0	0.0	0.0
1887	100.0	0.0	0.0	91.6	8.4	0.0
1892	100.0	0.0	0.0	84.4	13.3	2.3
1897	98.6	1.4	0.0	88.2	9.1	2.3
1902	72.0	8.2	19.8	60.5	35.7	3.8
1907	43.5	33.8	23.4	41.3	33.1	25.6
1912	31.1	51.8	17.1	30.4	36.0	33.6
1917	22.7	65.9	11.4	27.8	47.6	24.6
1922	23.7	65.3	11.0	26.7	57.6	15.7
1926	28.5	61.8	9.7	26.3	61.7	12.0

Note:
Author's estimate for missing data.

Sources:
Columns 1-3 – Byatt, *op. cit.*, p. 98. Hanna, *op. cit.*, pp. 425-426.
Columns 4-6 – *Electrical World*, Vol. 80, No. 11, September 9, 1922 (p. 546); Vol. 91, No. 1, January 7,
          1928, (p. 18).

Adding Up the Expenses

Producing electricity cost more in the UK than in the US, so utilities charged more. The UK's electric utilities had higher operating costs per kWh sold (Figure 4-14) despite having larger power plants (Figure 4-15) which normally run more efficiently than small ones. The British utilities did not utilize their plant as well either: they produced fewer kWh of output per kW of capacity (Figure 4-16). They, as a result, had to spend to maintain under utilized plant and to pay interest and dividends on money invested in that plant.

Figure 4-14. Utility Operating Costs (1902-1926) (¢ per kWh sold)

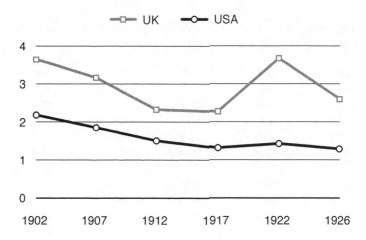

Source:
All data from Appendix, Table N.

Figure 4-15. Average Size of Utility Power Plant (1902-1926) (kW)

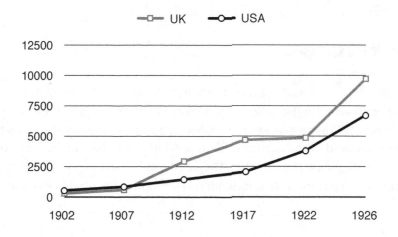

Source:
All data from Appendix, Table N.

Figure 4-16. Utility Average kWh Sold per Unit of Capacity (1892-1926) (kWh per kW of year-end capacity)

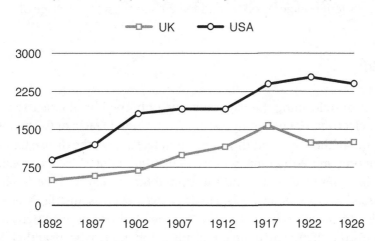

Source:
All data from Appendix, Table N.

In the first two decades of the twentieth century, British consumers paid (on a per kWh basis) about 4.4¢, the utility paid out 3.0¢ for operating costs and 1.4¢ remained to cover taxes, interest, depreciation and return to shareholders. American consumers paid about 3.2¢ for electricity, operating costs took around 1.6¢ and that left the utility with 1.6¢ for taxes, depreciation, interest and return to shareholders. The higher UK coal price accounted for about 0.4¢ of the difference in operating costs and the fact that US utilities could use water power and avoid fuel cost probably another 0.1¢. Thus the lack of inexpensive domestic energy sources, alone, made British electricity more expensive. Shipping in fuel would have cost money, and would have implications for domestic security, as well. (Imagine the consequences of German U-boats in World War I sinking coal carriers.)

British consumers also paid extra to maintain plant and equipment that sat idle because their utilities produced roughly 40% less electricity per kilowatt of installed plant than the Americans. They paid roughly another 0.3¢ per kWh for taxes, interest, profit and depreciation on investment in that idle plant. Thus, after subtracting the cost differential due to location (0.5¢), British consumers still paid 0.7¢ more per kWh, of which 0.3¢ was due to inadequate utilization of plant brought about, to a great extent, by government policies. Conceivably, part of the difference in operating costs, as well, could be attributed to the same policies that kept the utilities small, local and unable to standardize their services.

The British, at the time, were the leading manufacturers and financiers in the world and had invented much of the apparatus of the modern electric utility, so it is hard to swallow the notion that their managers did not know how to do a better job. Considering that they served a compact nation with manufacturing concentrated in a few regions and the population in big cities, they had the conditions needed to run efficient services if permitted to do so. Government policies probably added 20% to the average price of electricity.

Not that Bad

Activists intent on reforming the electric industry harped on the fact that the United Kingdom had electrified its economy less than the United States or Germany. They conveniently left out of the argument three pertinent facts. First, the British had also electrified their economy more than many other industrial nations. Second, wealthier countries tend to use more electricity and the United States had become wealthier. Third, countries blessed with abundant, cheap hydro power (such as Austria, Sweden and Switzerland) could lure in energy intensive industry, thereby boosting electricity usage within the economy. By the 1920s, when the criticisms of the industry were at their loudest, the United Kingdom sat in the same position as it had for decades, in the middle of the rankings for electrification, but significantly below those of the United States and Germany. (See Table 4-4.)

Table 4-4. International Comparisons (1902-1922) ($1990 and kWh per capita)

	GDP per Capita (1902)	GDP per Capita (1912)	GDP per Capita (1922)	Electricity Output per Capita (1902)	Electricity Output per Capita (1912)	Electricity Output per Capita (1922)
	[1]	[2]	[3]	[4]	[5]	[6]
UK	4,628	4,868	4,427	15	109	209
USA	4,426	5,207	5,546	75	260	554
Belgium	3,660	4,117	4,320			195
Switzerland	3,653	4,164	4,555			774
Netherlands	3,515	3,892	4,488			129
Germany	3,040	3,715	3,558	43	198	438
Austria	2,965	3,528	2,897			281
France	2,749	3,481	3,576	9	36	185
Sweden	2,496	3,064	2,906	19	218	447
Italy	1,782	2,411	2,574	6	48	119

Notes:

All data on European electricity from Mitchell for consistency. Mitchell's data and definitions may not align with other sources used in this book.

Sources (columns)

1-3. Maddison, op. cit.

4-6. Output per capita derived from Mitchell, European Historical Statistics, as well as authors' estimates based on available data.

Conclusion

The United Kingdom electrified slowly, hindered by government policies which focused more on market structure than consumer benefits. From 1882 through 1926, Parliament passed five major laws that set the terms for the structure of the electricity supply industry, and made many more decisions, as well. In that same time period, the federal government of the United States enacted one major law that affected the electric utility industry, the Federal Power Act of 1920, which set up a Federal agency to regulate dams on navigable waterways. The American industry, on its own, resolved the problem of multiple standards. Beyond that, the states made rules, within limits set by federal courts that concerned themselves solely with due process and constitutional provisions on the taking of property.

Ownership patterns affected industry development. Municipal governments owned electric utilities in major cities, shutting out more aggressive private power companies from rich markets. Municipal governments owned the competing gas works in many markets, so they had reason not to promote electricity sales. Private companies got the leftovers, and their strategy of agglomerating small and scattered leftovers did not produce the economies of scale that would encourage rapid electrification.

Forty years of parliamentary micromanagement had firmly established the role of the central government in the industry's decision making process, so it was no surprise that Lord Weir, an industrialist, would tell the government to solve the problems that he saw. Yet by the time Lord Weir got into the act, the industry had already begun to improve and, on an international scale, the United Kingdom had electrified its economy at least as much as many of its major competitors, but it still trailed far behind the United States or Germany, an odd position for one of the leading nations of the world.

Part Two:

The "Gridiron:"
Rationally Reorganizing the Industry

Chapter 5
Building the Grid (1926-1933)

... the salvation of the industry is not to be brought about by the waving of weird "gridirons" ... [70]

– The Electrical Review

After condoning a half-century of inefficiency, high prices and haphazard organization, British politicians began to fear that the supposedly backward state of the electricity suppliers might further weaken the country's failing industrial base. Britain needed low-cost energy. British electrical manufacturers needed a standardized home market that would support large scale production. The electricity suppliers required a reorganization.

The Central Electricity Board

Lord Weir, a Scottish industrialist, chaired the inevitable committee designated to review and reform the industry. His committee began work in January 1925. From the start he had no intention of presenting a menu of incremental improvements. The committee hired experts to propose reforms and then elicited an endorsement of the proposals from Samuel Insull, the English-born, uncrowned king of America's electric industry. In May, the committee finished its report, which touted the benefits of standardization and of a government-owned national transmission system. The report argued that 28 power stations produced 50% of output, 322 produced 11%, and the industry could operate with 60 of them. The 43% reserve margin showed, as well, that the industry had too many power stations. Somebody had to close down those unnecessary facilities, step on toes, and force an uncooperative industry to move ahead. The government did not publish the report when it was completed. Finally, a new Conservative government put forth a bill, which became the Electricity Supply Act of 1926. The law:

- Established the Central Electricity Board (CEB), a government-owned entity that would operate as a business, build and operate a nationwide transmission system and act as a power wholesaler that bought all power produced by designated power stations and sold that output to the electricity undertakings.

- Called for a standardization of the system.

- Set levels of profitability for industry participants.

---

70. Editorial about the Electricity Supply Act of 1926, quoted in Hennessey, *op. cit.*, p. 50.

During the debate, Conservatives reluctantly backed a bill that enlarged government's role in industry, Labor Party members saw it as a step in the right direction (toward socialism) and British manufacturers viewed it as necessary to create an efficient electric supply sector. The infamous Section 13 of the law, however, ordained that undertakings need not pay more for electricity from the CEB than what their costs would have been absent passage of the Act. That section forced the CEB to price against hypothetical and possibly fictitious costs, and deprived consumers of a significant portion of the savings generated by the Central Electricity Generating Board (CEGB).

The CEB was an unusual agency. It could invoke Treasury guarantees in order to raise funds yet it financed without guarantees. It paid commercial – rather than civil service – salaries. And the government could dismiss board members only for absence from the job.

The Electricity Commissioners, left over from previous legislation, still regulated the industry (including the CEB). Before the 1926 Act, they had little control over pricing. The original franchises often set maximum prices, relics of a period of higher costs. Some company franchises prohibited dividend increases until prices fell below prescribed levels, but prices had long ago fallen below those levels, so the companies effectively ran without price controls. The Act set a 5-6.5% range of return on capital invested in generation for companies, and a rate based on borrowing costs for municipal enterprises. At that time government bonds yielded about 4.5%, so the terms gave investors a return with a only slim premium over risk free bonds. But covering costs plus a return guaranteed by a government agency that exceeded the return on government bonds was not a bad deal.

The CEB built a 132,000 volt transmission system. Higher voltage lines already operated in Germany and the USA. The CEB standardized the network on three phase AC, 50 cycle (Herz) current, which left the only large, standardized system in the country out in the cold. NESCO had standardized at 40 cycles. Then the CEB "selected" the generating stations that would sell to the grid. The Weir Committee estimated that 60 power plants were required. The CEB instead selected power stations whose per unit variable costs of production were lower than the expected total costs at which the grid could supply electricity (costs of production from a new station plus cost of transmission). Wanting to assure cooperation from as many electricity suppliers as possible, the CEB selected 140 generating stations. The CEB also had to pay for conversion of power stations and electric motors to the new standards, which it did out of the proceeds from a tax on electricity sales. That way, every user paid for electricity modernization.

Building the grid took years. Initially it was less as a nationwide network for the transport of power than a collection of intra-regional connections that linked local power stations to distribution systems. Each region could operate independently, although weak connections linked the regions. The still incomplete National Grid commenced operation in 1933.

The CEB had a simple operating model: buy electricity from selected stations at a price that covered operating costs plus a return on the owner's capital, then sell that electricity to distributors at a price that covered the purchase price plus all other CEB costs. Distributors owning non-selected stations could generate their own electricity but often did not because the CEB offered them cheaper power.

As a result of the transmission interties, the electrical system became more reliable. Because only the most efficient plants generated, the overall system operated more economically. Because the local utility now had access to the power plants of others, it could run with less spare capacity, thereby reducing the burden of carrying excessive reserves.

Even with the CEB in business, the Labor government of 1929-1931 sought additional reforms. Minister of Transport Herbert Morrison put Sir John Snell to work. His report called for the formation of state-owned regional distribution boards to take over the distribution sector, or for larger undertakings to acquire smaller ones (including municipal takeover of companies and vice versa). The report remained unpublished because the Labor government collapsed.

The Electricity Commissioners, chaired by Snell, did not give up. They convened, in 1932, a private conference of industry leaders who worked out a new plan by mid-1933. It advocated formation of large distribution systems – companies, municipals agencies or state boards – that encompassed urban and surrounding rural areas. They targeted for takeover all undertakings with sales under three million kWh, and even considered undertakings with sales up to 25 million kWh as targets for consolidation. That proposal might have eliminated two-thirds of the undertakings. No legislation followed, but the conference's conclusions prompted a later investigation of the industry's structure. Morrison, incidentally, asked Lord Weir to chair a committee on railroad electrification whose conclusions did not please railway executives, who had no interest in electrifying without a government subsidy to pay for the process. The only concrete result of that report was a recommendation to permit the CEB to sell electricity to railways, which was incorporated in 1935 legislation.

Building Markets

First utilities had to connect customers, then persuade them to take more electricity, especially during non-peak periods in order to make use of utility plant for more than a few hours of the day. Utilities installed wiring for customers. They sold electrical appliances in their retail stores. They advertised. "Don't be old-fashioned Mother," the dapper 1920s flapper said to her blank-faced, grey-haired mother who carried a brush in one hand and a dustpan in the other, "For Health's Sake use Electricity."[71] Promotion of vacuum cleaners, electric cookers, clothing washers, and electric heaters added to the utilities' sales, although the multitude of voltages and frequencies must have hampered the sales efforts.

Utilities previously charged separate prices for electricity used to power different appliances, each operating on its own circuit. As appliances multiplied and customers wanted to move them from place to place within their homes, that pricing system became increasingly impractical. Utilities replaced those multiple tariffs with a block or two-part tariff. The block tariff charged a specified price per kWh for up to a given number of kWh (the first block), then a lower price for consumption in the next block, and an even lower price for consumption the following block. Block pricing encouraged consumption because the greater the usage the lower the cost.

For instance (in pence):

First 100 kWh	@ 6
Next 100 kWh	@ 5
Next 100 kWh	@ 4
All additional kWh	@ 3

The two-part tariff consisted of a fixed charge (to cover the cost of keeping equipment available to serve the customer) and a low price per kWh consumed, to encourage greater consumption.

Sales to industrial and transportation customers increased despite the deepening depression. Those big users began to rely more on the utilities and less on self-generation and put greater emphasis on electricity as an energy source. Horsepower count for electric motors rose 38% between 1924 and 1930.

Electricity suppliers expanded coverage. By 1928, electricity franchises covered 55% of the nation's area in which 95% of the population lived. Still, only one third of homes, overall, took electric service and less than one-tenth of homes in rural areas were connected to the local utility.

The Picture in 1930

The 1926 Act subsidized standardization. By 1930, four-fifths of generating plant produced three phase, 50 Hz power (the new standard), four-fifths of distribution systems delivered at 50 Hz frequency, and four-fifths of undertakings offered AC or AC and DC

---

71. Bowers, *op, cit.*, p. 22.

service, while one-fifth only offered DC. There were still too many power plants. Lord Weir, in 1925, said that the industry could operate with 60 power plants. In 1930, it still had over 500. (See Table 5-1.)

Table 5-1. Ownership of Generation (1930)

Owner	Number of Plants	% of Output	Average Generation per Plant (million kWh)
	[1]	[2]	[3]
Municipal	250	52.7	26.0
Company	245	35.8	18.0
Railway	28	6.7	29.5
Tramway	20	3.3	20.4
Other	16	1.5	11.3
Total	559	100.0	22.1

Source:
Hennessey, *op. cit.*, p. 178.

Between 1924 and 1930, British manufacturers employed more electricity in their processes. They raised the electrical percentage of factory power applied from 53% to 66% [72] but were still reluctant to rely on the local utility. The transit firms, on the other hand, just about gave up on self-generation. Yet the UK still lagged behind the US. In 1930 the average Briton earned about 85% the income of an American yet consumed only one third of the electricity. In the twenty-first century, policy makers might exclaim "Hooray, those Brits showed how efficient they could be versus their wastrel American cousins." Back in the 1930s, when electrification signified modernity and cleanliness, those numbers had a different message: British utilities either did not care to or had not found ways to get their product to the customers. One more symptom of the relative decline of the great imperial power? (See Table 5-2.)

---

72. Róstas, *op. cit.*, p. 59.

Table 5-2. Electrification in UK vs USA (1924-1930)

	UK (1924)	USA (1924)	UK (1930)	USA (1930)
Industry				
% of energy requirements provided by electricity	49	66	61	64
% of electricity purchased	58	65	60	66
% of electricity self generated	42	35	40	34
Transportation				
% of electricity purchased	72	38	95	67
% of electricity self generated	28	62	5	33
General				
Electricity generated (TWh)	8.9	75.9	14.1	114.6
Electricity sales (TWh)	5.2	45.2	9.2	74.7
Population (millions)	45	115	46	124
Real GDP ($1990 millions)	211,033	714,828	238,270	769,215
GDP per capita ($1990)	4,698	6,238	5,195	6,218
Utility sales per capita (kWh)	115	394	200	606
Electricity generated per $1000 real GDP (kWh)	42	106	59	149
Utility sales per $1000 real GDP (kWh)	25	63	39	97

Notes:
1. Estimates of electric include utility and non-utility sources. Estimates made by author based on government data are consistent with generation numbers shown elsewhere in text.
2. One terawatt hour (TWh) equals one trillion kWh.

Source:
See Tables 4-2, 4-3 and 4-4.

Conclusion

The 1926 Act reorganized the industry, produced standardization remarkably quickly and promoted electrification. It demonstrated how the government could cajole individual actors in a network industry to operate in a way that improved public welfare. The ultimate beneficiaries, the consumers, provided the funds, so nobody else could complain about the cost of compliance and those that did not care for the returns offered to generators did not have to build. The electricity commissioners knew how to herd cats. Still, in 1930, despite the progress, one-fifth of the industry offered non-standard service, and two-thirds of British homes lacked electricity. The 1926 Act would not make its full impact until completion of the National Grid

Chapter 6
The CEB Takes Charge (1933-1942)

The solution is such a simple one ... those concerns ... which have ... shown their ability to handle large areas ... should be entrusted with wider powers. [73]

– J. M. Donaldson

Once completed, the National Grid encouraged greater generating efficiency and more widespread electrification. The industry, though, faced further fragmentation due to the expiration of many franchises. Then World War II put a hold on corrective measures.

Finishing the Grid

In 1933, construction ended on the first stage of the Grid: 4,800 kilometers of 132 kilovolt (kV) transmission lines. The CEB then began to run as planned. By 1936 the Grid operated everywhere except in the northeast because the Grid and NESCO had different standards. Finally, in 1938, the northeast converted to the national standard. The original plan called for seven self-sufficient regions. Beginning in October 1938, the CEB ran the network as one.

Between the operation of the first Grid lines in 1933 to its conversion to a national network in 1938, Great Britain recovered from Great Depression lows. Electricity output grew faster than industrial production and the economy as a whole. Electricity prices and costs fell. So did the number of generating plants and undertakings. Reserve margin dropped. The Grid concentrated generation at efficient stations and lessened need for high reserves. Falling electricity prices encouraged consumption. Industrial usage rose twice as fast as industrial output, a trend encouraged by the CEB's willingness to help the local utility offer attractive prices to industrial customers. On the home front, the shortage of servants and the introduction of more home appliances raised consumption in affluent households while the introduction of prepaid meters encouraged low income consumers to sign up for electric service. By 1939, two-thirds of British homes had electricity. Just about everything that the UK's electricity industry should have done in the previous four decades it accomplished within a five year period after the National Grid opened up. (See Table 6-1.)

---

73. Presidential address to Institution of Electrical Engineers, 1931, quoted in Hannah, *op. cit.*, p. 251.

Table 6-1. Electrification and Economic Activity (1933-1938)

		1933 [1]	1938 [2]	% change [3]
1	Population (millions)	46.5	47.5	2.1
2	Real GDP ($1990 billions)	234.4	284.2	21.2
3	Industrial production (1933=1000)	100	133	33.3
4	Retail prices (1933=100)	100	112	12.0
5	Electricity generation (TWh)	17.4	31.2	79.3
6	Electricity sales (TWh)			
7	– All customers	11.3	20.4	80.5
8	– Industrial customers	6.1	10.3	68.9
9	Revenue per kWh sold (old pence)			
10	– All customers	1.26	1.04	-17.5
11	– Industrial customers	0.73	0.66	-9.6
12	Generating capacity (MW)	7,837	9,365	19.5
13	Maximum load (MW)	4,156	8,003	92.6
14	Number of generating stations	437	380	-13.0
15	Number of electricity undertakings	627	576	-9.1
16	Number of electricity customers (millions)	6.1	10.1	65.6
17	Working costs per kWh sold (old pence)	0.79	0.55	-30.4
18	Real price of electricity (1933 old pence /kWh)			
19	– All customers	1.26	0.93	-26.3
20	– Industrial customers	0.73	0.59	-19.2
21	Generation per capita (kWh)	374	657	75.7
22	Generation per $1000 real GDP	74.0	110.0	48.0
23	Usage per customer (kWh)	1,853	2,020	9.1
24	Reserve margin (%)	88.5	17.0	-80.8

Notes and Sources by Line:

1,2. Maddison, *op. cit.*, pp. 106, 107, 183.

3,4. London and Cambridge Economic Research Service, *The British Economy: Key Statistics 1900-1964* (London: Times Publishing, 1965), pp. 5,8.

5. Author's estimates based on government data.

6-12. Hannah, *op. cit.*, pp. 427-433.

10, 11, 17, 18. Before decimalization, the pound sterling was divided into 240 pence (d) refereed to in table as "old pence" to distinguish them from the "new pence" (100 to the pound).

13-17. *Garcke's Manual*, various editions.

18-20. Calculated using retail price deflator line 4.

21-23. Calculated from above data.

Reserve margin is defined as the difference between capacity and maximum load as a percentage of maximum load. For this calculation, it was assumed that year end capacity approximated capacity at time of maximum load because the maximum load usually took place in the winter.

The CEB did not achieve predicted savings. Section 13 of the 1926 Act required it to compete against imaginary costs. For political reasons, the CEB, did not close high cost generating stations so it had to pay higher generating costs. Some utilities objected to the return on investment that the CEB paid, so they resisted building large power plants. Still, the CEB's reforms reduced costs sharply, possibly lopping off as much as one-third off the average electric bill, but as economists James S. Foreman-Peck and Christopher J. Hammond concluded:

> ... the CEB's restructuring fell short of what was ideal. The extent to which it did so was one-half of the (one-third) industry cost reduction achieved by the CEB by 1937. [74]

That estimate may not have taken into account the costs associated with the Balfour clause.

The McGowan Report

National Grid enabled the industry to generate more efficiently and it created a much-needed transmission network. The distribution structure, however, remained parochial and was destined to fragment even more because, as franchises expired, municipalities could dismember regional systems by exercising their options to buy undertakings.

The nationwide holding companies owned scattered properties, controlled small percentages of the market and made limited progress in achieving economies or developing unified systems. One of the largest, Greater London and Counties Trust, was mired in financial difficulties caused in part by problems at its American parent. The holding companies had neither the resources nor the interest to act as integrators.

In 1934, the industry consisted of the CEB, three joint electricity authorities (structures authorized by the 1919 Act to integrate the industry), 635 undertakings and five joint boards of local authorities. Roughly half of those entities sold less than five million kWh. Half the undertakings combined sold less than 10% of the industry's output.

The secret conferences hosted by the Electricity Commissioners produced nothing. The Baldwin government in 1935 opted for the tried and true route to wisdom, the expert committee, this one chaired by Sir Harry McGowan of Imperial Chemical Industries (which, ironically, generated its own power rather than take it from the electric utilities). The McGowan Committee's 1936 report neither contained surprises nor pacified paro-

---

74. James S. Foreman-Peck and Christopher J. Hammond, "Variable Costs and the Visible Hand: the Re-Regulation of Electricity Supply, 1932-1937," *Economica*, Feb. 1997, p. 29.

chial interests. It advocated economies of scale in distribution, extension of franchises to 50 years, acquisition of small undertakings by companies rather than state boards, and giving the Electricity Commissioners power to force mergers of undertakings with sales of less than 10 million kWh.

In December 1936, Political and Economic Planning (PEP), a public policy think tank, produced its own study, *The Supply of Electricity in Great Britain.* The PEP report showed how the price of electricity rose as the size of the undertaking declined, with the very smallest undertakings charging almost four times more per kWh than the very largest ones. (See Table 6-2.)

Table 6-2. PEP Report: Utility Size vs Electricity Price (1934-1935)

Undertakings grouped by millions of kWh sold	Number of Utilities in the Group	Unweighted Average Price for Domestic or Lighting Purposes (d/kWh)
≥ 25	122	1.96
2.5-24.9	213	2.64
0.25-2.4	229	3.97
< 0.25	62	5.51

Source:
Hannah, *op. cit.,* p. 244.

The PEP report, however, said that size, alone, did not guarantee efficiency. PEP proposed that the government set up an agency to regroup the industry by having more the efficient undertakings take over the less efficient ones.

If it followed the McGowan recommendations, the government would have forced over 200 municipal power agencies out of business, with the bulk of them taken over by power companies. After much opposition, not to mention stalling, the government came up with a new proposal, which cut power companies out of the action and promoted the municipal undertakings as consolidators. By then, though, Stanley Baldwin had left office, with his most publicized accomplishment being management of Edward VIII's abdication.

Neville Chamberlain became prime minister. His proposed reform of the electricity sector, a continuation of Baldwin's efforts, unleashed a storm of protests. The small municipalities correctly accused the larger ones of secretly feeding the government damaging information about their smaller brethren in order to build up the case for takeovers of the small undertakings. The government shopped around its merger ideas, which seemed politically – rather than economically – motivated. The companies came out against proposed profit limitations and pro-municipal provisions. Government leaders decided not to hand this hot potato to Parliament, and that decision killed reforms.

World War II

Electricity suppliers prepared for war. They ran drills for operation under wartime conditions and set up emergency control centers and warehouses full of spare equipment. The government prepared, too. To prevent unnecessary reorganizations of ownership and operations that would have disrupted the war effort, Parliament passed the Special Enactments (Extension of Time) Act of 1940, which suspended the right of municipalities to purchase undertakings during the war. Those expected purchases could have created hundreds of new undertakings.

Thanks to the wartime blackouts, electricity sales fell off initially, so the CEB worried about having excess capacity after the war. Then demand from munitions factories built up, prompting concern about inadequate electric supply. The government located new war production factories as far as possible from German air bases. That shifted electricity demand to new regions, putting strains on the grid and on local power stations. Bombing damaged utility plant. Manufacturers that owned their own generators feared that they might not obtain fuel supplies so they began to buy more power from the electricity supply undertakings, thereby putting more pressure on them. Once the government saw that the electricity industry could meet the nation's needs, though, it refused requests for new generating equipment, a decision that would lead to electricity shortages in the early post-war years.

Sir William Beveridge, economist, former head of the London School of Economics, the man who devised rationing in World War I, attempted to introduce electricity rationing as part of a fuel saving scheme. Heated opposition put an end to that. The government decided to rely on voluntary conservation programs for electricity and gas, but it restricted coal deliveries, forcing those who could not purchase enough coal to take more electricity instead, and the electric companies, of course, burned coal. The government would neither let the utilities raise prices in order to dampen demand nor charge more in order to cover the rising cost of coal, because those actions would violate wartime price stabilization policies.

After the war, electricity demand rose. Electrical equipment manufacturers could not meet delivery schedules for previously ordered generators. The coal industry could not keep up with orders. Power companies feared adding to capacity because the just elected Labor government wanted to nationalize the industry, and the companies did not know how the government would compensate them for the new investment. Savoyards in the business must have thought of those classic lines from *The Mikado*:

Here's a pretty mess!

Here's a state of things![75]

Nanki Poo, of course, got his girl, and in a sense the Labor Party did, too, after the war, when it nationalized the industry.

Conclusion

The CEB prodded the British electric industry to unprecedented levels of efficiency. It created something close to a nationwide electricity management and control. Consumers reaped substantial benefits despite Balfour's extraction of undeserved profits for industry insiders and the CEB's hesitation in closing down inefficient stations and ventures. If the CEB had followed Weir's proposed path to its ultimate destination, Great Britain might have ended up with a modern, progressive and efficient electricity sector without the multiple post-war reorganizations. Politicians tried to avoid the obvious problem, the inevitable fragmentation of the sector. The war forced them to freeze the industry's structure. After the war, they would act.

---

75. W.S. Gilbert and Arthur Sullivan, *The Complete Plays of Gilbert and Sullivan* (NY: The Modern Library, no date), p. 378.

Chapter 7
Scotland and the TVA of the North (1885-1943)

I beseech hon. Members for the sake of the Highlands to give this industry a chance … A power station is not … unsightly … you cannot expect the people of Inverness to live forever on the Loch Ness monster. You must find work for the people .... [76]

– Frederick Macquisten, KC

Private and public power interests, sportsmen and landowners fought over development of Scotland's hydroelectric resources. Those debates led to actions that foreshadowed the nationalization of Britain's power industry after World War II.

Water and Aluminum

Northern Scotland, unlike most of Great Britain, had many sites suitable for hydroelectric development. The first project went up in 1885, erected by the Police Board of Greenock, Renfrewshire, supposedly to educate the public on the virtues of electric lighting. The operation worked, but the Police Board, which also owned the local gas works, closed it down two years later. In 1890, the Marquess of Ailsa considered a 2,000 kW facility on Loch Doon, but abandoned the idea because it would submerge Loch Doon Castle, ruin a salmon spawning ground and, perhaps more importantly, because no market existed for the electricity.

Rising demand for aluminum changed the picture. Aluminum production processes developed in the 1880s required vast quantities of power: 24,000 kWh per ton. The British Aluminium Company calculated that hydroelectricity was the cheapest source of power. It purchased an estate on the eastern shore of Loch Ness and constructed a 3,750 kW hydroelectric power facility and aluminum plant there. The British Aluminium power plant at the Falls of Foyer was approximately two-fifths the size of the Niagara Falls power station opened at roughly the same time. Operation began in 1896.

The Foyer plant raised worldwide aluminum production by 10%, the company had to stockpile its output, and the plant could take less than half of the hydro project's production. Demand for aluminum, however, eventually picked up. In 1909, a British Aluminium affiliate completed the Kinlochleven project with its astounding capacity of 25,725

---

76. Speech in the House of Common, 1936, quoted in Peter L. Payne, *The Hydro* (Aberdeen: Aberdeen University Press, 1988), p. 31.

kW. British Aluminium almost collapsed from burdens imposed by the facility, which would have quadrupled the nation's aluminum production. Demand created by World War I rescued the company. British Aluminium's management continued to think on a grand scale, though. With colossally bad timing, it finished the Loch Treig project (34,000 kW) in 1929 and then Loch Laggan (35,000 kW) in 1938. Rearmament needs before World War II bailed out the company.

Electric utilities planned less ambitious projects, many of which did not survive the antipathy of sportsmen, salmon enthusiasts and coal miners who fought the parliamentary acts required to license the dams. The Grampian Electric Supply Bill of 1922 did pass.[77] The company's promoters included prominent Scotsmen. The bill provided for watershed planning and guaranteed power for local use. The Grampian Electric Supply Company opened its doors in 1922. George Balfour's Power Securities holding company took control of Grampian in that same year. Balfour realized that the company could not make a profit simply serving its thinly populated service territory. It had to export power in order to thrive. In 1927, the CEB contracted to take the Grampian's output for the National Grid. Balfour now had a market. He could finance the projects. The company finished the first dam in 1930, the second in 1933, and more followed. The company had designed its projects so they could, ultimately, reach a capacity in excess of 82,000 kW. The bulk power contract with the CEB, signed in 1930, took 24,000 kW of that capacity. By 1940, Grampian had the largest territory of any undertaking in Great Britain. The company, however, served only 20% of the population in its service area and earned a low return on its investment.

The Clyde River Company completed two generating stations (15,000 kW total capacity) at the Falls of Clyde in 1924. William McLellan launched the Galloway Water Power Company in 1929. The company waited until 1931 for the CEB to agree to buy its output and then completed the five-station project in 1936. Galloway, with 99,000 kW of capacity, was the largest hydroelectric project in the United Kingdom. The Clyde and Galloway projects had required careful planning, attention to environmental issues, and political skill. Other projects fell victim to opposition from the tourist industry and landowners who wanted to protect pristine scenery and the coal mining interests that wanted to protect their markets. (See Table 7-1.)

---

77. The Grampians, a mountain range in central Scotland, separate the Highlands from the Lowlands.

Table 7-1. Major Scottish Hydro-Projects (before World War II)

Project	Developer	Capacity (kW)
Kinlochleven	British Aluminium Co.	22,000
Clyde Valley	Clyde Valley Electric Power Co.	15,520
Lochaber (Lochs Laggan and Treig)	British Aluminium Co.	85,200
Grampian	Grampian Electric Supply Co.	84,000
Galloway	Galloway Water Power Co.	103,250

Sources:

1. PEP, *The British Fuel and Power Industries* (London: PEP, 1947), p. 118.
2. Peter L. Payne, *The Hydro* (Aberdeen: Aberdeen University Press, 1988), pp. 3-35.

Notes:

Capacity figures differ between sources. The table uses PEP data.

Public Power

In 1938, Thomas Johnston, a Labor Member of Parliament, opposed a private hydroelectric project in Scotland because he did not want "to hand over to a private corporation, for purposes of gain, the great natural resources of our country".[78] Johnston became Secretary of State for Scotland in 1941 in Churchill's cabinet. He appointed a committee, headed by Lord Cooper, a Tory politician and Lord Justice Clerk (second highest judicial official in Scotland) to investigate the development of Scotland's hydroelectric resources. Among the members was Lord Weir, now serving in the war government as Director-General of Explosives. Cooper did not duplicate the work of John Snell's Water Power Resources Committee, which had already investigated the topic in 1918-1921. The Cooper Report of 1942, instead, analyzed opposition to development, which had left the Highlands scenic but economically depressed. It argued against accusations that developers exploited Highlanders because they sold power produced in the Highlands to outsiders. Electricity undertakings could not afford to supply the sparsely populated Highlands unless they could sell electricity elsewhere.

The Cooper Committee advocated the formation of a state-owned corporation, the North of Scotland Hydro-Electric Board, to develop new hydroelectric resources in the region, transmit and sell the output to existing utilities, and to distribute electricity to areas in which no other undertaking had a franchise. The Board should try to attract industry to the Highlands by offering low-cost power supplies as an inducement, build up the resources that serve existing consumers, and bring electricity service to isolated areas. The Committee declared that preserving the Highlands unchanged "for the benefit of ... holidaymakers" would "sterilize" an area "in which the dwindling remnants of the native

---

78. Payne, *op. cit.*, p. 32.

population could ... continue to reside until they eventually became extinct."[79] As Tom Johnston put it, even before the Cooper Committee had issued its report, "It is now or never for the Highlands."[80]

The North of Scotland Hydro-Electric Board became Britain's first state-owned generator and distributor, and the only British analog of the Tennessee Valley Authority (TVA) or the Bonneville Power Administration (BPA), New Deal agencies that attempted to develop the region as well as the hydro resources of the American river valleys they served.

Conclusion

As for results, a PEP report concluded that:

> Resources developed in Scotland prior, to the North of Scotland Hydro-Electric Development Act of 1943 totaled 310,000 kW .... The Board's programme, which was approved in May 1944, covered 102 projects ... equivalent to, say 3 million kW ... [81]

By 1946, Scotland produced 93% of Britain's hydroelectric power. The North of Scotland board was a new type of British electric utility, a precursor of the post-war nationalized industry.

---

79. Payne, *op. cit.*, p. 43.
80. Payne, *op. cit.*, p. 43.
81. Political and Economic Planning, *The British Fuel and Power Industries* (London: PEP, October 1947), p. 11.

Chapter 8
The Road to Nationalization (1942-1947)

Nationalization thus became a central feature in the socialist programme, and it was nationalization rather than public ownership. For public ownership includes ownership by local authorities. [82]

– H. A. Clegg and T. E. Chester

During the war, the government fretted over the structure of the electric industry. According to policy analysts H.A. Clegg and T.E. Chester:

By 1945 the industry had added to its difficulties a vast expansion of demand during the war years in which no complementary expansion of … facilities was possible – when, in fact, even maintenance and repairs had often to be skimped.[83]

Once the Labor Party took power, it solved the problem in the same way as it had intended to solve all previous electricity problems: by nationalizing the electric industry.

Stirrings

In 1942, the government formed the Ministry of Fuel and Power, headed by Gwilym Lloyd George. It began post-war energy planning. Sir William Jowitt, Paymaster-General and a veteran Laborite, investigated the industry. He pointed out that the hypothetical costs of the infamous Section 13 of the 1926 Act victimized the CEB. He advocated that the CEB take over all power plants, create a nationwide price for generation, and consolidate local activities into fewer than 30 regional boards under central control. The government solicited comments. Predictably, the municipal vs. private, large vs. small, generating vs. purchasing, and holding vs. operating company divisions prevented the industry from developing its own consensus for reform.

In the middle of the debate over structure, Parliament established the North of Scotland Hydro-Electric Board and the London & Home Counties Joint Electricity Authority tried to take over all 75 undertakings in the area. The Electric Power Engineers Association proposed a National Electricity Supply Board to own all utility generating stations, the National Grid and the distribution systems. George Wansbrough, Labor MP and

---

82.   H. A. Clegg and T. E. Chester, *The Future of Nationalization* (Oxford: Basil Blackwell, 1953), p. 5.
83.   Clegg and Chester, *op. cit.*, p. 28.

utility executive, commented during the 1943 debates that the problem was not the "untidy" nature of the industry, but rather "the imminent maturity of local authorities' purchase rights ...." [84]

The Plank in the Labor Platform

Legendary Fabian socialist Sidney Webb and his wife Beatrice ejected the anarchists from the movement and "presented Socialism in the form of a series of parliamentary measures, thus making it possible", as George Bernard Shaw put it, "for an ordinary respectable religious citizen to profess Socialism...".[85] In 1917, Webb issued his manifesto, *Labour and the New Social Order.* He advocated a cooperative, planned economy with scientifically organized industry, and nationalized railway, steamship, coal mining and electricity industries. The Labor Party adopted Webb's ideas. Clause IV, Part 4 of its 1918 constitution called for the party:

> To secure for the workers... the full fruits of their industry... upon the basis of common ownership of the means of production... [86]

That clause became Labor dogma.

The party's election manifesto for the 1945 election spelled out its intentions:

> The nation needs a tremendous overhaul... Production must be raised to the highest level... Britain needs an industry organized to enable it to yield the best that human knowledge and skill can provide... each industry must have applied to it the test of national service...

> The Labour Party is a Socialist Party... There are basic industries ripe and over-ripe for public ownership and management in the direct service of the nation... [87]

Moving from the abstract to the concrete, the manifesto urged:

> Public ownership of the fuel and power industries. For a quarter of a century the coal industry... has been floundering... Amalgamation under public ownership will bring great economies... and... raise safety standards... Public ownership of gas and electricity undertakings will lower charges, prevent competitive

84. Hannah, *op. cit.*, p. 341.
85. George Bernard Shaw, *The Intelligent Woman's Guide to Socialism and Capitalism* (NY: Brentano's, 1928), p. 220.
86. Labour Party, *Labour Party Rule Book 2013* (London: Labour Party, 2013), p.3.
87. "1945 Labour Party Election Manifesto," http:// *www.labour-party.org.uk/manifestos/1945/1945-labour-manifesto.shtml,* retrieved on 2007-08-21.

waste, open the way for coordinated research and development, and lead to the reforming of uneconomic areas of distribution. Other industries will benefit. [88]

Labor Wins

In 1944, Labor reaffirmed that it would nationalize the electricity and fuel industries if it returned to power. Its 1945 election manifesto repeated that point. The Labor Party had written "common ownership of the means of production" into its 1918 constitution. It advocated national – not local – public ownership. It believed that nationalized industries should run on a non-profit basis, or that any profit earned should go directly to the government. In 1945, Labor won the general election. Clement Attlee, ex-member of the Stepney Council Electricity Committee and ex-vice president of the London Joint Electricity Authority, became Prime Minister.

First, the government nationalized the coal industry. Herbert Morrison, Lord President of the Council, indicated that electricity was next on the list. Emanuel Shinwell, Minister of Fuel and Power, commented in 1946 that:

> While great things have been achieved in electrical development ... I feel that there is little doubt we have reached a stage at which the existing boundaries are retarding progress and the full play of enterprise, and I can see no way of securing orderly national development over the whole country apart from unified ownership and control. [89]

The only debate within government was whether or not to centralize industry operations.

In December 1946, the government introduced its bill. Shinwell recited the history of the industry and all the attempts to reorganize it. The municipalities, he noted, regarded electric service as a local affair, while the companies argued that economic development of the service required larger franchise territories:

> Moreover, if the local authorities were to exercise their purchase rights during the coming ten years ... local authorities undertakings would increase to 878 and the companies decline to 148 .... Most of these additional ... undertakings would consist of uncoordinated fragments taken from the companies. [90]

Herbert Morrison noted that government boards already:

---

88. *Ibid.*
89. *Garcke's Manual*, 1946-47 Edition, pp. 16-17.
90. *Garcke's Manual*, 1947-48 Edition, pp. 2-3.

... have been set up to run socialized industries on business lines on behalf of the community .... A large degree of independence for the boards ... is vital to their efficiency as commercial undertakings. [91]

Shinwell did not get to do the deed. The government made him the fall guy for the coal shortages of 1947 and ejected him from the cabinet. By the time that Hugh Gaitskell had taken on the job of Minister of Fuel and Power, the debate had wound down, and he could comment that the bill "may not be very sensational, dramatic or even revolutionary ...." [92]

The Electricity Act of 1947 created a British Electric Authority (BEA) to own transmission and generation facilities, and to coordinate the activities of 14 local area boards that would distribute electricity. Thanks to the doggedness and skill of Tom Johnston, now its chairman, the North of Scotland Hydro-Electric Board not only escaped from control of the British Electric Authority but also took over all undertakings within its territory. The government paid market value for securities of companies, but paid municipalities an amount equal to the municipal utility's debt outstanding, meaning that municipalities recovered a fraction of the value of their electricity undertakings. Some annoyed municipal governments reacted by running their undertakings at a loss or stripping assets out of the utility in order to give local citizens some benefits before they had to give up their utilities to the BEA. Transfer of assets ("vesting") took place in April 1948.

Shortages

At war's end, the electric industry faced organizational uncertainty in the midst of an operational crisis. Who would own or operate assets? Would owners receive adequate compensation? Managers who did not know the answers to those questions had to figure out how to deal with inadequate supply, surging demand, fuel shortages and bitterly cold weather as well.

Coal output fell in 1945. While rising in 1946 and 1947, it still remained below pre-war levels. Electric generators and British manufacturers needed more coal. Nationalization produced no productivity miracles in the coal mines. The electric industry had inadequate generating capacity in place. Delivery of electric plant on order was behind schedule. And electricity prices no longer tracked costs, thereby encouraging wasteful consumption.

By late 1946, the government had to divert coal from electricity suppliers to manufacturing industry in order to boost economic output. Consumers unable to secure coal turned

---

91. *Garcke's Manual*, 1947-48 Edition, p. 4.
92. *Garcke's Manual*, 1947-48 Edition, p. 1.

to public electricity supply. Electricity undertakings needed coal to meet demand. Generators warned that they had low fuel inventories, so a cold winter would bring on electricity shortages. Shinwell ignored the warnings. In January 1947, snow fell. Coal piles froze. Electric utilities shed load and reduced voltage and frequency. At the time of peak demand on January 29, the CEB estimated that 16% of customer demand remained unmet. The government diverted coal from manufacturing industry, turned off street lights, cancelled trains and fired Shinwell.

In the fall of 1947, just before Gaitskell's appointment to replace Shinwell, a member of Parliament asked foreign minister Ernest Bevin, "Well, Ernie and when are we going to get out of this mess," to which Bevin replied, "When Gaitskell produces the forty million tons extra coal,"[93] that is, when coal production returned to the 1930s level. That would not happen. The country needed higher coal production to supply industrial manufacturers, electric generators and the export market to earn scarce foreign exchange. In 1939, the mines produced 231.3 million tons and exported coal worth £38.3 million. In 1947, they produced 197.5 million tons and exported only £2.5 million worth. Gaitskell commented that "I made it clear that if he was to get his exports, I must have some means of stopping domestic electricity demand."[94] The fuel prospects for the coming winter looked bleak. The new Fuel and Power minister told a political gathering in November, "It means getting up and going to bed in cold bedrooms. It may mean fewer baths."[95]

Gaitskell and his permanent secretary, Donald Fergusson, worked out plans for the industry. Gaitskell believed that he had to let the boards, which had statutory powers, manage the businesses, as long as they served the public interest. Fergusson prepared a memorandum for him, setting forth these policies and suggestions:

- Let National Coal Board do its own negotiations with labor, and run its business.
- Quickly nationalize the gas industry, to achieve a "coordinated fuel and power policy."[96]
- Bring in the right managers to run the electric industry.
- Work out a "scientific set-up" for fuel policy.[97]
- Focus on two big items: "... the main issue ... is the need to control and restrict domestic consumption of electricity. The other main problem likely to arise is trouble in the pits.."[98]

---

93. Hugh Gaitskell, *The Diary of Hugh Gaitskell 1945-1958*, Philip M. Williams, ed., (London: Jonathan Cape, 1983), p. 39.
94. *Ibid.*
95. Gaitskell, *op. cit.*, p. 44.
96. _____, *op. cit.*, p. 42.
97. _____, *op. cit.*, p. 43.
98. *Ibid.*

- Give the Ministry of Fuel and Power time to organize.

Gaitskell sponsored a tax on electric heaters, and then limited their production. Early in 1948 the electricity supply industry prepared for nationalization by asking consumers to use less electricity and by limiting the number of new electric connections.

Conclusion

The CEB/National Grid combination created the central production and control structure that made full consolidation of the industry so much easier. The hypothetical pricing requirements of Section 13 of the 1926 Act, though, worked against its efforts. The distribution sector, in contrast, not only remained fragmented, but faced disintegration. Consumers did gain from the reforms. Additional restructuring would have added to benefits. Unfortunately, restructuring efforts affected groups within the industry in different ways. Reaching consensus among them seemed impossible. Restructuring by nationalization of an industry more than half-owned by governmental entities anyway scarcely seemed a radical measure, especially with a government committed to a business-like, scientific and coordinated approach.

Chapter 9

The CEB Delivered the Goods: UK vs USA (1926-1946)

In 1925, when the problem of accelerated electric development was tackled, the domestic electricity load was in an unsatisfactory state... [99]

– PEP

Electric industry reforms worked. Utility sales grew faster than the economy. Real electricity prices declined. The industry earned a profit in the worst of times. By time of nationalization, the industry had made remarkable progress in improving operations and in the electrifying the economy, although it still faced organizational challenges.

Catching Up

With the CEB in place, the electricity industry moved into high gear, not catching up with the United States, but rapidly narrowing the US lead. Electricity generation in the UK grew twice as fast as in the US (Figures 9-1 and 9-2). Electricity generated per capita (Figure 9-3) and per unit of real GDP (Figure 9-4) both shot up, with the latter measure almost reaching reached American levels, despite wartime and post-war shortages.

Figure 9-1. Electricity Generation (1927-1946) (billion kWh)

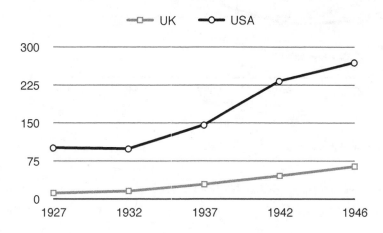

Sources:
All data from Appendix, Table B.

---

99. Political and Economic Planning (PEP), *Report on the Gas Industry in Great Britain*, (London: PEP, March 1939), p. 104.

Figure 9-2. Electricity Generation (1927-1946) (billion kWh) (Semilog scale)

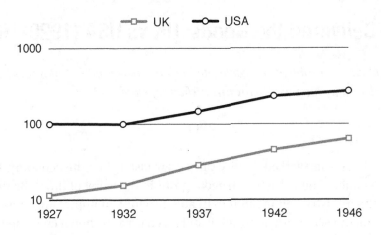

Source:
See Table 9-1.

Figure 9-3. Electric Generation per Capita (kWh)

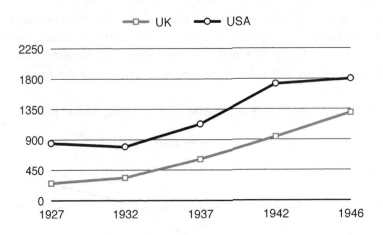

Source:
All data from Appendix, Table C.

Figure 9-4. Electric Generation per $ Real GDP (million kWh per $1990 billion)

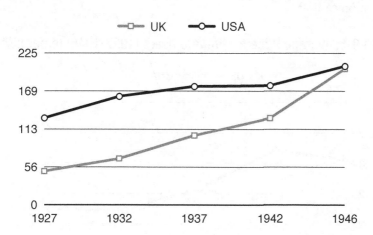

Source:
See Figure 9-3.

Pricing Down

The National Grid had two aims: make the electricity industry more efficient and transfer the benefits of those efficiencies to consumers in the form of lower prices. It delivered on both counts. Electricity prices fell in nominal (Figure 9-5) and in real terms (Figure 9-6) faster than in the United States. Prices to industrial customer, though, dropped less than the average price in both countries, though, and the British continued to charge industrial customers more relative to residential ones than did the Americans.

Doing a back of the envelope calculation, on a per kWh basis in the 1927-1946 period, UK customers paid roughly 2.2¢, from which the electricity supplier laid out 1.5¢ on operating costs, leaving about 0.7¢ for depreciation, taxes and return on capital. Industrial customers paid roughly 1.4¢, slightly below the average operating cost. American customers paid about 2.3¢, of which 1.1¢ went to pay operating costs, leaving 1.2¢ for depreciation, taxes and capital. Industrial customers paid about 1.2¢, slightly more than operating costs.

As before National Grid, UK utilities paid more for fuel than their American counterparts, a difference that added about 0.5¢ per kWh to British costs. Taking that difference into account, then, the British had managed to bring all other operating costs close to American levels. They did not, however, bring production per unit of capacity up to the American level, and that poorer utilization may have added about 0.1¢ to the price. Under National Grid management the UK electric industry reduced operating costs to American levels, to the extent possible given the coal cost differential, and the electric companies ate the higher coal costs, in a sense, by taking a smaller portion of the bill to cover depreciation, taxes and return on capital.

Figure 9-5. Average Price to Ultimate Users (1927-1946) (¢ per kWh)

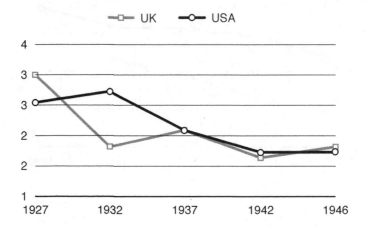

Source:
All data from Appendix, Table G.

Figure 9-6. Real Price to Ultimate Users (1990 ¢ or p per kWh) (semilog scale)

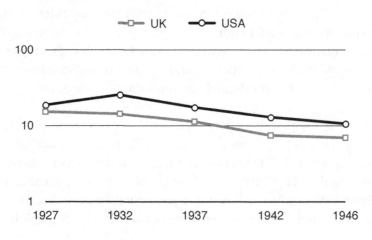

Source:
All data from Appendix , Table H.

Gas Loses Steam

Gas sales grew more slowly than electric. Gas had limited uses. In America, gas companies competed against a vigorous, declining-cost electricity sector but they had an advantage over the British gas companies. They also sold natural gas cheaper than manufactured gas and plentiful in some of the fastest growing states. By the 1920s, American electric utilities faced a revitalized gas sector. In the United States, gas consumption in total (Figure 9-7), per capita (Figure 9-8) and per unit of real GDP (Figure 9-9) grew faster than in the United Kingdom. The British gas industry did not so much fade way as move at a gerontological pace.

Figure 9-7. Gas Consumption Excluding Sales to Electric Generators (1927-1946) (billion therms)

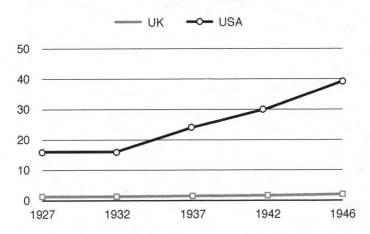

Source:
All data from Appendix, Table I.

Figure 9-8. Gas Consumption Excluding Sales to Electric Generators per Capita (therms)

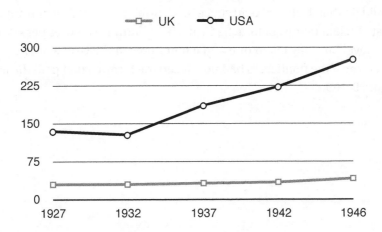

Source:
All data from Appendix, Table K.

Figure 9-9. Gas Consumption Excluding Sales to Electric Generators per $ Real GDP (therms per thousand $1990)

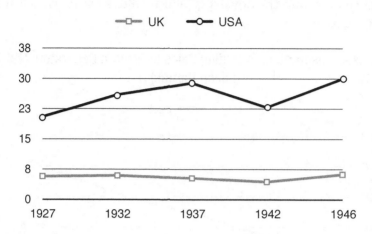

Source:
All data from Appendix, Table J.

Operating Better

National Grid, rather than forcing the industry to operate more efficiently, instead offered the utilities the opportunity to take advantage of the grid's lower prices and subsidies. National Grid paid to convert consumers to a standard electricity service but grid legislation had separated the generator from the customer. National Grid operated only in the wholesale market. It could not pitch its services to ultimate customers. Only the local utility could do that. National Grid succeeded, principally, by concentrating generation at the most efficient plants. Before the Grid, UK operating expenses per kWh sold ran at twice the US level but they fell to one and a half times in the Grid era. The gap would have narrowed more if the politicians had not rigged the formula that paid the power stations. (See Figure 9-10.)

Figure 9-10. Operating Costs (1927-1946) (¢ / kWh sold)

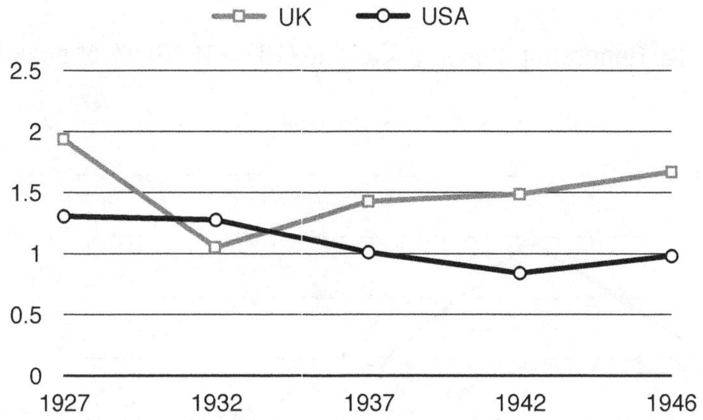

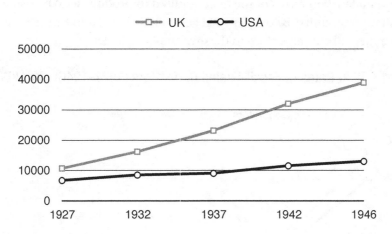

Source:
All data from Appendix, Table N.

The Grid enlarged the market for large, efficient power stations. And British ones grew larger relative to American ones, from roughly one and a half times to over two times larger. Opening larger stations and closing down smaller ones raised operating efficiency (Figure 9-11).

Figure 9-11. Average Size of Power Station (1927-1946) (kW)

Source:
All data from Appendix, Table N.

Electric utilities with inadequate interconnections maintain extra generating reserves to meet emergency needs because they cannot get back up power from neighboring utilities. Those reserves add to the electric bill. In the United States, utilities voluntarily built transmission networks to provide needed interconnections. In the UK, National Grid did it. In both countries, margin of reserve in excess of peak load declined, although by the end of World War II, the reserve margin looked too thin for comfort (Figure 9-12).

Figure 9-12. Generating Reserve Margins (1927-1946) (% of peak load)

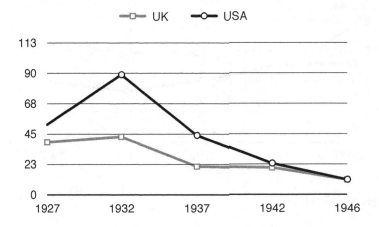

Source:
All data from Appendix, Table O.

British generators operated at a disadvantage because they paid a high price for fuel. So they attempted to use as little fuel as possible per kWh generated. The industry judges generating fuel efficiency in terms of the heat required to produce a kWh , usually measured in British thermal units (BTUs) per kWh generated. As seen in Figure 9-13, British generators operated at heat rates close to those of the Americans.

Figure 9-13. Heat Rates of Fossil-Fueled Generators (1927-1946) (Btu/ kWh)

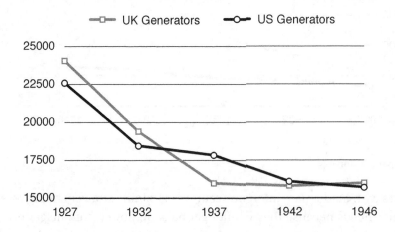

Source:
All data from Appendix, Table O.

Adding bigger power stations to the mix, concentrating output at those stations and eliminating the need for excess capacity improved reliability and reduced costs. The industry, though, could have made better use of those stations. In the pre-grid days, Americans sold almost twice the electricity per unit of capacity as the British. After National Grid took over, Americans sold only 40% more per unit of capacity. (See Figure 9-14.)

Load factor measures average utility plant capacity utilized during the year as a percentage of capacity used at peak periods. A low load factor indicates that most plants sit idle during most of the year, a situation to be avoided because the utility has to pay to maintain that excess plant. But too high a load factor indicates that the utility has little time to take the plant out of service for routine maintenance. The load factor in the UK remained consistently below US levels. (Figure 9-15.) National Grid and the CEB could only do so much.

Figure 9-14. Utility Average kWh Sold per Unit of Capacity (1927-1946) (kWh per kW of year end capacity)

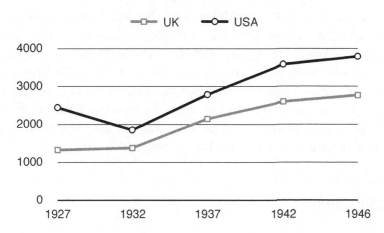

Source:
All data from Appendix, Table N.

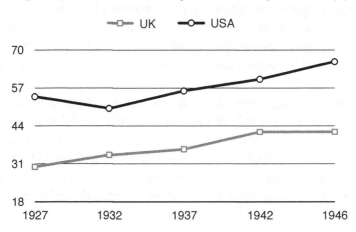

Figure 9-15. Electric Industry Load Factor (1927-1946) (%)

Sources:
All data from Appendix, Table O.

Despite the similar sales mix and the milder weather that should have raised the UK load factor, it stayed well below that of the US. Encouraging consumers to take electricity at off-peak periods was a key to raising the load factor and industrial and transportation users were the ones most likely to act if offered reasons to do so. They might not, though, if electricity suppliers, despite reform of the wholesale market, kept doing what it did before on the retail level, that is, treat large customers indifferently. Utility managers, with no idea whether their firms would stay in business due to impending franchise expirations and the threat of nationalization, had little incentive to act differently.

Industry Organization

Market reforms reduced the power station count faster than the utility count. (See Table 9-1.) Power companies grew faster than municipal undertakings, took market share and consolidated into larger entities. (See Table 9-2.) They overtook municipal utilities in almost every respect except size of power plant. (See Table 9-3.) Larger companies, however, faced an uncertain future. As franchises expired, municipal authorities could exercise their options to take over the investor-owned supplier, and thereby fragment the market again.

Table 9-1. Census of Power Stations and Electricity Suppliers (1922-1946)

	Number of Municipal Generation Stations	Number of Company Generation Stations	Number of Municipal Electricity Suppliers	Number of Company Electricity Suppliers
	[1]	[2]	[3]	[4]
1922	277	196	NA	NA
1926	272	210	339	248
1927	267	223	355	258
1932	243	208	379	261
1937	214	175	374	213
1942	205	160	370	200
1946	NA	NA	NA	NA

Source:
Garcke's Manual, various issues.

Table 9-2. Municipal (Local Authority) vs Company (1925-1942)

	1925	1926	1931	1936	1942
	[1]	[2]	[3]	[4]	[5]
Municipal utility % of kWh sales	64	65	64	63	60
Average revenue per kWh sold (old pence)					
Municipal	1.59	1.68	1.19	1.00	0.95
Company	1.54	1.52	1.02	0.99	1.00
Average operating cost per kWh sold (old pence)					
Municipal	0.91	1.11	0.69	0.74	0.79
Company	0.84	1.01	0.57	0.63	0.73
Average sales per utility (MWh)					
Municipal	11,280	11,312	17,174	32,409	48,635
Company	8,149	7,912	13,362	32,042	59,422
Average number customers per utility					
Municipal	4,375	4,720	10,160	16,965	18,969
Company	2,240	2,530	5,365	13,634	16,344
Average utility maximum load (kW)					
Municipal	5,852	5,803	8,150	12,759	15,005
Company	4,191	4,951	7,406	15,647	22,645
Average capacity per utility (kW)					
Municipal	9,278	9,313	12,309	14,281	19,070
Company	5,927	5,938	9,545	16,162	23,315
Average capacity per generating station (kW)					
Municipal	10,718	11,607	19,432	24,958	34,420
Company	6,785	7,262	12,712	20,647	28,888
Average utility load factor (%)					
Municipal	22	22	24	29	37
Company	22	18	21	23	30

115

Source

Garcke's Manual, various issues.

Notes:

All data based on Electricity Commissioners' statistics.
Average number of customers estimated by author for 1925, 1926.
Years are fiscal 1925-1926, 1926-1927, 1931-1932 1936-1937 and 1942-43 for municipals and calendar 1925, 1926, 1931, 1936 and 1942 for companies.

Profitability

While legislation capped profitability, to the public distress of company executives, the cap did not discourage companies from expanding more rapidly than government owned enterprises. The UK industry earned returns on invested capital above prevailing government bond (consols) rates but lower than returns of investor-owned utilities in the United States, where government bonds had lower yields. Since government-owned utilities (which earned a lower return on capital because they relied more on debt financing) made up the bulk of the industry, and many of the privately owned utilities sold output to a government-backed entity, National Grid, they operated a low risk business that deserved a low return. (See Table 9-3.)

Table 9-3. Profitability (1922-1947) (%)

Year	Return Earned UK Utilities	Bond Yield UK Consols	Return Earned USA Utilities	Bond Yield US Treasuries
	[1]	[2]	[3]	[4]
1922	7.7	4.4	7.5	4.3
1926	6.0	4.6	7.4	3.7
1927	7.2	4.6	7.4	3.3
1932	5.6	4.4	6.3	3.7
1937	5.1	3.3	5.8	2.7
1942	4.9	3.0	5.8	2.5
1946	4.8	2.9	6.7	2.2
1947	NA	2.6	6.4	2.3

Sources (by column):

1. *Garcke's Manuals*, various dates.
2. London and Cambridge Economic Service, *The British Economy: Key Statistics 1900-1964*.
3. Edison Electric Institute, *EEI Pocketbook of Electric Utility Statistics*, various years.
   U. S. Department of Commerce, *Long-Term Economic Growth 1860-1970* (Washington,DC: U.S. Government Printing Office, June 1973), p. 273.

Notes:
Author's estimates, 1946. (Col. 1), 1926 and 1942 (col. 3).
Return in cols. 1 and 3 defined as operating income as percentage of total capitalization. Accounts use all modern business deductions including depreciation.

International Comparisons

From 1926 to 1937, the UK's economy grew faster than elsewhere and that growth showed in electricity production. Economic recovery, then, as well as the industry reforms, helped to raise electrification levels. But despite the UK's high level of income, its per capita usage remained below the norm, except for right after World War II, when the Continental nations still suffered from wartime destruction. (See Table 9-4.)

Table 9-4. International Comparisons (1926-1947) ($1990 and kWh per Capita)

Country	Real GDP per Capita 1926	Real GDP per Capita 1937	Real GDP per Capita 1947	Electricity Output per Capita 1926	Electricity Output per Capita 1937	Electricity Output per Capita 1947
	[1]	[2]	[3]	[4]	[5]	[6]
UK	4,713	5,937	6,306	282	664	1,026
USA	6,610	6,438	8,896	799	1,131	2,125
Switzerland	5,549	6,087	8,926	1,119	1,636	2,160
Netherlands	5,528	5,301	4,925	199	406	482
Belgium	4,682	4,856	4,699	344	646	834
France	4,209	4,444	4,099	304	479	634
Germany	3,592	4,809	2,763	532	1,151	894
Austria	3,436	3,177	2,181	332	428	589
Sweden	3,404	4,664	6,176	661	1,272	1,979
Italy	2,862	3,247	2,856	184	317	380
Average for:						
All 10 countries	4,429	4,896	5,183	476	814	1,107
Continental 8	4,120	4,573	4,578	459	792	990
UK as % of:						
All 10 countries	106	121	122	59	82	90
Continental 8	114	130	138	61	85	104
USA	71	92	71	35	59	48

Note:
Col. 6 – Author's estimate of East and West Germany combined, 1947.
Sources:
Col. 1-3 – Maddison, *op. cit.*, pp. 104-106, 194-196.

Col. 4-6 – Mitchell, *European Historical Statistics*, pp. 479-482. USA from Dept. of Commerce, *Historical Statistics*, 1945 and 1970 editions.

Conclusion

Building the National Grid and then concentrating production in the most efficient facilities reduced costs and made electricity economically attractive and geographically available to most of the population. Profitability declined, so customers collected some benefits of the reform, but not all, and the imminent expiration of franchises threatened to undo benefits by tearing apart larger firms. How much the drop in profitability affected investment is unclear because the upcoming franchise expirations would also have affected investment decisions, too. Investors did not know how much they could recover when the franchise expired.

Scientific socialism, as exemplified by the professionally run government firm (National Grid) had already produced results in the electricity sector so why not extend the concept to the rest of the industry? By the end of the war the industry had run out of time to further improve efficiency and organization because it would soon fragment as a result of franchise expirations. Time to nationalize completely.

Part Three:

Scientific Socialists and Nuclear Conservatives: Nationalized Electricity

Chapter 10
The Citrine Era (1947–1957)

... Citrine is an intellectual ... He has the integrity and loyalty characteristic of the better type of British mechanic... [100]

– Beatrice Webb

The first head of Britain's nationalized electricity industry melded a disparate hodge-podge into the largest electric utility operating under one ownership. That new industry spread the benefits of electrification, while following a cautious engineering policy and always emphasizing the virtues of trade union membership.

Organizing in the Midst of Shortages

In the new era, the British Electric Authority (later called the Central Authority or CA), owned all generating stations and transmission lines and the fourteen local area boards that distributed electricity within their own service territories. The enlarged North of Scotland Hydro had an ambiguous and often tense relationship with the BEA, having a seat on the board, but reporting to the Secretary of State for Scotland, while the BEA reported to the Ministry of Fuel and Power.

Outside of the North of Scotland territory, the BEA provided electricity to the area boards and to special customers, controlled the area boards' finances and set general policy for the industry. The Minister appointed the chairmen of the area boards, thereby assuring their autonomy from the BEA. Four board chairmen sat on the BEA's board on a rotating basis, thereby assuring coordination within the industry. Consultative Councils made up of local citizens looked after the interests of consumers. Those councils had no power. They simply expressed their views to the Area Board, to the BEA or to the Minister.

The Labor government chose Lord Citrine to head the BEA. Walter Citrine had worked his way up from apprentice electrician in Liverpool to general secretary of the Trades Union Congress, the top spot in the labor movement. Early, in his chairmanship, Citrine confided to Gaitskell that "he did not feel that other people on the Board respected him enough" so he memorized the Electricity Act because "he felt that if he could quote the Act on any occasion ... it would impress them."[101] Citrine introduced a consultative, committee-

---

100. Hannah, *op. cit.*, pp. 355-356.
101. Gaitskell *op. cit.*, p. 50.

ridden style of management, forced disparate elements in the industry to work together, and maintained the industry's independence from the Minister. He encouraged unionization to such an extent that by 1949 all but 150 out of 155,000 employees belonged to a union. The industry negotiated national pay scales and agreed to restrictive work rules. In return, unions worked amicably with the industry and did not tolerate unofficial strikes, that bane of postwar British labor relations.

Emanuel Shinwell wanted the industry to meet demand for electricity by 1951: impossible due to shortages that forced the government to ration electrical machinery. To move ahead, the industry standardized in order to facilitate production, settling for small, conservatively designed 30 and 60 MW generators. Consumers endured power shortages. They wanted more electricity and they had new uses for it. The fraction of customers with electric heating ("fires"), for instance, rose from one-quarter pre-war to three-quarters post-war. The industry had to come up with the electricity to power them.

The BEA advocated pricing policies that exacerbated the shortages, to the chagrin of some Laborites. Right before nationalization, many municipal utilities deliberately underpriced, which delighted local consumers but produced operating losses at the soonto-be-nationalized undertakings. The nationalized industry had to raise prices in order to cover costs reported on its books but still underpriced services in relation to long run incremental costs (the costs of new facilities), a policy that encouraged uneconomic demand for electricity. Shinwell told the industry to raise prices enough to cover costs without defining costs. The law, after all, only required break-even operation.

The government said that nationalization would lead to lower prices. High profits would create embarrassment as well as making a dent in the budgets of already suffering working class families. Anyway, the BEA and the area boards liked low prices, which gave them a competitive edge against the gas industry and helped them meet what Leslie Hannah described as "a long-run obligation to sell as much electricity as possible at as low a price as possible."[102] Uninhibited by rational pricing, demand surged. The industry did not have the equipment to meet the demand. It responded by limiting power supply and reducing voltage, banning the sale of electric appliances and pleading with the public to reduce consumption. Industrialists then warned about the economic consequences of inadequate electric supplies.

Hugh Gaitskell realized what was going on. In February 1948 he appointed a committee chaired by Sir Andrew Clow to examine pricing. Clow, former governor of the Indian state of Assam, now chaired the Scottish Gas Board. The committee concluded that the industry should introduce a winter peak surcharge in order to discourage peak demand, and should consider time-of-day meters and load control measures. The recommendations

---

102. Leslie Hannah, *Engineers, Managers and Politicians* (Baltimore: The Johns Hopkins University press, 1982), p. 35. [Hannah, *Engineers*].

made sense theoretically. Unfortunately, area boards read meters quarterly, so consumers would not know the cost of their peak period consumption until after the end of the winter, which meant that they would not see price signals in time to do anything about their consumption in the peak season.

Unsettled by this seeming government interference with management prerogatives, industry leaders furiously attacked the proposals. Peak load pricing, they said, would not dampen peak demand, would affect the summer lull in a way to interfere with maintenance, and would reduce sales enough to cause operating losses. In other words, the price increase in the winter would not reduce demand, but the price decrease in summer would increase demand, and the reduction in demand which was not supposed to take place would be so far out of proportion to the price increase that the industry would lose money.

Gaitskell took the Clow report to the cabinet and received approval for the pricing plan. Citrine signed on to it, then reneged because "all his experts were violently opposed to it."[103] Then he said he expected no problem with the area board chairmen. So, in July 1948, Gaitskell entered the lions' den, meeting the board chairmen. Gaitskell, afterwards, wrote:

> The meeting … was an uproar… I did not mind the opposition but the unbelievably stupid arguments they put forth … The trouble, of course, is that they are all madly keen to sell electricity and just cannot get used to the idea that at the moment they should stop people from buying it."[104]

Citrine the negotiator worked out a compromise: an inadequate summer/winter price differential so poorly and belatedly publicized that customers only discovered the higher prices after the heating system had ended. A few days after the meeting, Citrine, with his advisors present, met at the ministry, and after a long session, he blew up, "accusing the Ministry of trying to dominate the Board and put things across them."[105] Citrine afterwards explained to Gaitskell that he had to do it in order to build up the chairmen's confidence in him. The industry, however, had asserted its independence, sabotaged an attempt to introduce rational pricing and won the battle.

This difficulty set Gaitskell to thinking, again, about the relationship between the boards and the ministry. He concluded that the ministry would need more power over the boards:

---

103. Gaitskell, *op. cit.*, p. 79.
104. *Ibid.*
105. *Ibid.*

It is really very unsatisfactory having to deal with the Boards as though they were independent authorities with no special obligations to the Government. [106]

A few months before this incident, Labor ministers discussed the problem of (or need for) ministerial control over boards. Aneurin Bevan, a left winger, advocated full control. Gaitskell admitted that it was "irritating not to be able to keep them on the right lines" but being "committed … to the principle of the semi-independent Board," they had to "give this particular … relationship a fair trial."[107] At that point, Sir Stafford Cripps, Chancellor of the Exchequer, commented: "All I can say is we must remember it is an experiment."[108] Future electricity policymakers might have benefitted from such humility.

Operating Conservatively

During a period of fuel, equipment and capital shortages, the British Electric Authority had to build new generating capacity, operate existing power stations plus the transmission network, improve existing facilities, standardize the industry, cope with rising demand, extend electricity to those without it, keep prices at politically acceptable levels and maintain financial stability. The BEA's staff knew the transmission business from the days of running the National Grid. Generating experience resided in the field with the employees of the former undertakings.

In 1950, the BEA announced the Supergrid project, a ten-year long construction effort to erect a 1,150 mile transmission network operating at 275 kV. (The National Grid operated at 132 kV. The first American 230 kV line went into service in 1923 and the first 287 kV line, from Hoover Dam to Los Angeles, in the early 1930s.) The Supergrid, a north-south interregional network, would permit the BEA to build power stations in northern coalfields and transmit electricity to southern load centers, and to site power stations to increase system efficiency.

On the generating side, engineers initially chose 30 and 60 MW units with relatively low steam pressures and thermal efficiencies. That satisfied British electrical equipment manufacturers, comfortable in the low technology rearguard of the world's industry. (American utilities installed more advanced units in the 1930s.) In 1953, a new engineering management pushed the size of generators to 100 MW, then 120 MW, then 200 MW, and by 1956, they had a 275 MW unit on the drawing board. Even under this new regime, the British were at least 10 years behind the Americans in terms of size, steam pressure and temperature, and they took longer to complete their facilities, too.

---

106. Gaitskell, , *op. cit.*, p. 80.
107. _____, *op. cit.*, p. 72.
108. *Ibid.*

Although standardization of electric supply at the distribution level was one of the original reasons for nationalization, that goal had to take a back seat as the BEA spent its scarce resources on desperately needed generating equipment and on extension of electrification to rural areas. Not until the mid-1950s could the industry put resources into standardization of distribution.

Meanwhile, despite shortages, the industry promoted electrification through sale of appliances at industry-owned stores, electrical contracting operations and, beginning in 1954, with nationwide advertising. Area boards offered incentives to builders to install wiring, sockets and appliances. The British Electric Authority's own research staff, headed by Paul Schiller, showed that electric heating accounted for over half of peak load, and that cost of serving the heating load exceeded revenue collected from it. The industry downplayed Schiller's conclusions.

The BEA developed a bulk tariff rate (price of electricity sold to area boards) that reflected cost of capacity at peak plus operating costs, with adjustments for regional fuel costs. But area board chairmen opposed off-peak pricing for the bulk tariff. Did they believe that a lower price would not encourage more off peak or that bulk pricing signals represented an interference in their domain? Only the Eastern board, chaired by Cecil Melling, offered off-peak tariffs. Eastern also encourage sale of storage heaters that could charge up during low-tariff periods. Other boards sold off-peak electricity on separate wiring circuits, a throwback to old practices, until persuaded to use price as an inducement instead, in 1955.

Tories Return

Conservatives won the 1951 election. They did not denationalize the industry but rather tinkered with the model, in 1952, transferring the South West and South East Scotland boards and local generation and transmission assets in those areas from the jurisdiction of the Ministry of Fuel and Power to that of the Scottish Office. In 1955, those two boards merged into the South of Scotland Electricity Board. That move, effectively, put the Scottish industry in a separate category from that of England and Wales. The new Macmillan government also launched an export drive that diverted electric equipment from domestic utilities that badly needed it to catch up with demand. With Scotland out of the picture, the government changed the name of the British Electric Authority to the Central Electric Authority (CA).

Viscount Ridley, of a prominent Northumberland Tory family, chaired a committee in 1952 that recommended pricing that would ensure fair competition between nationalized industries. That, in turn, led to still another committee, made up of industry chairmen, which buried Ridley's proposals in even more committees, but gave some support to the views of Paul Schiller. Then, in 1954, the government put Sir Edwin Herbert, an experienced lawyer with an interest in commercial issues, in charge of a committee whose report criticized the industry's over-centralization, budgeting procedures and lack of research. The report asked, specifically, whether the electric industry made the best possible use of capital and manpower, charged the correct prices, and was organized and staffed in a manner to allow it to move forward. The report concluded that "This is not … an inefficient industry but rather one … that could be improved…" [109] In 1956, with a new minister in charge, the government decided to reorganize the industry along lines unattractive to the still presiding Lord Citrine.

Even before that, Tories interfered with operations. Fearing a fuel shortage, the government pressured the industry to convert generating stations to burn oil, an uneconomic decision. In the same year, 1954, the government decided to launch a nuclear power program that would produce plutonium for the military as well as electricity for civilian purposes. It did so without consulting the CA. Based on experience with military reactors, the government planned two reactors for 1957, two more for 1959, and eight more by 1965. Engineers at the Atomic Energy Authority (AEA) calculated that, even after adding in the value of the military plutonium, nuclear power would cost more than coal-fired electricity. No matter. The government's working party on nuclear energy, chaired by a Treasury official, Burke Trend, spoke of "modest investment" and "considerable return,"[110] projections that proved wildly in error. Early in 1955, with little consideration, the CA took on the task of building the reactors.

The government produced a rhapsodic White Paper singing praises of nuclear power. A nuclear Britain would "lead another industrial revolution in the second half of the century."[111] The government selected a British, gas-cooled design whose advantages were primarily military: the Magnox (named after the magnesium alloy used for the fuel containers). Would another design better suit civilian purposes? That did not matter. The CA had no other options. The government was in a hurry. The experimental reactor at Calder Hall, run by the AEA, opened in 1956. The AEA wanted to double the nuclear program. The CA hesitated, but knuckled under in October 1956, during the Suez Canal crisis. In December, worried about dependence on foreign oil, the supply minister proposed tripling the nuclear effort. Macmillan, back as prime minister, took charge of the nuclear program himself. The AEA called for 6,000 MW of nuclear capacity by 1965. The cabinet approved, rejected the CA's call for caution and the British electric industry plunged into the nuclear abyss. Construction bids on the first stations were three-and-a-half times those of conventional coal-fired units on a per kW basis, the CA's projections showed the nuclear stations earning a negative rate of return, and a coal shortage that had worried the cabinet had turned into a coal surplus. But the nuclear program moved ahead.

---

109. A.H. Hanson, "Electricity Reviewed: The Herbert Report" *Public Administration*, Vol. 34, June 1956, p. 211.
110. Hannah, *Engineers*, p. 173.
111. Hannah, *Engineers*, p. 174.

Having interfered with decisions about fuel and pushed on the CA an inherently centralized production process, nuclear power, the Conservative government decided to circle back to complaints about the industry's over-centralization and bureaucracy aired in the Herbert report. The government passed the Electricity Act of 1957 which created an Electricity Council (on which all board chairmen would serve) to act as a toothless coordinating body, a forum for whatever ideas members chose to share, serving, too, as a financial supervisor in case any of the boards failed to meet financial obligations and a Central Electricity Generating Board (CEGB) to own the power stations and transmission lines. The chairman of the Electricity Council had no power base. Whoever headed the giant CEGB would become the industry's leading figure.

The government wanted a new leader to replace Citrine, an outsider, a nuclear enthusiast.

Conclusion

The British Electric Authority united a fragmented industry that had difficulty meeting the public's demands. At nationalization, one-quarter of British households had no electricity. A decade later, the percentage had fallen to one tenth. In 1948, 32% of farms had electricity. By 1958, 72% did. Hundreds of electric undertakings consolidated into 15. Perhaps the BEA was too bureaucratic and cautious, but it did rescue the industry from postwar fragmentation and solved problems caused by hypothetically priced electricity (section 13 of the 1926 Act).

Lord Citrine first defended the industry's independence from political interference, a dubious accomplishment given that it came at the expense of a forward thinking pricing policy that might have revolutionized the industry and made it a real world leader. Later, he compromised the industry's independence by acquiescing to ill-conceived oil and nuclear schemes.

Ironically, the Labor Party, having nationalized the industry, seemed to take seriously the ideal of the independently run nationalized industry, even if that meant allowing it to pursue less than ideal pricing policies. The Conservatives, on the other hand, interfered with the industry's operations, foisted uneconomic policies on it, and enlarged the government's role in the economy through a huge nuclear program, but claimed, at the same time, that they disliked centralization.

Chapter 11

The Nuclear Push (1957–1970)

In twentieth-century, Britain Socialism has been the only creed and the Conservatives have had to accept the greater part of it. [112]

– C. Northcote Parkinson

The electricity industry needed to catch up with demand, wanted to increase consumption per customer and had to satisfy the orders of its owner to promote the role of nuclear power.

Changing the Guard

The government wanted the changes at the top. Ronald Edwards of the London School of Economics, for instance, had participated in studies critical of the industry. He could have been the fresh face to chair the toothless Electricity Council, but he was too controversial for that spot until 1964. The government, instead, chose Sir Henry Self, civil servant, and former deputy to Citrine. For the big job, CEGB chair, the government picked Sir Christopher Hinton an AEA official. Leslie Hannah, historian of the nationalized industry, characterized Hinton as "a difficult person" nicknamed "Sir Christ" back at the AEA, but the government saw a nuclear future for the electricity sector, and wanted a "nuclear knight" to lead the way. [113]

The area boards held an odd position in the newly reorganized industry. They operated independently, except for budget review by the Council, but they had to buy all their electricity from the CEGB, although in theory they could generate electricity if the minister approved. Since purchased power represented their biggest cost, whatever the CEGB did affected them. When the CEGB finally put together a bulk tariff that made sense, they had to respond by changing their business practices. From 1960 to 1962, the CEGB increased the capital component of the bulk tariff, gave discounts to interruptible loads, and instituted a night vs. day differential for the running charge, with the backing of Ronald Edwards, not yet chairman. The bulk tariff reflected costs of production. It signaled consumers to use electricity during off-peak periods when prices were lower. Once the area boards faced realistic bulk tariffs for the electricity they bought, they designed marketing

---

112. C. Northcote Parkinson, *Left Luggage* (Boston: Houghton Mifflin, 1967), p. 173.
113. Hannah, *Engineers*, pp. 186-187.

programs to encourage customers to take electricity during periods of low demand. The industry succeeded in selling more appliances and raising its load factor.

About this time, the Scottish Office, every ready to tidy up the electricity industry north of the border, saw its chance when Tom Johnston retired from chairing North of Scotland Hydro. The Office gave the chairmanship of the board to Lord Strathclyde, figuring that he was more amenable to merging the two Scottish systems than Johnston. Strathclyde had retired from a long career in the Royal Navy to become a chartered accountant, served as a minister of state in the Scottish Office, and he was well beyond normal retirement age already, so, perhaps, not likely to put up much of a fuss when the government dealt the knockout blow to Johnston's baby.

North of Scotland Hydro-Electric Board at time of nationalization in 1948 acquired all electricity undertakings in northern Scotland. The government formed South of Scotland Electricity Board in 1954 to consolidate the South West and South East Scotland Electricity Boards and to acquire local generating and transmission assets from the British Electricity Authority. North of Scotland served a thinly populated region largely with hydro-electric power. It charged consumers a uniform tariff, which equalized prices between high cost rural and low cost urban areas, in order to spread the benefits of electrification throughout its territory. South of Scotland served an urbanized, industrial region, largely with fossil fueled power stations.

The 1962 Mackenzie report argued that merger would improve the Scottish industry's efficiency. The North of Scotland board, led by Strathclyde, resisted. The government had to compromise: the two Scottish boards would remain independent, but would cooperate in generation. The arrangement kept the Hydro from launching uneconomic waterpower projects and let the Scots take a more adventurous approach to design than the CEGB, but the joint arrangement also tied the Hydro into an expensive nuclear power station.

After the 1965 Joint Generation Agreement (JGA), South of Scotland dispatched generating units of both utilities on a merit order basis (lowest cost units go into service first), reducing overall generating costs throughout Scotland. The boards pooled generation, transmission and some distribution costs. North of Scotland contributed low-cost hydro generation and in turn, enjoyed the economies of scale of South of Scotland's large stations. The southern board built and operated the nuclear power plants and contracted to buy fuel from British Coal for the thermal units. The northern board operated most of the hydro facilities, put the Peterhead station (which burned oil and gas) into service in 1980 and signed contracts for North Sea gas for that station. The JGA stayed in force until March 31, 1989.

The government had other goals for the entire nationalized industry beyond providing better electric service, such as manipulating the industry's capital budget in order to protect the value of the pound, helping depressed areas, maintaining price stability or reducing government spending. It imposed price freezes in 1957 and 1958. The law required the industry to operate at least on a break-even basis but the Tories wanted profits. In 1958, the industry changed its method of calculating depreciation, which reduced reported profits but not cash flow, thereby disguising profitability. The CEGB introduced a discounted cash flow test to improve investment decision making but did not apply that test to its biggest investment, nuclear power. The government wanted more nuclear investment, greater financing of expenditures from internal sources and stable prices. All at the same time.

Riding the Nuclear Wave

Generation accounted for over half of electricity cost and capital budget. During winters in the early 1960s, due to inadequate generating capacity the CEGB had to reduce voltage and interrupt service. The government already had a solution: a 6,000 MW nuclear program (equivalent to 25% of 1957 generating capacity) scheduled for 1965 completion. CEGB's engineers opposed it as uneconomic. The only buyer to believe the story about economic nuclear power was the South of Scotland Electricity Board, whose members were miffed by being left out of early nuclear planning. That board ordered a Magnox reactor in 1956. The Hunterston station took twice as long to complete and cost twice as much as expected. The Queen Mother opened it in 1964. Hannah noted that "This was the only Magnox order placed in the expectation that it would be economic...,"[114] and that, although it qualified "as the most incompetently executed Magnox construction project... operationally it was to prove highly satisfactory."[115]

Industry engineers soon calculated that nuclear stations would produce at a higher cost than coal generators. The AEA intended to unload its high-cost nuclear fuel inventories on the industry, on top of which it backed off from plans to buy back spent fuel from the generators at a high price. Then, as a result of an accident at the AEA's Windscale facility, the Magnox units required a redesign. On top of those problems, the government-owned National Coal Board (NCB) began to pressure the CEGB to take more coal. The NCB feared a coal surplus. CEGB's engineers had managed to reduce fossil fuel generating costs, too. They calculated total cost per kWh of electricity produced by Magnox reactors was twice that of modern coal stations.

---

114. Hannah, *Engineers*, p. 270.
115. Hannah, *Engineers*, p. 272.

Even Hinton became cautious about nuclear power, which disappointed government officials who dreamed of exporting Magnox reactors to the world. Hinton delayed the nuclear program but would not back the latest fossil fuel technology, either. He feared that the CEGB would end up with experimental nuclear and coal units, neither of which would work well. So, the CEGB opted for an uneconomic nuclear technology and missed the economies that became available from modern, conventional power stations. Eventually, the CEGB did build large, fossil fuel power stations with several 400-500 MW units on the sites located in rural areas. That policy required a stronger transmission grid, the 400 kV Supergrid.

Hinton left the CEGB in 1964. F.H.S. Brown, the long-time engineering chief, took over.

Operations and Markets

From the late 1950s to 1970 – before discoveries in the North Sea rejuvenated the gas industry – the electric industry thrived. It boosted plant size, employed staff more effectively, improved generating efficiency and raised transmission voltages. Load factor rose. The industry produced and priced more efficiently. Customers bought more electricity, installed new appliances and even responded to off-peak pricing. (See Table 11-1.)

Table 11-1. Operations and Markets (1948-1969)

	1948	1956	1969
	[1]	[2]	[3]
Generation			
Largest plant in service (MW)	105	600	2,000
Employees/MW	3.86	2.53	1.56
Thermal efficiency-conventional steam (%)	21.2	24.9	26.3
Load factor (%)	46.9	47.9	54.7
Nuclear % of generation	0	0	10
Transmission (% of circuit kilometers)			
65 kV and lower	28.9	8.1	1.0
132 kV	71.1	79.8	10.3
275 kV	0.0	12.1	41.1
400 kV	0.0	0.0	47.6
Distribution			
Employees/GWh sold	3.00	2.00	0.77
Consumption/customer (kWh)	3,201	4,650	9,210
Off peak consumption/customer (kWh)	NA	21	728

	1948	1956	1969
Estimated home appliance saturation (5)			
Electric oven	18	27	39
Kettle	20	32	55
Refrigerator	3	10	53
Washing machine	4	21	63
Water heater	11	26	56

Notes:

1. NA – not available.
2. Fiscal years 1948-1949, 1956-1957 and 1969-1970.
3. Column 1 – Largest plant in service is author's estimate based on Hannah, op. cit., p. 114.
4. Column 2 and 3 – Transfer of lower voltage transmission lines to area boards from CEGB reduces percentage of lower voltage lines owned by CEGB.
5. Off peak consumption defined as sales on restricted hour and day/night tariffs. Column 2 estimated by author.
6. All home appliance saturation extrapolated from surveyed years.

Sources:

1. The Electricity Council, *Handbook of Electricity Supply Statistics 1987* (London: The Electricity Council, 1987).
2. Hannah, *Engineers, Managers and Politicians*, p. 114.

Conclusion

The nationalized electricity suppliers had to take on government-imposed burdens: development of nuclear power, support of chosen vendors, boosting a technology that nobody else wanted. It had to operate with stop-start budgets determined by overall governmental needs, rather than by what the industry required, and pricing that served political as well as business purposes. It exhibited caution when dealing with the conventional, and a politically pressured daring when dealing with the experimental, ending up with an uneconomic nuclear program in place of an economic fossil fuel program. Yet the nationalized industry in its first 20 years almost quintupled output, tripled sales per customer and increased its operating efficiency. In those respects, it did what a commercially driven industry would have done, in spite of the obstacles.

Chapter 12

Slowdown, Strike and Privatization (1970 – 1990)

Such arguments embody a common and basic misconception – that centralized management and size can be equated with efficiency. [116]

– Allen Sykes and Colin Robinson

The electricity industry had to compete with a reinvigorated gas industry, deal with an expensive nuclear program and survive a strike that almost closed down the country. Finally after four decades of socialism, the government decided to exit the electric business.

Competition Comes

In 1965, British Petroleum found natural gas in the North Sea. More discoveries followed. British Gas' reluctance to pay enough for the gas initially slowed development. But natural gas found a market and changed the country's energy mix, accounting for less than 1% of the nation's energy in 1970, 22% in 1980 and 24% in 1990. The nationalized gas industry could now compete vigorously against nationalized electric and coal industries, which were raising their prices at a faster pace. The average gas customer took two and a half times as much gas in 1979 as in 1969. (See Table 12-1.)

Table 12-1. Electricity vs Gas (1969 to 1979)

	1969 [1]	1979 [2]	Annual increase (%) [3]
Volume			
Gas (million therms)	5,235	16,736	12.3
Electric (million kWh)	168,230	204,762	2.0
Revenue / unit sold			
Gas (pence/therm)	9.44	17.61	5.2
Electric (pence/kWh)	0.79	2.8	13.5
Customers (millions)			
Gas	13.35	15.26	1.3
Electric	18.27	20.33	1.1

---

116. Allen Sykes and Colin Robinson, *Current Choices: good ways and bad to privatise electricity* (London: Centre for Policy Studies, 1987), p. 13.

Notes:

1. Fiscal years 1969-1970 and 1979-1980.
2. Electric data for England and Wales only.

Source:

Electricity Council.

Costs and Coal

Margaret Thatcher's Conservatives fervently believed in the superiority and even necessity of private over state ownership of industry, with the subtext that privately owned
industries would run more efficiently. But in the 1970s the nationalized electric industry
had improved operating ratios, reduced staffing and held salaries in check. Prices rose
less than inflation despite the huge hike in coal bills imposed by a nationalized coal industry not subject to normal market forces because its main customer was stuck with it.
(See Table 12-2.)

Table 12-2. Electricity Revenues and Cost Indices (1969 - 1979)

	1969	1979	Annual Rate of Change (5)
	[1]	[2]	[3]
Income statement (p/kWh)			
Revenue	0.82	2.92	13.5
Costs			
Fuel	0.25	1.46	19.3
Salaries	0.13	0.44	13.0
Depreciation	0.16	0.31	6.8
Other operating expenses	0.09	0.45	17.5
Total operating costs	0.63	2.66	15.5
Operating income	0.19	0.26	3.1
Indices (1969=100)			
Coal price	100	582	19.3
Weekly earnings			
Electric	100	454	16.3
Manufacturing	100	488	17.3
Retail prices	100	499	17.4
Electricity prices	100	360	13.7

Notes:

1. Fiscal years 1969-1970 and 1979-1980.
2. Fiscal years ending March 31.
3. England and Wales only for revenues and expenses.
4. Revenues include miscellaneous revenues less costs.

5. Other expenses includes rates (local property taxes), supplemental depreciation and sales and administrative expenses.
6. Operating income is interest expense plus net profit. Most of interest expense is paid to the government which also owns the industry.

Sources:
1. Electricity Council.
2. Government statistics.

The industry expanded the high voltage network, opened large power stations and raised thermal efficiency. Sales growth slowed, though, and the industry could not convince customers to take more electricity at off-peak times, which limited its ability to improve asset utilization. The picture would have looked better had the government not spent so much time and effort to develop a nuclear industry at the expense of the electricity customer. (See Table 12-3.)

Table 12-3. Operations (1969 - 1979)

	1969 [1]	1979 [2]
Generation		
Largest plant in service (MW)	2,000	2,640
Employees/ MW	1.56	1.06
Employees / GWh generated	0.36	0.26
Thermal efficiency – conventional steam (%)	28.3	33.3
Load factor (%)	54.7	57.7
Nuclear % of generation	10.0	11.4
Transmission (% of circuit kms)		
Under 275 kV	11.3	5.9
275 kV	41.1	30.1
400 kV	47.6	64.0
Distribution		
Employees / GWh sold	0.77	0.48
Consumption / customer (kWh)	9,210	10,070
Off peak consumption/ customer (kWh)	728	764

Note:
1. England and Wales.
2. Fiscal years 1969-1970 and 1979-1980 ending March 31.

Source:
The Electricity Council.

Labor relations deteriorated. This was post-war Britain, after all, land of work stoppages, strikes, tea breaks, shunning, low productivity and class warfare, the world celebrated in the film classic *I'm All Right Jack*. In late 1970, a work-to-rule protest (slow down) caused by a fight over wages led to a 30% reduction in power output, parliamentary debates by candlelight and warnings that the elderly could freeze to death. In late 1977, another work-to-rule slowdown spread through the industry. This time, newspapers headlined surgery by candlelight and obligatory items about Parliament operating in the dark. Almost every electricity employee was a union member, so not even management would cross picket lines and operate plants, a common practice during the occasional American electric utility strike.

In the nuclear sector, however, the CEGB touted ambitious goals while demonstrating an inability to deliver on them. In 1972, the CEGB needed two nuclear plants by 1980 and 18 thereafter. In 1973 it planned to order 18 by 1980 and 18 thereafter. In 1979 it would require one per year or maybe 15 in the decade. Meanwhile, the advanced gas-cooled reactor (AGR), the government designated successor to the Magnox, turned out to cost more than the old design.

Thatcher Takes Charge

Margaret Thatcher took office in May 1979. Unlike previous Tory leaders, she actually believed in competition and sale of state-owned businesses to the private sector. In 1980, Secretary of State for Energy, David Howell, said that the government intended to repeal the statutory prohibition on any entity "generating electricity as a main business."[117] At that time, business organizations could generate their own electricity but could sell their excess output only to the CEGB, not to other business organizations.

In March 1982, Energy Secretary Nigel Lawson declared "There was no case for a state monopoly in electricity. There should be increasing scope for private generation of electricity."[118] The Energy Act of 1983 permitted independent electricity generation. But with no published tariff for purchases from independent producers, and with the nationalized industry firmly in control of transmission and distribution, independent generating never got off the ground. By the late 1980s, non-industry sources supplied less than 2% of the electric industry's needs, the bulk of which came from two government agencies, British Nuclear Fuels and the Atomic Energy Authority, not from independent entrepreneurs.

---

117. Hugh Noyes, "Government relaxes hold on three nationalized industries," *The Times*, July 22, 1980, p. 1.
118. Peter Riddell, "Lawson plans to end state monopoly of electricity generation," *Financial Times*, March 29, 1982, p. 1.

The CEGB and National Coal Board (NCB) began the decade by signing a Joint Under-standing to run through March 1985. The CEGB would limit purchases of foreign or pri-vately mined coal and take up to 75 million metric tons per year from the NCB as long as prices rose no faster than the retail price index. Later, the CEGB altered the Joint Un-derstanding. It would buy 95% of its coal from the National Coal Board (later renamed British Coal) with price adjustments tied to oil prices. Still, prices for British-mined coal exceeded world prices.

The Joint Understanding saddled CEGB with high-cost fuel and assured the coal board of a market and price for 70% of its output. The understanding raised fuel costs and elec-tricity prices, and provoked complaints from industrial electricity buyers. CEGB and NCB responded by introducing the Qualifying Industrial Consumers Scheme (QUICS). NCB provided CEGB with four million metric tons per year at close to world market prices and CEGB passed on those savings to QUICS customers. The rest of the customers could subsidize NCB.

Meanwhile, the Mergers and Monopolies Commission (MMC) report on nuclear policy released in May 1981, denounced the 15,000 MW nuclear expansion strategy of Decem-ber 1979 as based on "appraisals which are seriously defective," doubted that nuclear power offered a cost advantage, asserted that the CEGB could have made better deals for coal transport and coal purchases from British Railways (another government-owned corporation) and NCB, said that CEGB should import more coal, and denounced it for its "seriously inaccurate" forecasts that had led it to order too many power stations.[119]

In 1982, Lawson chose Dr. Walter Marshall, a mathematician, to chair the CEGB. Tony Benn, left wing Labor minister and former aristocrat, had fired Marshall from the posi-tion of chief scientist at the Department of Energy, said the *Financial Times*, "for advo-cating nuclear policy too passionately," after which Marshall headed the Atomic Energy Authority, and chaired a "task force to salvage plans for the first big British pressurized water reactor (PWR)."[120] Marshall headed the CEGB and its principal successor, almost to privatization.

Despite slowing sales growth, CEGB put mammoth facilities into service. Drax station began operating with 2,000 MW of capacity in 1983. The CEGB later added another 2,000 MW, making Drax one of the largest coal-fired generating stations in the world. In 1984, CEGB laid an undersea cable to France, opening the UK market to France's nuclear power giant, Électricité de France (EDF). The CEGB pressed ahead with its nuclear program, this

---

119. "Britain's nuclear foes get ammunition for their fight against atomic energy," *World Business Weekly*, June 8, 1981, p. 15.
120. David Fishlock, "A crusader for the nuclear future," *Financial Times*, May 28, 1982, p. 10. Other reports say that Benn fired Marshall in a dispute about how the poor would pay their energy bills. Either way, the firing brought Marshall to the attention of Margaret Thatcher and her crew.

time using the American-developed PWR. In order to achieve economies of scale and standardization, CEGB planned four plants for service by 2000, beginning with Sizewell B whose pre-construction public hearings stretched out 27 months in 1983-1985 before construction began. The 1,175 MW plant was completed in 1994. The other three plants were cancelled in 1989.

Putting Walter Marshall in charge of the CEGB did not solve its biggest non-nuclear problem: a joined-at-the-hip relationship with NCB. Someone had to shape up that agency. Mrs. Thatcher picked Ian MacGregor, a Scottish engineer who, after a successful metals and mining career in the United States, returned to the United Kingdom in 1977 to straighten out British Leyland, the troubled automobile maker. In 1980, Thatcher put him in charge of nationalized British Steel, where he sacked staff and slashed losses. He took over NCB in 1983. Given his contentious labor relations record, nobody should have had any illusions about what might come.

His opposite number, Arthur Scargill, the Marxist leader of the National Union of Miners (NUM) opposed closure of any mine that still had coal in it. He took over the top spot in the union in 1982, succeeding Joe Gormley. At the time, the left and right wings of the labor movement were engaged in a bitter fight for control of the unions and the Labor Party. Scargill was on the left and Gormley on the right.

CEGB prepared for a strike, stockpiled coal and readied power plants to burn oil. To Nigel Lawson, it was "just like rearming to face the threat of Hitler in the 1930s."[121] In March 1984, the NUM struck to protest mine closings. Labor organizer Ken Smith described the "… great miners' strike of 1984-1985" as "the longest lasting, most bitter industrial dispute of the second half of the 20th century in Britain…"[122] and a "battle between the workers and the ruling class."[123] The strike lasted for a year, but many miners stayed on the job, encouraged by a mysterious figure called the "Silver Birch" and aided in part by David Hart, a real estate developer, libertarian acolyte of Margaret Thatcher, and, in the words of his *Financial Times* obituary, a "shadowy fixer" who helped to defeat the strike "by bankrolling strike-breakers from his suite at Claridge's."[124] With mines open, coal supply reached power plants.

The CEGB had close calls but kept power flowing by switching fuels, from 77.2 million metric tons of coal and 2.7 million of oil in fiscal 1983-1984 to 40.5 million metric tons of coal and 22.0 million of oil in fiscal 1984-1985. It shifted generation to units with access

121. Ken Smith, *A Civil War Without Guns – 20 Years On* (London: Socialist Publications, 2004), chapter 1 (no pagination in web edition).
122. _____, *op. cit.*, chapter 1.
123. _____, *op. cit.*, introduction.
124. Brian Groom, "Flamboyant libertarian who helped Thatcher defeat the miners," *Financial Times*, January 8/9, 2011, p. 5.

to non-strike coal. Power production rose in the strike year, but at a cost: electric industry net profit fell from £1.4 billion in the previous year to a loss of £0.6 billion. The strike failed, broke the NUM and focused the Thatcher government on security of electricity supply, which really meant freeing the country from dependence on coal (and coal miners) and promoting nuclear power.

Despite labor drama and little sales growth during the Thatcher years, the industry managed to eke out incremental improvements. (See Tables 12-4 and 12-5.) Its most intractable problems, ironically, centered around the unsatisfactory operating record and construction delays of the nuclear stations that constituted a key ingredient of Thatcherite energy policy.

Table 12-4. Markets and Prices (Fiscal Years 1979 - 1989)

	1979	1989	Annual Rate of Change (%)
	[1]	[2]	[3]
Sales volume			
Gas (million therms)	16,736	18,552	1.0
Electric (million kWh)	204,762	230,493	1.2
Customers (millions)			
Gas	15.26	17.72	1.5
Electric	20.33	22.18	0.9
Indices (1979=100)			
Coal price	100	174	5.7
Weekly earnings			
Electric	100	193	6.8
Manufacturing	100	196	7.0
Retail prices	100	192	6.7
Electricity prices	100	184	6.0
Gas prices	100	208	7.6

Sources:
1. Electricity Council, *Handbook of Electricity Supply Statistics*, 1987.
2. Department of Trade and Industry, *Energy Statistics 2001* (London: HMSO, 2001).

Table 12-5. Operations (1979 - 1989)

	1979 [1]	1989 [2]
Generation		
Employees/ MW	1.08	0.81
Thermal efficiency-conventional steam (%)	33.3	35.5
Load factor (%)	57.7	60.8
Nuclear % generation	11.4	16.2
Transmission (% of circuit kms)		
Under 275 kV	5.9	5.5
275 kV	30.1	28.0
400 kV	64.0	66.5
Distribution		
Employees/ GWh	0.48	0.35
Consumption/ customer (kWh)	10,070	10,393

Sources:

1. Electricity Council, *op. cit.*
2. Department of Trade and Industry, *op. cit.*

The Sales Pitch

The Tories sold state-owned enterprises one after another: British Aerospace and Cable & Wireless in 1981, Amersham International and Britoil in 1982, Associated British Ports in 1983. In 1984 after selling Enterprise Oil and Jaguar, they put a traditional, monopolistic utility, British Telecommunications (BT), on the block. That sale demonstrated the contradictions of the process, the desire to roll up into one package a company profitable enough to fetch a high price, a light-handed regulatory scheme to increase efficiency at an ex-Post Office department and a competitive telecommunications market. They could have split BT into competing firms, but that would have complicated the sale, and investors would have paid less for smaller firms that had to compete so they kept BT intact but licensed Cable & Wireless as a competitor, which guaranteed only limited competition near term. The government sold 51% of BT's stock initially and maintained a "golden share" that gave it a veto over key corporate decisions.

Regulation of BT followed along lines developed by Prof. Stephen Littlechild of the University of Birmingham. Price of service would rise annually at the rate of inflation (the retail price index or RPI) less a productivity factor (X) for a five year period. This framework gave BT an incentive to reduce costs, modernize, and sell more services because it could retain benefits derived from those actions until the end of the five years. The concept needed work, however, because BT managed to earn high profits while still struggling to provide better service.

British Gas (BG) emerged from state ownership in December 1986 as a nationwide gas monopoly. BG operated with an RPI-X formula plus a pass-through of natural gas purchase costs. The regulatory incentives to increase efficiency, then, did not extend to the company's biggest expense, the gas it bought.

The government converted ten regional water authorities in England and Wales into companies and sold them in November 1989. The X in the RPI-X formula was heavily influenced by the need to finance huge capital expenditures for water quality improvement. Functionaries talked up the wonders of competition in the water business, which at the time looked like window dressing to disguise the fact that the pro-competition government had just created a classic regulated monopoly.

Shares of privatized companies rose dramatically after the initial public offering (IPO). Critics accused the government of underpricing shares, depriving the Treasury of billions of pounds. But, if the goals were to get the government out of business and to encourage the public to invest, the sales succeeded.

UK vs US Regulation

Tories touted the superiority of RPI-X over American regulation because it encouraged efficiency, required a small regulatory agency and interfered less with managerial decisions. American utilities charged customers for all prudently incurred costs plus an allowed rate of return. The British set price and let the utility earn more or less than the allowed rate of return, depending on managerial prowess. American managers, they implied, ignored costs because they collected a set profit no matter their efficiency while British managers would watch costs carefully because they could retain any extra profit derived from stringent cost control. (See Table 12-6 for a simplified example of the American regulatory regime.)

Table 12-6. American Regulation: An Example

	Year 1	Year 2	Year 3	Year 4	Year 5
Sales (kWh)	1,000	1,000	1,000	1,000	1,000
Price per kWh (¢)	11	11	11.5	11.5	11
Revenues ($)	110	110	115	115	110
Expenses	100	105	105	100	100
Profit	10	5	10	15	10
Investment ($)	100	100	100	100	100
Return on investment (%)	10	5	10	15	10

US Utility invests $100 in rate base. Regulators find that Utility should earn 10% on investment in order to provide a fair return to investors, which means a $10 profit on the $100 invested. Utility produces 1,000 kWh, at an operating cost of $100. Utility charges customers 11¢ per kWh in order to collect $110 ($100 to cover operating costs and $10 for profit). In the next year, operating expenses rise to $105 and profit falls to $5, only a 5% return on investment. Utility seeks to raise revenues $5, in order to restore profit to the allowed $10. Regulators agree, and raise price per kWh to 11.5¢. In the third year, with the new prices in effect, Utility collects $115 of revenue, and shows a $10 profit (10% return on investment). In the fourth year, expenses fall to $100, thanks to a cost cutting drive. Profit rises to $15 (15% return on investment), which gets the attention of regulators, who find that Utility is making too much money. In the fifth year, they order Utility to reduce price back to 11¢ per kWh, revenue declines to $110, and profit to $10.

American utilities, the British argued, pass costs to customers as quickly as allowed and keep savings only until the regulator orders a price cut to reduce profits to the allowed level. In reality, though, regulatory proceedings drag on ("regulatory lag") and regulators neither force utilities to disgorge excessive earnings nor make them whole for earnings lost during the lag. Regulators in some states fix rates for several years at a time, which forces the utilities to watch costs. They may, also, benchmark costs (or determine their prudence) before allowing recovery. American utilities have practical rather than theoretical reasons to run efficiently.

American regulation, said critics, encourages excessive investment because profit increases (assuming a fixed rate of return) only when rate base does. Excess investment raises bills because consumers pay a return on the excess. The incentive to over invest, though, occurs when expected rate of return on new investment exceeds cost of capital. A utility would not voluntarily invest to earn a 10% allowed return if capital cost 11%, but it would if capital cost 9%. The problem arises when the regulator sets the return too high.

The UK's regulator caps price rather than profit. At the beginning of a multiyear (usually five-year) period the regulator sets an initial price per kWh to produce revenue needed to

cover expenses plus return on investment. So far, this procedure looks like American regulation in disguise, stretched over five years with one difference. In the UK prices rise or
fall automatically each year at the rate of inflation plus or minus an X factor designed to
encourage efficiency and cover costs of expansion, a procedure that eliminates frequent
hearings to adjust prices. The utility retains savings that exceed X but eats the deficiency
if it cannot make the efficiency target.[125] (See Table 12-7 for a simplified example of UK
regulation.) At the end of the five year pricing period, the regulator resets prices and X
factor for the next five years.

Table 12-7. UK Regulation: An Example

	Year 1	Year 2	Year 3	Year 4	Year 5
Sales (kWh)	1,000	1,000	1,000	1,000	1,000
Rate of inflation (RPI) (%)	1	2	3	2	1
Productivity adjustment (X) (%)	-1	-1	-1	-1	-1
Price change from previous year (%)	0	1	2	1	0
Price/kWh (p)	11.00	11.11	11.33	11.44	11.44
Revenues (£)	110.0	111.1	113.3	114.4	114.4
Expenses (£)	100.0	105.0	105,.0	100.0	100.0
Profit (£)	10.0	6.1	8.3	14.4	14.4
Investment (£)	100	100	100	100	100
Return on investment (%)	10.0	6.1	8.3	14.4	14.4

UK Utility has 1,000 kWh sales, £100 invested, and operating costs of £100. Regulator sets the
opening price at 11p (£0.11) per kWh, based on a 10% return on investment plus operating costs,
and orders X factor annual price reductions of 1% to encourage greater productivity. Inflation price
adjustments vary from year to year. In the first year, Utility charges the opening price, the rate of
inflation at 1% offsets the productivity factor of -1%, revenues reach £110, expenses £100 and
profit is £10, all as expected. In the second year, price is raised for a 2% inflation rate less a 1%
productivity factor, for a net increase of 1%, which brings price to 11.11p per kWh and revenue to
£111.1, but expenses, rise to £105, and profit falls to £6.1. In the third year, inflation rises to 3%, so
the price increase less the productivity factor is 2%, bringing the price to 11.33p. and revenue to
£113.3. Expenses remain at £105, though, and profit rises to £8.3, better than last year, but still far
from the anticipated number. In the fourth year, inflation is 2%, the price adjustment nets out to 1%
after subtracting the productivity price cut, and price reaches 11.44p. Revenue rises to £114.4, but
management has reduced operating expenses to £100 and profit shoots up to £14.4, far better than
expected. In the fifth year, inflation is only 1%, which is offset by the 1% productivity price reduction,
price of electricity remains unchanged at 11.44p, and revenues remain unchanged at £114.4, as do
expenses at £100. As a result, UK Utility again earns £14.4.

---

125. If X factor reduction is £10 while productivity improvement is £20, the utility earns an extra £10. If the utility
    reduces costs only £5, it earns £5 less than the base line earnings.

Because they can file for rate relief when needed and because constitutional rules circumscribe regulatory actions, American utilities should show steady growth around a trend line. (See Figure 12-1.) They rarely cut the dividend. Predictability reduces cost of capital.

Figure 12-1. Earnings Pattern of US Utility

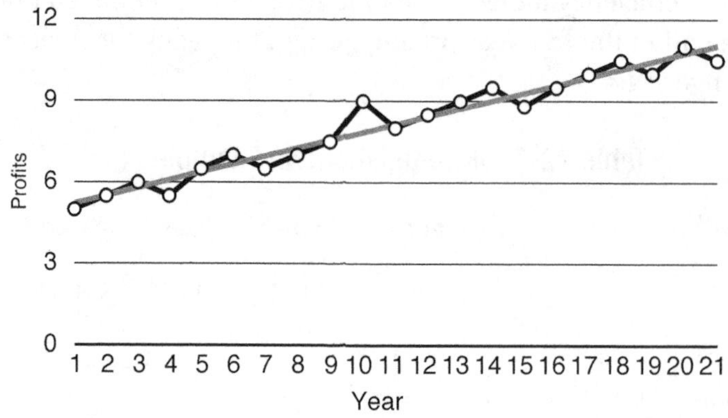

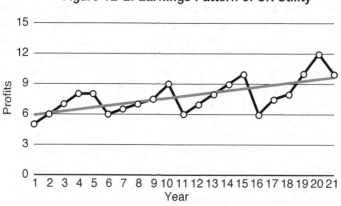

The price resets every five years in the UK cause jagged earnings and dividend patterns, add to uncertainty and make the utilities more volatile and therefore riskier investments (Figure 12-2). Compared to American utilities British ones have tended to sell at lower stock market valuations in spite of higher returns, indicating that UK regulation raises cost of capital.[126] Deregulators – fascinated by the lure of incentives and efficiencies – ignored cost of capital on both sides of the Atlantic. But for deregulation to pay off over the long term, the savings derived from competition and greater efficiency have to exceed the higher cost of capital.

Figure 12-2. Earnings Pattern of UK Utility

---

126. A Citi brokerage report written with benefit of twenty years of experience, claimed the existence of an "additional premium on cost of equity demanded from utilities operating under price cap/benchmarking regulation ... in relation to cost plus systems..." Marcelo Brito, Alexandre Kogake and Kaique Vasconcellos, "CPFL (Company Update)", Citi Research, 19 December 2012, p. 2.

The UK's regulatory framework has advantages:

- It provides real price certainty to consumers.
- It encourages efficient operations.
- Setting price every five years reduces regulatory expense.
- It protects the utility from the effect of inflation.

The UK formula has disadvantages:

- A regulatory miscalculation could persist for five years.
- It discourages unplanned investments because it would provide no return on them.
- It encourages the utility to hide information about potential cost savings from the regulator.
- It raises investor uncertainty about the size of price and dividend changes every five years.

If conditions diverge materially from expectations, the regulator could reopen the price order, but that would weaken incentives inherent in the arrangement.

The UK system supposedly gives utilities greater scope and incentive to innovate and cut costs because they can retain the benefits of those actions for a predictable period of time. But after utilities squeeze out excess costs, the regulator has less room to reduce prices in anticipation of productivity gains, and the regulatory framework reverts to the equivalent of American rate of return regulation cloaked in UK terminology.

Electricity Privatization

Electricity supply, with fifteen parts and more on the way, was the most complicated privatization of all.

The Conservative Party Manifesto of May 1987 advocated privatization of electricity supply and promoted nuclear power, which really meant independence from British coal. Allen Sykes, an executive of the mining group Consolidated Gold Fields and a world-renown expert on the development and financing of giant projects, and Colin Robinson, an economics professor, wrote in a study for The Centre for Policy Studies, a Conservative think tank, that:

> ... the Conservative election manifesto ... is almost silent on the reasons for electricity privatization. The single sentence on the subject states 'following the

success of gas privatization, with the benefits it brought to employees and millions of consumers, we will bring forward proposals for privatizing the electricity industry subject to proper regulation.' The benefits of electricity privatization seem to be regarded as self evident....

The manifesto ... supports nuclear energy as a supplier of low-cost electricity, stating that to depend on coal alone '... would be short-sighted and irresponsible...'[127]

The Thatcher government believed that privatization, *per se*, produced public benefits, the country should reduce its dependance on coal and increase the role of nuclear power, but it sidestepped the question of what to do with the nationalized coal industry. In fiscal 1986-1987, fuel cost made up about 40% of the electric bill. Coal generated over 80% of electricity. The industry paid roughly 50% more than world prices for NCB's coal. Privatizing NCB might shake it up and reduce its operating costs. Sykes and Robinson emphasized that:

> ... electricity consumers stand to gain a great deal from coal privatization. Similarly, if coal is not privatized a large part of the potential gains from electricity privatization will not be realized. Electricity customers should benefit from better management and reductions in ... costs under ... electricity privatization; but they might well forego the reductions ... which coal privatization would stimulate. [128]

Bringing coal prices down to international levels would lower electric bills 15%. Privatizing electricity without dealing with coal left consumer benefits on the table. Nuclear power had not met expectations but with the 1984-1985 miners' strike in mind, Tories pushed for more nuclear. Breaking up NCB and importing foreign coal might have just as effectively reduced the power of one union to close down the electric industry.

In determining industry structure, the government decided to aim for more competition, do something simple and easily sold within the life of the existing Parliament, make the industry more efficient, get a good price for the shares and promote nuclear power. Mrs. Thatcher selected Cecil Parkinson as Secretary of State for Energy, as the man in charge. Parkinson was a businessman, accountant, ex-chairman of the Conservative Party and, early on, one of the prime minister's key advisors. The government, though, underestimated the complexity of the choices and the time required to complete the process.

Most economists and consultants had concluded that generation could operate competitively. But how many generators constituted a competitive market? A Centre for Policy Studies paper by Alex Henney, an engineer and industry consultant, concluded that:

---

127. Sykes and Robinson, *op. cit.*, pp. 39-40.
128. _____, *op. cit.*, p. 27.

The CEGB should be broken up into nine or ten separate generating companies, all of similar capacities .... Most importantly, in order to maximize competition, the generating capacity should ... be dispersed throughout the country. [129]

Breaking the CEGB into ten parts would have left no firm large with the scale required to manage the nuclear power sector, so the government chose to divide CEGB generation into two. The larger company, National Power, would retain nuclear operations in England and Wales. Eventually, bankers convinced the government that they could not sell a generator with the CEGB's nuclear assets, construction plans and liabilities, so it took the nuclear plants from National Power and put them in a state-owned Nuclear Electric. If the Tories had planned to retain nuclear assets within the public sector, they could have split the remaining generating sector into more parts in order to create a competitive market. But time was running out, and still another reorganization might have pushed the sale beyond the next election.

The government reconstituted 12 area boards into 12 regional electric companies (RECs), with an average of 1.8 million customers each. The RECs varied in size and prospects, but restructuring them would not have been worth the disruption. The government put transmission into a new company, aptly named National Grid, owned indirectly by the RECs. The Scottish Office, also taking the path of least resistance, retained two integrated utilities.

The nuclear security blanket turned into a can of worms. Nuclear cost more than conventional power. The government had to force RECs to purchase it. Aged Magnox reactors ran well, but were scheduled for near term retirement and would require large sums for decommissioning. Advanced gas reactors (AGRs) had longer life expectancies but poorer performance records. Pressurized water reactors (PWRs), according to a parliamentary report, would produce electricity at "two or three times the price of fossil-fuel generation."[130] British Nuclear Fuels, the firm that treated spent nuclear material, raised reprocessing fees when its own decommissioning estimate rose eleven times over (by over £4 billion) between 1988 and 1989. The CEGB raised its decommissioning cost estimates for Magnox units by £2 billion. The future liabilities of the nuclear establishment exceeded the value of the operating assets. [131]

---

129. Alex Henney, *Privatise Power: restructuring the electricity supply industry* (London: Centre for Policy Studies, 1987), pp. 39-40.

130. "The Cost of Nuclear Power," Fourth Report, Energy Committee, House of Commons, June 7, 1990 (London: HMSO, 1990), p. xxiii.

131. Press reports at the time intimated that the CEGB's managers had put something over on the government, had hidden the nuclear financial liabilities. Yet the CEGB's annual reports (public documents) in the late 1980s, before the privatization, hid nothing. The 1986/1987 report, for instance, showed that the CEGB had accrued £2.6 billion in provisions for future nuclear liabilities on the liability side of the balance sheet, but there was no corresponding fund on the asset side of the balance sheet into which the CEGB had set aside the funds to pay for the nuclear decommissioning. In effect, the CEGB had reduced its annual earnings with an allowance for future nuclear expenditures, but it spent the money. The "surprised" politicians and civil servants could have done what ordinary shareholders in ordinary companies do all the time: read the annual reports.

Nervous investment bankers claimed they could not sell nuclear facilities to investors unless somebody else took responsibility for spent fuel disposal and plant decommissioning costs. Privatized nuclear firms would have insufficient funds to pay those costs, no means to charge customers for past deficiencies or future needs, and no call on the government to pick up the tab. Bankers advised: remove nuclear assets from the privatization. The Energy Department seemed genuinely "surprised"[132] according to testimony to the Energy Committee of Parliament.

In July 1989, Cecil Parkinson, who had assured Parliament that nuclear power was economical and that the industry had made "full provision" for "the anticipated costs dealing with spent nuclear fuel and retired nuclear plant"[133] removed Magnox reactors from National Power's generating fleet, leaving them in government hands. He then moved on to the Transport ministry. John Wakeham, an accountant and businessman who earned the soubriquet of "Mr. Fixit" for his problem-solving ability, became Energy Secretary. On the day in November 1989 when the East Germans opened the Berlin Wall's gates, the government decided to retain all nuclear plants and obligations that it previously refused to guarantee. Now the government had no reason left to divide privatized generation into only one big and one small part, but no time to reorganize again before the stock offering. Lord Marshall, proponent of nuclear power, the man who kept the lights on during the coal strike, tipped to head National Power, resigned as chairman of the CEGB shortly thereafter.

The government had to design a competitive market that assured generators enough revenue to contract for overpriced British coal and still make a predictable profit. It created a daily market for electricity which sets generation prices, but also required the RECs to sign multiyear contracts to buy electricity from generators at specified prices. Large customers could bypass the REC and make deals directly with a generator, but too much bypass would detract from the value of the RECs so the government limited bypass opportunities.

A predictable formula set prices for services at the local level. Two factors would control prices: the rate of inflation (measured by changes in the Retail Price Index or RPI) and potential for productivity increases (the X factor).

The Electricity Act of 1989 outlined the framework for privatization. Over a year of bargaining and planning ensued before the actual sales took place. In November 1990, the month in which the Tory leadership ejected Margaret Thatcher from office, the government sold the 12 RECs. The two generators went public in February 1991, and the two Scottish utilities in May 1991. All shares shot far over the issue prices.

---

132. *Ibid.*
133. Alf Young, "Delicious ironies in nuclear power U-turn," *Glasgow Herald*, Nov. 13, 1989, p. 13.

Conclusion

The nationalized electric industry improved its performance even as politicians turned against state ownership. The Thatcher government wanted to sell off monopolistic, nationalized utilities and institute competition in the electricity marketplace. Yet it showed an extraordinary ignorance of the industry that it owned and supposedly supervised and imposed policies that tied the industry to uneconomic nuclear power and British coal, hampered development of competitive markets and prevented consumers from realizing the full benefits of restructuring. The privatization process, in this case, involved too many contradictions: competition desired but duopoly instituted, freedom of choice but nuclear required, and rewards to the generator for cutting costs but its biggest cost (coal) fixed. The industry and the regulators spent years, thereafter, working their way through those issues.

Chapter 13

Nationalization's Semi Success: UK vs USA (1947-1990)

The leader of Albania, Enver Hoxha, warned before his death in 1985 that departures from the true path of Stalinism would end up as a 'bucket of crabs'. If only someone had given a similar warning to Mrs. Thatcher... [134]

– John Rentoul

Nationalization began with high hopes. An independent, unified, government-owned industry operating on businesslike principles would run better than a fragmented one. The new managers adhered to a conservative course when operating independently of government interference. They favored timidity and bureaucracy over technological leadership.

Government policies seemed as focused on supporting British coal and nuclear establishments as on producing electricity. The nationalized industry did not act in the independent manner envisioned by its founders, and government policy makers deserved more blame for the industry's egregious failures than utility managers who chalked up a record of improvements. The government – not the managers – made the daring decision, the push for nuclear power, and it manipulated the industry's investment programs in order to fit them into the budget.

Going Forward and Backward

In 1947-1970, electricity sales outpaced the economy. After that, electric sales growth fell off as natural gas reached the market, the energy crisis of the 1970s raised prices, and heavy industry declined. At nationalization, the UK produced one-fifth as much electricity as the USA and at the tail end of the nationalized era, only one-tenth as much. (See Figure 13-1.) Before 1970, UK and US generation grew at close to the same rate. Afterwards, UK electricity growth slowed down relatively and absolutely. (See Figure 13-2.)

---

134. John Rentoul, "Privatisation: The Case Against," in Julia Neuberger, ed., *Privatisation: fair shares for all or selling the family silver?* (London: Papermac, 1987), p. 1.

Figure 13-1. Electric Generation (1947-1989) (billion kWh)

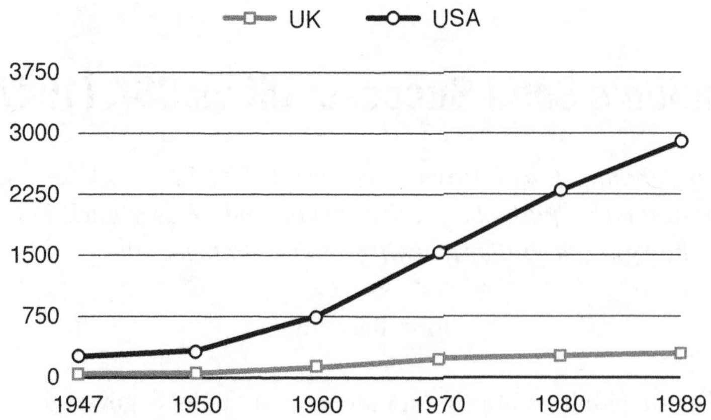

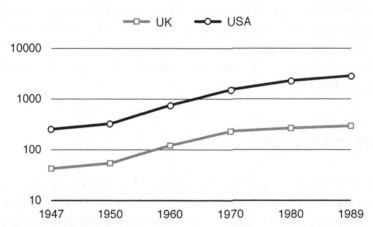

Source:
All data from Appendix, Table B.

Figure 13-2. Electric Generation (1947-1989) (billion kWh) (semilog scale)

Source:
See Figure 13-1.

Post 1970, electric generation per capita (Figure 13-3) grew at a declining rate and generation per unit of real gross domestic product (Figure 13-4) actually fell, trends that emerged sooner in the UK than in the USA. Generation per unit of GDP peaked in the UK in the 1970s and in the USA in the 1980s. Optimistic electric industry planners always predicted growth. They had a hard time adjusting to the slowdown well after it should have been apparent.

Figure 13-3. Electric Generation per Capita (1947-1989) (kWh)

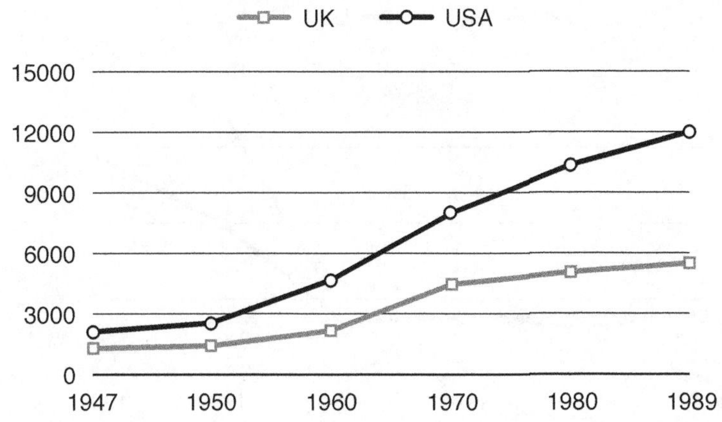

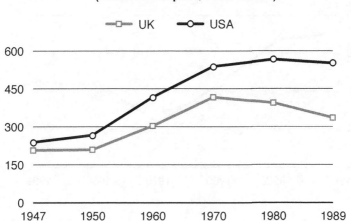

Source:

All data from Appendix, Table C.

Figure 13-4. Electric Generation per $ Real GDP (1947-1989) (million kWh per $1990 billion)

Source:

See Figure 13-3.

Pricing Up

Electricity prices declined for nine decades to the 1970s, when global energy shortages raised fuel prices, economies of scale in generation ended and mismanaged nuclear programs raised fixed costs. British and American prices stayed close until the 1970s, after which UK prices rose more. (See Figures 13-5 and 13-6.)

Figure 13-5. Average Price to Ultimate Users (1947-1989) (¢ per kWh)

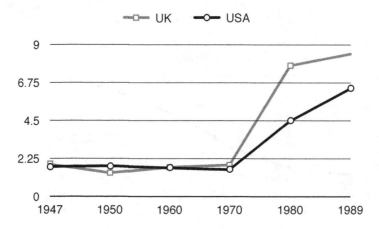

Source:

All data from Appendix, Table G.

Figure 13-6. Real Price to Ultimate Users (1947-1989) (1990 p or ¢ per kWh)

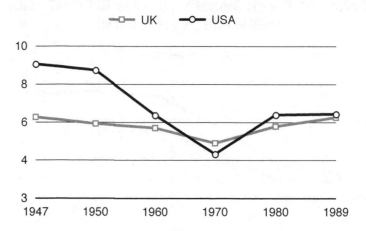

Source:
See Figure 13-5.

In the first half of the nationalized era, UK and US customers paid (all on a per kWh basis) about 1.8¢. Coal, the primary fuel, cost the same in both countries. Industrial customers in the US got a better deal, paying about 1.0¢, vs 1.4¢ in the UK. In the second half of the nationalized period, UK customers paid an average of 6.1¢ vs. 4.2¢ in the US. Higher coal costs added about 0.8¢ to the UK electricity price. American industrial customers paid about 3.1¢ and the British 4.8¢. Expensive British fuel raised bills, and put energy-intensive British manufacturers at a competitive disadvantage.

Gas Revived

The manufactured (town) gas industry peaked by the 1960s, when North Sea natural gas fields began production and the gas industry started taking market from electricity. Natural and town gas combined accounted for 10% of UK energy consumption in 1970 and 30% in 1980. Rising electricity prices helped sell gas. The British, though, still paid more for and consumed less gas per capita than Americans.

UK gas consumption (excluding sales to electric generators) doubled in the 1960s, tripled in the 1970s, and then growth slowed. (See Figure 13.7.) In the USA, gas consumption almost doubled in the 1950s and rose 60% in the 1960s. On a per capita basis, gas usage continued to increase in the UK while it peaked in the United States in the 1970s (Figure 13-8). Gas consumption per unit of real GDP topped out in the 1980s in the UK and in the 1970s in the US (Figure 13-9). Basically, other than selling gas to electric generators, the gas industry could not find additional uses for the product, and the fact that new appliances were more efficient than old ones made it harder to make the business grow.

Figure 13-7. Gas Consumption (Excluding Electric Generators) (1947-1989) (billion therms)

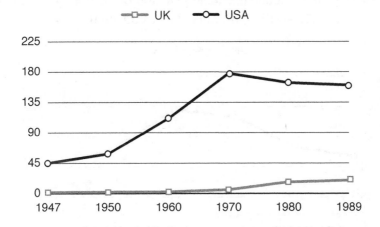

Source:
All data from Appendix, Table I.

Figure 13-8. Gas Consumption (Excluding Electric Generators) per Capita (1947-1989) (therms per capita)

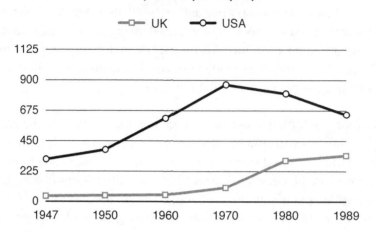

Source:
All data from Appendix, Table K.

Figure 13-9. Gas Consumption (Excluding Electric Generators) per $ Real GDP (1947-1989) (therms per $1,000 1990)

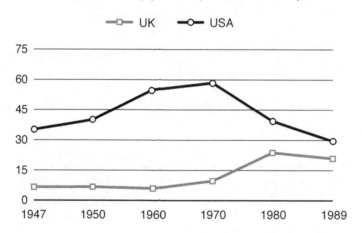

Source:
All data from Appendix, Table J.

Operating on Par

Managers can control expenses by choosing low-cost fuels (not an option for the CEGB), burning fuels more efficiently (more electricity per unit of fuel), minimizing wage bills (fewer employees and lower wages) and utilizing plant better (producing more with existing facilities). Costs of financing and maintaining fixed assets add up to at least 20% of the electric bill, so making existing assets more productive benefits consumers.

The UK electric industry increased the volume of electricity sold per unit of generating capacity (Figure 13-10), the load factor (Figure 13-11) and lowered the reserve margin (Figure 13-12) to American levels during the nationalized period. Unfortunately, although the UK industry improved its asset utilization, it also chose the wrong assets, saddling consumers with bills for overly expensive nuclear facilities that never met expectations. Then again, so did utilities elsewhere. America ended up with roughly the same proportion of nuclear power as the British, generated by some extraordinarily expensive and cranky power stations. AGR units, however, made up over half of UK nuclear capacity. Cumulative capacity factors for AGRs, by the mid 1980s, were only 29%, versus about 60% for Magnox reactors and standard reactors used elsewhere. [135] Americans picked the right kind of nuclear station and could not blame the government for strong arming them into going nuclear. They did it voluntarily. They thought the numbers would work out. The British did it for other reasons. At least UK consumers could find some solace because that expensive nuclear power saved them from buying expensive local coal.

Figure 13-10. Electric Utility Average kWh Sold per Unit of Capacity (1947-1989) (kWh per kW of Year End Capacity)

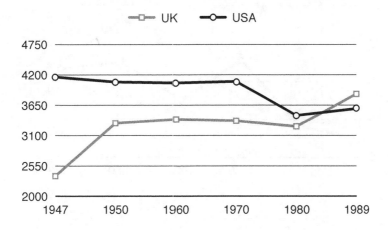

Source:
All data from Appendix, Table N.

---

135. L.R. Howles, "Nuclear Station Achievement: 1984 Annual Review," *Nuclear Engineering International*, May 1985.

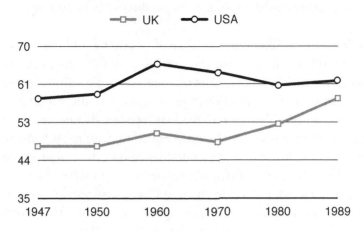

Figure 13-11. Load Factor (1947-1989) (%)

Source:
All data from Appendix, Table O.

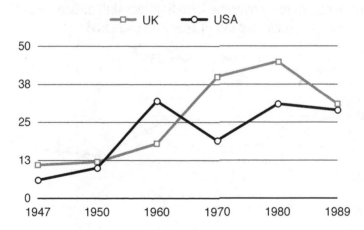

Figure 13-12. Reserve Margin (1947-1989) (%)

Source:
See Figure 13-11.

In the old days, British generators burned fossil fuels almost as efficiently as the Americans. For a good part of the post-war period, though, British efficiency fell below American standards. Strikes, unwillingness to build the most efficient type of station and preoccupation with the nuclear solution all hurt. But, by the end of the nationalized era, the two industries were again neck and neck for fuel efficiency, as measured by the heat rate. (See Figure 13-13.)

160

Figure 13-13. Heat Rate of Fossil-Fueled Generators (1947-1989) (BTUs per kWh)

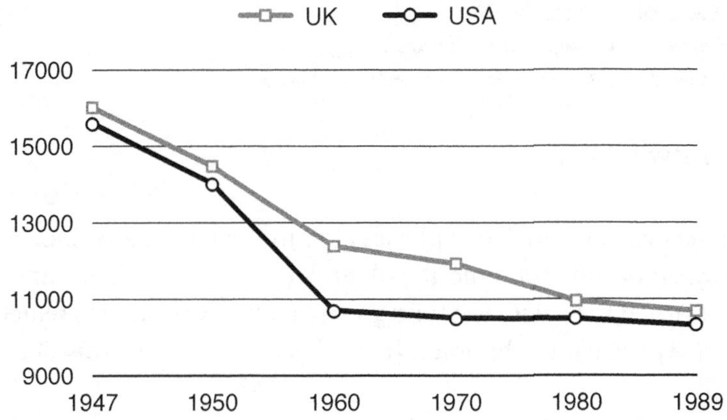

Source:
All data from Appendix, Table O.

Employment data show a different picture. The British required roughly 50-100% more employees than Americans for the same level of output. Employment costs generally ran at around 15% of the electric bill, so bringing staffing down to American levels could have cut UK electric bills 5-7%: the average price per kWh in the first half of the nationalized era would have been 1.7¢ as opposed to 1.8¢, and in the second half 5.7¢ versus 6.1¢. The Labor Party found jobs for labor, and the Tories went along with that policy. (See Table 13-1.)

Table 13-1. Operating Efficiency Measures (1947-1989)

Year	UK Customers per Employee	USA Customers per Employee	UK GWh sold per Employee	USA GWh sold per Employee
	[1]	[2]	[3]	[4]
1947	83	125	250	809
1950	75	139	264	1,071
1960	84	134	498	1,589
1970	93	148	950	3,072
1980	129	149	1,312	3,631
1989	196	161	2,113	4,144

Notes:
Columns 1 – 1947 and 1989 estimated by author.
Column 1 – England and Wales only.
Columns 3 and 4 – GWh is abbreviation for gigaWatt-hour (one million kWh).
Columns 2 and 4 – Investor-owned utilities.

Sources:
Column 1 – *Digest of UK Energy Statistics,* various issues.

Columns 2, 4 – Edison Electric Institute, *Statistical Yearbook*, various issues.
_____, *Pocketbook of Electric Utility Statistics*, various issues.
Columns 1 and 3 – Electricity Council, *op. cit.*
Hannah, *Engineers, Managers and Politicians*, p. 294.
Government reports, prospectuses and brokerage reports.

Hurry Up and Slow Down

The electricity market went through two phases after the war. For more than two decades output rose rapidly, outpaced economic growth and real price fell. Then came the energy crisis, consumers learned to do with less, and growth stalled. Consumers reluctant to buy high-priced electricity turned to abundant natural gas whose price was declining. (See Table 13-2.)

Table 13-2. Annual Rates of Change (1947-1989) (%)

	1947-1960	1960-1970	1970-1980	1980-1989
	[1]	[2]	[3]	[4]
Real GDP	2.8	2.9	1.9	2.9
Electric generation (all)	5.9	6.2	1.3	1.1
Electricity sales (utility)	8.4	6.5	1.5	2.1
Gas sales	2.1	7.9	11.5	1.4
Real price				
Electricity (all)	-0.5	-1.4	1.5	-1.0
Electricity (residential)	-2.0	-1.5	-0.3	2.0
Gas (all)	2.4	-4.0	-4.5	-0.1

Notes:
Real GDP in 1990 £.
Electric generation and sales in kWh.
Gas sales in therms.
UK GDP deflator used to calculated real price changes.

Sources:
See Appendix, Table A for real GDP, Table B for electric generation, Table D for electric sales, Table I for gas sales, Table H for real price of electricity (all and residential), and Table L for real price of gas.

UK vs the World

Nationalized industry managers set out to build sales, no matter the adverse consequences. They outsold their Continental peers. Later, electricity sales growth slowed more in the UK than on the Continent or in the USA, not because British prices were higher than on the Continent (they were not) nor because of competition from natural gas (which became competitive in many European countries at roughly the same time as in Great

Britain). The anemic state of the British industrial sector undoubtedly contributed to the results but, conceivably, so did the UK's energy executives who finally espoused and preached the virtues of using electricity more efficiently. Maybe they got their message across to consumers. (See Table 13-3.)

Table 13-3. International Comparisons (1947-1989) ($1990 and kWh per Capita)

	Real GDP per Capita ($1990)			Electricity Generated per Capita (kWh)		
	1947 [1]	1970 [2]	1989 [3]	1947 [4]	1970 [5]	1989 [6]
UK	6,306	10,694	16,288	1,026	4,494	5,697
USA	8,896	14,854	21,783	2,125	7,797	12,114
Continent						
Switzerland	8,926	16,671	21,381	2,160	5,656	7,748
Sweden	6,176	12,717	17,593	1,979	7,540	17,023
Netherlands	4,925	11,670	16,024	482	3,133	5,249
Belgium	4,699	10,410	16,299	834	3,167	6,532
France	4,099	11,558	17,457	634	2,894	6,511
Italy	2,856	9,508	15,650	380	2,187	4,223
Germany	2,763	8,463	18,015	894	3,988	7,165
Austria	2,181	9,813	16,305	559	4,022	6,312
Averages						
Continent	4,578	11,351	17,341	990	4,080	7,598
All	5,183	11,636	17,680	1,107	4,513	7,859
UK as % of						
Continent	138	94	94	104	110	75
All	122	92	91	93	100	72
USA	71	72	75	48	56	47

Sources (by columns):
1-6: Maddison, *op. cit.*
4-6: International Energy Agency, *Energy Statistics of OECD Countries 1970-1979, volume 2* (Paris: OECD, 1991).
Mitchell, *European Historical Statistics*, pp. 481-482.
United Nations, *1990 Energy Statistics Yearbook* (NY: UN, 1992).
Edison Electric Institute, *Statistical Yearbook*, various issues.

Notes:
Combined East and West Germany.

Conclusion

Nationalization saved the electric industry from fragmentation, a goal that could have been as easily accomplished by extending the franchises. The nationalized industry's executives ran a conservative, overstaffed operation whose efficiency improved over time. They followed government policies that embraced nuclear power and overpriced British coal. The government put its budgetary needs above those of the industry, too, which interfered with long term planning. A poisonous combination of unreasonable labor demands and political maneuvering provoked electricity stoppages that periodically endangered public welfare. Costs and prices rose above American but not Continental levels. Overall, the nationalized industry made economically unjustified decisions, but so did private- and government-owned utility industries elsewhere. In Britain, politicians – not bumbling managers lacking economic incentives to work hard – made the really big, bad decisions. The managers, on the whole, did what utility managers did elsewhere. They could have run a tighter ship, as demonstrated by savings achieved after privatization, but that would have required the government to permit what it previously would not countenance.

John Kay, the economist, described the UK's experience with its electricity system before 1990 as "dreadful." [136] But, dreadful compared to what? The work stoppages certainly stood out, for instance, but were typical of overall British labor problems, not just those of government-owned industry. Electricity network managers elsewhere did not cover themselves with glory either.

---

136. John Kay, "Governing by announcement leaves us all out in the cold," *Financial Times*, 1 October 2014, p. 9.

Part Four:

Capitalism Triumphant:
The Return to
Private Ownership

Chapter 14
Packaging and Selling an Industry (1990-1991)

Brutality, ignorance, poverty, corruption and waste – the constant qualities of socialism – can and will be vanquished. [137]

– Peter Clarke

After selling ports, bus lines, airports, manufacturers, a savings bank, telecommunications, natural gas and water utilities, the Tories got to electricity. They left the most difficult for last. They restructured the industry to encourage competition and made deals to subsidize the coal sector, all on a tight timetable. The old industry expired on March 31, 1990.

New Structure

In England and Wales, they reassembled the electric industry into six parts:

1. Generation – Power stations sold into an unregulated wholesale market called the "Pool." Generators had no public service obligation to sell electricity.

2. Transmission – National Grid owned and operated high voltage transmission lines connecting generators to load centers and ran the system control center. The regulators set prices.

3. Distribution – Twelve regional electric companies (RECs) distributed electricity over their wires to customers at regulated prices. RECs could own generation and act as suppliers.

4. Supply – Suppliers bought electricity at wholesale in the Pool to resell to consumers. Generators and RECs owned suppliers. Initially small customer supply was regulated.

5. Pool – The Pool operated a computerized market that selected power plants on the basis of price offers to generate electricity on the following day.

6. Regulator – The Electricity Act licensed companies. The Director General of Electricity Supply (DGES) ensured that licensees followed the rules. The Office of Electricity Regulation (OFFER), a part of DGES, set prices.

---

137. Peter Clarke, "The Argument for Privatisation," in Julia Neuberger, ed., *Privatisation ... Fair Shares or Selling the Family Silver* (London: PAPERMAC 1987), p. 91.

Scottish utilities generated, transmitted, distributed and supplied electricity. The Scottish Office of OFFER set prices, initially, for all services except supply to non-franchise customers.

Northern Ireland Electricity, which owned the regulated transmission and distribution facilities, purchased power from generating stations on a contract basis.

How It Operated

The Pool calculated the next day's electricity needs. Generators submitted bids to meet the need. The Pool selected the lowest bids. Next day, the system operator ordered generators to produce the electricity. Suppliers bought from the Pool at Pool price plus costs, then sold to consumers at a higher price. Large consumers could buy directly from the Pool.

Generators injected power into the high-voltage transmission network for delivery to load centers where voltage was lowered to level suitable for consumer use and then sent over small lines (the distribution network) to consumers.

Buyers and sellers protected themselves from price fluctuations by signing contracts for differences (CFDs). If Pool price exceeded the "strike" price, the generator paid the excess to the supplier and if it fell below strike price, the supplier paid the difference to the generator. The CFD gave price certainty to both parties.

See Figures 14-1, 14-2 and 14-3.

Figure 14-1. Electrical Distribution Boundaries in the UK

Key:

NIE- Northern Ireland Electricity, SH- Scottish Hydro-Electric, SP- Scottish Power, NW- NORWEB, NE- Northern Electric, YE- Yorkshire Electricity, EM- East Midlands Electricity, ME-Midlands Electricity, MW- Manweb, SW- South Wales Electricity, EE- Eastern Electricity, LE- London Electricity, SEE- SEEBOARD, SE- Southern Electric, SWE - South Western Electricity. The two Scottish companies had limited connections to England and Wales. Northern Ireland had no connection to the UK networks. England and Wales had a limited connection to France.

Figure 14-2. Flow of Funds in the Electricity Market

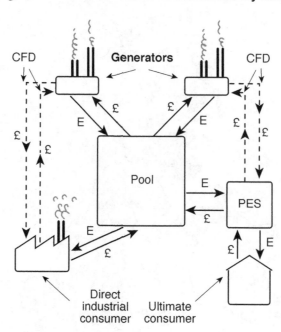

Key:
E- flow of electricity, £- flow of money, CFD- contract for differences, PES- public electricity supplier.

The generators sold electricity in the Pool and were paid Pool prices set every day. Public electricity suppliers and big industrial firms (direct users) bought that power in the Pool at Pool price plus a markup. The public electricity suppliers resold the electricity to consumers for what they paid plus an additional markup. In order to obtain a predictable price for the power, the generators signed contracts for differences with public electricity suppliers and direct industrial customers that set a strike price. When the Pool price exceeded that price, the generators paid the difference back to the buyers. When the Pool price fell below the strike price, the buyers paid the difference to the generators. Thus, no matter the Pool price, buyers and sellers knew at what price they could conduct business.

Figure 14-3. Flow of Electricity from Generator to Customer (England and Wales)

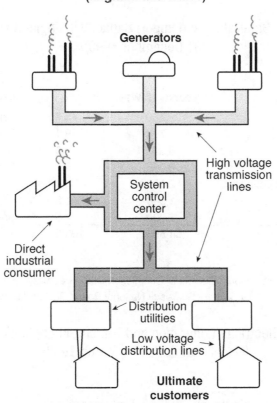

The generators injected power into the transmission network. The system control center gave operating orders to generators in order to maintain a reliable flow of electricity that met customer demand. Many big industrial customers took electricity directly from the transmission network. Distribution networks took electricity from the transmission lines, lowered the voltage and sent that electricity over their lines to their customers. All generation goes into one pot, so to speak, and the customer cannot specify electricity from a specific generator. The system control operator cannot separate the input. Everyone gets the same electricity. For that reason, all participants in the market must adhere to rules, because any violation would affect all users.

Generation

If the government had gone through with its plan to privatize the nuclear plants, National Power, the designated nuclear owner, would have controlled roughly three-fifths of the England and Wales market. After withdrawing nuclear plants from privatization, the government did not have time to split up the fossil generating sector before the share offering, so three companies ended up dominating the market. Two owned three quarters of generating capacity, produced over 70% of output and set market prices. (See Tables 14-1 and 14-2.)

Table 14-1. Generating and Import Capacity in England and Wales
(31 December 1990)

Company	Capacity (MW)	%	Comment
National Power	29,486	45.7	Largely coal
PowerGen	18,764	29.1	Largely coal
Nuclear Electric	8,357	12.9	Nuclear
National Grid	2,088	3.2	Pumped storage
Électricité de France Interconnect	2,000	3.1	DC line
Scottish Interconnect	1,000	1.5	Often limited to 850 MW
Others	2,900	4.5	Some for private use
Total	64,595	100.0	

Source:
Preliminary Placement Memorandum Subject to Completion Dated February 1, 1991, Rule 144A.
Package Offering of American Depositary Shares of National Power PLC and PowerGen plc, p. 58.

Table 14-2. Electricity Generation and Transfer in England and Wales
(Fiscal Year Ended 31 March 1989)

Company	% of Generation
National Power	43
PowerGen	28
Nuclear Electric	16
France	5
Scotland	1
All other	7
	100

Source:
Nigel Burton and Mark Loveland, *Electricity: The New Beginning* (London: S. G. Warburg Securities, 14 May 1990), p. 49.

The Big Two looked like Tweedledum and Tweedledee on paper. PowerGen, led by an aggressive Ed Wallis, who had directed CEGB operations during the coal strike and had already sent off his own chairman and the corporate raider Lord Hanson, moved faster. It secured gas supplies first. National Power, directed by a well-spoken, deliberate John Baker, who took over after Walter Marshall quit, followed. In the end they did the same things.

The Big Two owned two-thirds of base load plants and had an even stronger grip on other generation through which they could control Pool prices (Table 14-3). The government, which touted the virtues of competition, delivered the electricity consumers to a duopoly.

Table 14-3. Generating and Import Capacity in England and Wales (31 March 1991)

	(GW)			(%)		
	Total	Base Load	Other	Total	Base Load	Other
National Power	28.3	14.0	14.3	47	42	53
PowerGen	18.8	8.0	10.8	31	24	40
Nuclear Electric	8.4	8.4	0.0	14	25	0
National Grid	2.1	0.0	2.1	4	0	7
Scottish Interconnect	0.8	0.8	0.0	1	3	0
Électricité de France Interconnect	2.0	2.0	0.0	3	6	0
Total	60.4	33.2	27.2	100	100	100

Note:
Peak demand in year was 47.0 GW. Reserve margin was 29% (excluding private generation).
Base load units are large, low cost stations that the utilities attempt to keep running at a constant rate.

Source:
Leonard S. Hyman, "National Power," Merrill Lynch, December 20, 1991, p. 12.

The three largest generators, however, had an aged fleet of plants (Table 14-4). Maybe they would retire the old clunkers to make room for new, competitive power stations.

Table 14-4. Age of Power Stations (31 December 1990)

Age (Years)	(% of capacity in operation)		
	National Power	PowerGen	Nuclear Electric
Over 30	12	16	0
21-30	51	50	29
11-20	29	27	23
10 and under	8	8	48
	100	100	100

Notes:
Years from commissioning of first generator in station. Nuclear only for Nuclear Electric.

Sources:
Preliminary Placement Memorandum Subject to Completion Dated February 1, 1991, pp. 96, 168.
Burton and Loveland, *op. cit.,* p. 283.

The government downplayed the small number of competitors. After all, capacity was "substantially in excess of that which would have been considered necessary by the CEGB..," [138] evasively implying that excess would prevent generators from manipulating the market, although, of course, the Big Two controlled the excess capacity. Barclays de Zoete Wedd, the broker, though, predicted that reserve margins would fall from 20% to the high teens within a few years, so "excess" capacity would decline to normal industry levels. [139] The government banked on competition from new generators – combined cycle gas turbines (CCGTs) built by the RECs, the plant Nuclear Electric would open in 1995 – and the expansion of the Scottish interconnection to England. Perhaps there was hope for a competitive generation market. The prognosis for new builds looked good, anyway.

By Spring 1990, incumbents had about 6,000 MW of new generating capacity in planning or under construction and others had 12,000 MW in the works, adding up to a 30% increase in capacity. [140] In early 1991, the Big Two still planned to install about 5,500 MW of new capacity and close about 3,500 MW of old plant. The other entrants had about 6,500 MW of solid proposals in the pipeline. Netting out additions and closures, capacity might rise 15%, lower than the previous estimate but perhaps enough to reduce the Big Two's market power. The government, however, had other problems to fix and the solutions entailed locking in customers for generators, not creating more competition.

Packaging Generators

The government had to attract buyers for companies operating in an untested market. REC investors wanted predictable earnings and dividends and generator investors assurances about sales volumes and prices, both unlikely if the companies faced daily price changes.

Then there was the coal problem. Coal accounted for over 80% of the Big Two's output. They burned about three quarters of UK coal production. British Coal (BC), the dominant miner, charged more than foreign suppliers. As private firms trying to maximize profit, the generators would look for alternatives to BC's coal, thereby threatening the viability of the mines.

There was a nuclear problem, too. At projected Pool prices, Nuclear Electric would lose money. The government would not subsidize nuclear power. Somebody had to.

---

138. *Preliminary Placement Memorandum February 1, 1991, Rule 144A Package Offering of American Depositary Shares of National Power PLC and PowerGen plc*, p. 59.
139. Chris Rowland and Alex Milne, *Issues for Investors* (London: Barclays de Zoete Wedd, September 1990), p. 94.
140. Nigel Burton and Mark Loveland, *Electricity: The New Beginning* (London: S.G. Warburg Securities, 14 May 1990), p. 64.

Everything depended on solving the coal problem. The CEGB's successors bought 80 million metric tons of coal annually, 90% from BC, which, as of March 1990, charged 50% more than private mines and 65% more than imported coal. The government persuaded the generators to sign a three-year contract with BC for 70 million metric tons in the first and second years and 65 million in the third. British Coal, in turn, reduced prices 9% from March 1990 levels, with future prices tied to an inflation formula. The gap between BC and world prices after the deal still added about 7% to the average electric bill.

Having agreed to buy enough coal to fuel about 70% of their expected output, the generators needed to sell enough appropriately priced electricity to cover the costs. The new market offered neither assured buyers nor assured prices.

The government solved that problem. RECs would purchase a prescribed volume of electricity at a guaranteed price from the Big Two for resale to the RECs' own franchised (first tier) supply customers. The generators also signed fixed term contracts to sell electricity to large industrial customers and to RECs for resale to non-franchise customers. The contracts had three-year terms with escalation clauses. Thus, as long as the generators could keep their plants running efficiently, investors knew that the companies could earn predictable profits.

How, though, could RECs and generators fix prices if they had to buy and sell in the Pool at prices determined by the Pool, not by a contract? The solution: generators and electricity buyers did not sign ordinary contracts. They signed "contracts for differences." [141]

Next, the government estimated the gap between Nuclear Electric's costs and the smaller sum it would collect selling power in the Pool. To cover the shortfall, the government

---

141. The principal contracts with RECs were "two way." For instance, Generator and REC make a deal for 100 kWhs at a price of 5p per kWh. Generator produces the 100 kWhs. On days when the Pool price reaches 7p, Generator collects 700p from the Pool, REC pays the Pool 700p for the electricity, and then Generator refunds to REC the difference between the Pool price and the agreed price, times the volume produced, or 200p. After the refund, Generator has 500p left (5p per kWh). On days when the Pool price falls to 3p, Generator collects 300p from the Pool, REC pays the Pool 300p, but then REC pays Generator the difference between the Pool and the agreed price, times volume purchased, or 200p, and Generator ends up with 500p for its output, and REC has a cost of 500p for its purchase. Thus Generator and REC fix price for a set volume of electricity, despite oscillations in the Pool price. Generator, however, must produce the 100 kWh in order to collect.

The "one way contract" does not provide such symmetrical results. For that contract, REC pays a fee to Generator which, in return, agrees to pay REC a sum of money that equals the difference between the Pool price and an agreed upon ("strike") price times the volume contracted for. For example, REC pays 100p for a one way contract involving 100 kWh, with a strike price of 5p. If the Pool price rises to 8p, Generator collects 800p from the Pool, and REC has to pay 800p for its purchase of 100 kWh, but Generator then has to refund 300p to REC, making REC's effective purchase cost 600p (500p for the electricity plus 100p for the contract fee). If, however, the Pool price falls to 2p, Generator collects 200p for the electricity plus 100p for the fee, for an effective sale price of 300p. The one way contract for differences does not provide full protection from price swings, but acts, rather as an insurance policy against a particular event. With the two contracts, buyers and sellers could either protect themselves from all or some price fluctuations in the market.

imposed a Nuclear Non-fossil Fuel Obligation ("Nuclear Levy") on all suppliers for eight years. Suppliers transferred funds to Nuclear Electric by means of CFDs. The agreement covered about 90% of Nuclear Electric's expected output in the first year of the levy. This "tax" added about 10% to electric bills initially but would decline if Nuclear Electric hit operating targets.

To float generators, buy time to reorganize British Coal, keep nuclear plants running, and assure predictable prices, the government bundled together interlocking arrangements of three years' duration, with electricity consumers paying all costs. Although consumers paid more than in a truly competitive market, the carefully controlled process protected them from the sort of abuses and disasters that befell the California victims of its poorly deregulated market.

All generators of over 10 MW capacity (except for those dedicated to on site consumption) had to obtain a license and participate in the Pool. Units of 100 MW or larger had to submit to central dispatch, provide ancillary (grid reliability) services when requested, comply with use-of-network rules, and not discriminate in favor of one customer (contracting party) over another. The generating license neither regulated price nor obligated the generator "to invest in needed generation capacity, to maintain existing capacity or, except in certain circumstances, to declare generating sets available to generate electricity."[142] The generator became a strictly private profit-maximizing operator with no social responsibility.

The regulator also issued generators a second tier supply license (to make sales contracts with competitive market customers other than RECs) that did not regulate price. For the first four years, National Power and PowerGen together could not supply over 15% of demand in a REC territory (excluding direct contracts with large customers already implemented by the CEGB) and no more than 25% of demand in the next four years.

Although generators could not promise the same steady returns as RECs, the returns offered implicitly in the prospectus (Table 14-5) exceeded those projected for the RECs by only two percentage points, not a high differential for businesses that could face stiff competition once the contracts ended, and far below returns projected for the fully regulated National Grid.

---

142. *Preliminary Placement Memorandum Dated February 1, 1991*, p. 37.

Table 14-5. Projected Combined Financial Data for National Power and PowerGen (Fiscal Year Ended 31 March 1991) (Million £ or %)

	Pro-forma	Pro-forma excluding extraordinary items
	[1]	[2]
Invested capital (end of period) (£)		
Debt	550	550
Equity	3,296	3,344
Total	3,846	3,894
Valuation of assets (£)		
Net plant (end of period)	3,865	3,865
Market value of equity (offering price)	3,601	3,601
Enterprise value (offering price)	4,151	4,151
Income for invested capital (£)		
Interest	103	103
Net profit	466	514
Total	569	617
Rates of return on (%)		
Invested capital	14.8	15.8
Enterprise value	13.7	14.9
Equity	14.1	15.4
Net plant	14.7	16.0

Notes:

1. Enterprise value equals common equity at market value plus debt.

2. Numerator for returns on invested capital, enterprise value and net plant is income for invested capital.

3. Calculations by author based on information in prospectus.

Sources:

1. *Preliminary Placement Memorandum dated February 1, 1991.*

2. Alex Henney, *A Study of the Privatisation of the Electric Supply Industry in England & Wales* (London: EEE Limited, 1994), p. 196.

The government freed generators from the old public utility obligation to serve, espoused the virtues of competition, set up a market structure, launched too few competitors into the market, and then tied up sales in contracts that gave key players little room to maneuver.

Transmission

National Grid owned and operated high voltage transmission lines in England and Wales, operated interconnections to Scotland and France and ran the network control center that dispatched power plants for daily operation. It managed ancillary services (such as reactive power, standby equipment and control of frequency) that maintained grid reliability. It operated the settlement system through which Pool buyers paid sellers. NG had to issue a seven-year forward plan each year to help generators site their plants. National Grid also owned two pumped storage hydroelectric stations (used during peak periods and emergencies). National Grid, a common carrier, had to treat all users equally.

Rather than float National Grid, the government made it a subsidiary of National Grid Holdings (NGH), owned by the RECs, thereby making RECs more attractive to shareholders by making them less parochial. As the prospectus for the REC's said, "the dividends from NG ... should provide a material and stable contribution ... to profit..." [143] Those dividends provided about 15% of projected REC earnings. On a consolidated basis, NG accounted for about 35% of REC earnings. National Grid was the industry's crown jewel.

The RECs appointed NGH's directors who could not interfere with NG other than to prevent it from entering extraneous fields of business or from taking actions that threatened the company's viability. National Grid's transmission license, not its owners, guided company activities.

NG operated ancillary services for costs plus a fee, the settlement process for costs plus a markup and other functions under the RPI-X formula. The government set initial X at zero, so prices would rise with cost of living, and, presumably, revenues would increase by rate of inflation plus growth in demand for service, except the actual formula was more complicated, using weather adjusted data and trailing four year averages. [144]

---

143. *Placement Memorandum Dated December 11, 1990,* Eastern Electricity plc, East Midlands Electricity plc, Manweb plc, Midlands Electricity plc, Northern Electric plc, NORWEB plc, SEEBOARD plc, Southern Electric plc, South Wales Electricity plc, South Western Electricity plc, Yorkshire Electricity plc, 12,306,000 ADS Package Units, Rule 144A Offering of American Depositary Shares in ADS Package Units only, each ADS Package Unit being equivalent to 1/100th of a U.K. Ordinary SharePackage Unit, p. 194.

144. National Grid pricing for services complicated a simple process. According to the prospectus, "In setting its charges for each year, NG estimates the level of inflation, the level of demand on the total system and ... the level of revenue over- or under-recovered by the end of the previous year." In addition, "The price control formula ... adjusts the permitted maximum average charge per kW ... to allow for changes in the system maximum ACS [average cold spell] demand... compared with the average of such demand in that year and in the previous four years." (*Placement Memorandum*, p. 90.)

    The regulator, then, not only set price but also volume (by forcing use of an averaging formula). Price times permitted volume equals permitted revenue. Actual revenue might differ. The difference could result from a deliberate misestimation or from any number of small projection errors. The regulatory rules, however, did not require automatic adjustment next year to correct for past errors. Nor did they guarantee recovery of large shortfalls. If the overages or shortfalls were small, the regulator would do nothing. If large, the regulator might act.

The regulatory formula had to support new grid investments. In the prospectus, National Grid projected capital spending that probably justified 5-6% annual revenue increases (in real terms) to cover its costs, no doubt more than the formula would produce. NG could, however, levy additional charges for "new connections to the transmission system and extensions to existing connections" and charge "the appropriate costs directly or indirectly incurred in providing such connections, together with a reasonable rate of return on the capital employed." [145]

National Grid collected revenues through charges for use if system and for connection to it.[146]

Despite National Grid's placement as a subsidiary of a holding company owned by RECs, privatization restored transmission to a central position in the industry.

Distribution

The biggest REC was three times the size of the smallest, measured by customers or revenues and thirty times larger by area served. Their percentage of sales to industrial customers ranged from 11% to 55%. Bankers feared that investors would shun small RECs because of illiquid markets for their shares (hard to trade) and RECs dependent on big users (customers might leave in search of better prices). The farther from London

---

Presumably, given the regulatory uncertainty, National Grid would craft projections carefully, trying not to beat the formula by more than a small margin.

For example, in the base year, ACS peak hit 1,000 MW. NG charged 10p per MW, and revenue reached 10,000p. NG predicted that the cost of living would rise 3% in the coming year, so the new price per MW in that year was set at 10.3p. The company expected peak to rise to 1,030 MW, but the five year average peak formula produced 1,010 MW, so NG should have expected to collect 10.3p x 1,010 = 10,403p in the year. Unfortunately, NG miscalculated. Inflation hit 4% and actual peak reached 1,050 MW. The company collected 10.3p x 1,050 = 10,813p. It should have collected 10.4p x 1,010 = 10,504p. That excess of 309p exceeded allowed revenue by less than 3%, so the company squeaked in under the 3% excess, so NG did not have to assure the regulator that it would not mess up again.

In case of substantial shortfall, the regulator might only permit a partial recovery in order to limit price increases. In case of excess revenues, the regulator might or might not require NG to adjust future revenues for the excess. Although the formula gave NG an incentive to reduce costs (the company could retain the savings), it provided little incentive to develop new products and services (different or more efficient ways to use the grid). If the new services increased revenues the company could not keep the increase and if they lowered demand at peak, that would eventually reduce NG's revenues.

145. *Op. cit.*, p. 122.
146. Use of system charges – Suppliers and generators paid a system service charge to maintain the network based on peak demand (in kW) and an infrastructure charge based on peak demand, generating plant capacity and kWh output to cover costs of delivering bulk power and maintaining reliability. The infrastructure charge varied by region to reflect cost differences.

Connection charges – Generators paid entry charges, and RECs and consumers connected directly to the grid (large users such as industrial facilities) paid exit charges, all based on current value of assets used to connect them to the network.

the REC, the more likely that it served a poor region with a big industrial load. (See Table 14-6.)

Table 14-6. Regional Electric Companies in 1990

REC	Customers (millions)	Sales (TWh)	Industrial % Total Sales	Revenues (£ millions)	Service Area (thou. km²)	Xd Factor (%)	LV3 Price (p/kWh)	HV Price (p/kWh)
	[1]	[2]	[3]	[4]	[5]	[6]	[7]	[8]
Eastern	2.9	22.7	28	1,616	20.3	0.25	1.34	0.46
East Midlands	2.1	22.1	4.3	1.263	16.0	1.25	1.57	0.64
London	1.9	18.3	11	1.148	0.7	0.00	1.59	0.59
Manweb	1.3	17.3	55	891	12.2	2.50	1.76	0.51
Midlands	2.1	22.4	41	1,295	13.3	1.15	1.61	0.60
Northern	1.4	15.2	50	820	14.4	1.55	1.93	0.47
NORWEB	2.1	20.6	38	1,232	12.5	1.40	1.63	0.53
SEEBOARD	1.9	16.8	24	982	8.2	0.75	1.41	0.56
Southern	2.5	24.4	23	1,457	16.9	0.65	1.48	0.59
South Wales	0.9	11.6	57	604	11.8	2.50	2.20	0.73
South Western	1.2	12.1	27	748	14.4	2.25	1.87	0.51
Yorkshire	2.0	23.1	53	1,258	10.7	1.30	1.67	0.58

Notes:

Column 1 – Customers as of July 1, 1990.

Columns 2-4 – For fiscal year ended March 31, 1990.

Column 5 – As of date of prospectus.

Column 6 – X factor for distribution for period beginning April 1, 1990.

Column 7 – Predominant tariff category for low voltage service.

Column 8 – Predominant tariff category for high voltage service.

Columns 7-8 – Revenues for LV3 and HV generally make up over two thirds of REC revenue.

Source:

Placement Memorandum Dated December 11, 1990, Eastern Electricity plc, East Midlands Electricity plc, London Electricity plc, Man web plc, Midlands Electricity plc, Northern Electric plc, NORWEB plc, SEEBOARD plc, Southern Electric plc, South Wales Electricity plc, Yorkshire Group plc, Rule 144A Offering of American Depositary Shares.

The RECs had four lines of business:

- Distribution, the wires that delivered the electricity to the local customer, accounted for almost all assets and profits. The Director General of Electricity Supply set prices.

- RECs, as first tier license suppliers, sold to all customers in their territory with a load of less than 1 MW, at regulated prices. As second tier suppliers they could sell to any consumers in the UK taking 1 MW or more, at unregulated prices. Before privatization, half the RECs planned to aggressively pursue of second tier customers, while the rest viewed the business as a defensive measure to keep local customers in the fold. [147]

- A subsidiary could invest in generation up to a capacity of about 15% of the REC's own load. At time of flotation, eight RECs had committed to generation investments (several in one large project sponsored by Enron) and four indicated interest in doing so.

- Each REC owned electrical appliance stores that served as local payment offices and an electrical contractor. Appliances and contracting contributed 6% and 2% of sales respectively. These businesses were unregulated.

First tier supply was regulated with prices set by a variant of the RPI-X formula. The REC could make adjustments in the following year, in case it misestimated components of the formula, but actual recovery of lost revenue was not guaranteed.[148] Since the RECs had committed to keeping supply prices (excluding the nuclear levy) stable in real terms, full recovery of first tier supply costs was more something to wish for than to expect.

An $RPI + X_d$ formula set distribution prices. RPI was the percentage change in the Retail Price Index from October of the previous year to October of the current year and X_d the productivity factor for distribution. With X a positive number, prices would rise faster than the rate of inflation.[149] The regulator then added to the price increase an electric loss factor, a bonus for bringing electrical losses below a historical average and a penalty if the REC's losses rose above the that average. The REC could rearrange prices between customer groups as long as it did not cross subsidize one class of costumer as against another, and the average price stayed within the limit set by the formula.

---

147. Dr. Simon Williams, *The Electricity Company Notebook* (London: Kleinwort Benson, Sept. 1990).

148. The regulator used an $RPI + X_s + Y$ formula. RPI was the retail price index, X_s a productivity factor for the supply function and Y the cost of purchased power passed on to customers (about 95% of the supply bill). The regulator already told the RECs to buy power in the most economical manner so he would not question the power purchase costs unless the RECs acted in a questionable manner.

    Assume that in the base year, the consumer paid 5p for the service element and 95p for the purchase element of the supply bill. The regulator sets a productivity factor of -2% ($X_s = -2$). The REC expected cost of living to rise 5% (RPI = 5) and power costs 4% (Y = 4). In the following year, then the service component of the bill would rise 3% ($RPI - X_s = 5-2=3$) from 5p to 5.15p (5.00 x 1.03). the purchase component would rise 4% from 95 to 98.8p (95p x 1.04) and the total bill would increase from 100p to 103.95p.

    However, there was a catch. The REC could make adjustments in the following year in case it had misestimated components of the formula but actual collection of lost revenues was not guaranteed, but "will depend on … the extent of competition with other suppliers in the non-franchise market, competition with other energy sources where relevant… and the judgement of … management as to the desirability of seeking to recover these amounts… In future years, application of the correction factor should normally allow a substantial proportion of any deficit to be recovered in the next financial year or years." (*Placement Memorandum Dated December 11, 1990.*, p. 71.)

149. Considering that the RECs operated on an April 1 - March 30 fiscal year, RPI effectively measured the price change from the average consumer price level of the previous fiscal year to that of the current fiscal year.

The X_d factor supposedly took into account expected changes in productivity, benefits from growth in volume and need to raise money to refurbish or expand the network. Planned capital expenditures, though, would have increased tangible assets devoted to serving customers by 15% in the first year, with high spending to continue in subsequent years. Revenues could not have risen enough to pay the costs of the new facilities without price increases.

The X_d factor varied from REC to REC. It was, like the opening price of electricity, a number that the government could manage in order to equalize the attraction of the RECs. Generally speaking, small RECs charged general (LV3) customers more and could raise prices faster (higher X_d) and RECs with the highest industrial percentage of sales also had the highest LV3 rates and X_d factors. All RECs had roughly the same HV (industrial) rates.

Perhaps the numbers just happened to work out that way, charging poorer customers more and raising their prices faster than in more affluent parts of the country. More plausibly, the numbers resulted from hard bargaining by REC managements playing on the government's need to move fast and its fears that the offering might fail due to investor reluctance to buy some of the RECs. It was as if the government set the numbers to make the sale more successful and in the process inadvertently assured that already expensive services in some regions would become relatively more expensive. But distribution charges made up less than one-fifth of the electric bill, so maybe customers would not notice. (See Table 14-7.)

Table 14-7. X_d, and Prices by Size and Industrial Sales

Tercile	Average Revenues (£ millions)	Average Industrial sales	Average X_d (%)	Average LV3 (p/kWh)	Average HV (p/kWh)
	[1]	[2]	[3]	[4]	[5]
By size					
1	1,408	–	0.82	1.50	0.57
2	1,155	–	0.86	1.58	0.57
3	766	–	2.20	1.94	0.56
By industrial % sales					
1	–	54	1.96	1.89	0.59
2	–	38	1.01	1.46	0.56
3	–	21	0.91	1.59	0.56

Sources and Notes:
See Table 14-6.

At year end, RECs had to account for price forecast accuracy. The regulator might excuse 3-4% overcollection if not persistent, but if a REC, "for two successive years ... under-charges by more than 10 percent of the maximum permitted average charge, the DGES may ... limit the amount by which such undercharging may be recovered." [150] RECs could get away with small errors in their favor but would lose badly for making an egregious error in the customer's favor.

RPI-X provided the RECs with incentives to reduce costs, boost sales, cut electrical losses, delay capital spending and improve profits.

Supply

Suppliers purchased electricity wholesale at the Pool and sold it at retail to ultimate users.

The RECs, as first tier suppliers, had exclusive right to sell electricity to all customers with demand under 1 MW in the first four years, and to all customers with demand under 0.1 MW in the following four years. That reservation for first tier suppliers removed 70% of the market from the competitive arena for the first four years. At the end of the restricted period, all customers would buy in the competitive market. (See Table 14-8.)

Table 14-8. Distribution of Customers and Sales Before Privatization

Size of Demand	Number of Customers	% of kWh Consumed
Over 10 MW	400	13
1 to 10 MW	3,600	17
0.1 to 1 MW	45,000	20
Below 0.1 MW	22,200,000	<u>50</u>
		100

Source:
Burton and Loveland, *op. cit.*, p. 112.

RECs and generators dominated the supply markets. Entry into supply, an S.G. Warburg report said, "... involves a degree of risk which, other than for a generator or distributor, is probably not justified by the likely returns." [151] Suppliers with margins of a few percent on sales, contending with volatile Pool prices, could end up paying more for electricity than their customers paid them. Their thin margins left no room for error.

---

150. *Placement Memorandum Dated December 11, 1990*, p. 38.
151. Nigel Burton and Mark Loveland, *Electricity: The New Beginning* (London: S.G. Warburg Securities, 14 May 1990), p. 113.

Suppliers had to meet nuclear and renewable requirements, belong to the Pool, comply with the grid code and could not cross subsidize. The supply business looked like an effort contrived to give the appearance of competition where little existed that also provided minimal profit in return for taking great risk. Could supply exist as a free standing enterprise?

The Pool

The Pooling and Settlement Agreement ("the Pool") went into force on March 30, 1990. Members elected an executive committee to supervise the Pool. A National Grid subsidiary administered the settlements – payments for transactions between members. The government declared that "Sales and purchases ... are made between participating generators and suppliers according to a set of rules (the pool rules) ... The pool does not itself buy or sell electricity."[152]

All but the smallest licensed generators and all suppliers transacted through the Pool. Generators had no obligation to make power available to the Pool, only to trade through it. Unlike other organized markets, the Pool did not make good on a transaction if one of the parties defaulted. The Pool was like a club whose members did business with each other on uniform terms.

The Pool worked this way. At 10:00 AM, all electricity generators in England and Wales declared which power plants they would make available for each half-hour period of the following day. Generators in Scotland and Électricité de France, did the same for electricity they would deliver over the transmission interconnectors into England and Wales. They specified their minimum price to operate the generating unit, with conditions attached (such as price adjustments for start-up costs and output requirements for that unit), and their prices to provide standby capacity (keeping the facility ready to operate) and to operate the unit above normal levels.

National Grid, as system operator, made a schedule of demand for each half-hour of the next day. It then assigned generators to each half-hour slot so to minimize costs to consumers. In doing so, it considered transmission bottlenecks, costs of operating power stations that had to run continuously and stability of the network (which required operation of certain stations to protect network stability rather than to sell power to consumers).

With that information, the computer sorted out which generators would sell into the Pool during each half-hour slot, starting with the lowest priced offering and adding on

---

152. *Preliminary Placement Memorandum Dated February 1, 1991.*, p. 23.

generators until it filled the quota. For instance, the Pool required 100 kWh for the half-hour period 10:30-11:00 AM on the following day and seven generators put in their offers:

Generator	kWh Offered	Offering Price (p per kWh)
A	30	0
B	10	2
C	30	3
D	30	6
E	10	7
F	20	7
G	30	8

The computer first selected the unconstrained schedule: offers of A, B, C and D. All selected generators received 6p per kWh, the price required to bring in enough generation to serve the expected demand. Generator A bid 0 p in order to assure selection, fully expecting that price paid would exceed its costs. The price paid to those selected was the System Marginal Price (SMP). Then the operator revised the list to account for problems on the grid that would preclude certain plants from operating as planned. Those changes produced the revised unconstrained schedule, which determined the next day's production plan.

The Pool paid a bonus over SMP to assure capacity available at time of greatest need.[153] It divided the day into Table A (high demand) and Table B (low demand) periods. In the A period, few generators are available to replace a unit that goes down, so a power outage might ensue if spare capacity were not available. In the B period idle generators could replace out-of-service plant, so an outage would not cause difficulty. Therefore the Pool paid a bonus only to generators that made units available during Table A periods. It calculated that payment based on value of lost load (VOLL), a hypothetical number measuring the damage done when a generator could not produce the promised electricity.[154]

---

153. As Alex Henney put it, the price of electricity had to "be both economically efficient" and allow the generator to "recover the investment cost of capacity." (Alex Henney, *A Study of the Privatisation of the Electric Supply Industry in England & Wales* [London: EEE Limited], 1994, p. 111.)

154. Value of lost load (VOLL) was cost to supplier (buyer) of losing load because generator could not fulfill its obligation. To simplify, the Pool set VOLL at £2 per kWh for all Table A periods, increasing every year at the rate of inflation. Then the system operator calculated the likelihood in every half hour that generation could not meet demand, the loss of load probability (LOLP). Finally, in order to pay the generator for availability, as opposed to paying it to generate electricity, the payment had to exclude the payment for electric output, the SMP.

Price paid to generators, the Pool Input Price (PIP), in Table A periods was:

$$PIP = SMP + LOLP (VOLL- SMP)$$

For example, SMP was 5p per kWh, the chance of losing the load was 1% (0.01), and value of lost load was set at £2 (200p) per kWh. Therefore, the generator, during Table A hours, should have collected:

$$PIP = 5 + (0.01) (200 - 5) = 5 + 1.95 = 6.95p$$

The generator collected 5p per kWh to provide electricity plus 1.95p per kWh to be available at a time when loss of a generator would cause inconvenience and expense to consumers.

Back in CEGB days, the system coordinator selected plants for daily operation by cost (the "merit order") to produce the lowest aggregate cost of operation. In the privatized industry, unregulated generators had no obligation to sell electricity or divulge costs. The system operator selected plants for operation based on voluntary offers that had prices attached to them.

Competition forces generators to reduce costs in order to bid low enough for selection and customers reap the benefits. That was the theory, anyway.

Why should all selected generators collect the highest price when some seemed willing to accept less? In reality, low bids were part of a positioning process to assure that the Pool selected the generator for next day operation. The generator expected the final price to exceed its bid. If that generator expected payment based on its bid it would have raised that bid to a level that would, at least, have covered operating costs.[155]

The Pool was not an ideal market. Too few generators competed. They operated similar plants. Ex CEGB staffers running the facilities knew the other plants, and could figure out how their competitors bid into the market and how to bid collusively without even talking to each other.

During Table B periods when capacity was ample and plant outages could cause little inconvenience:

$$PIP = SMP$$

Winning generators could not count on collecting revenue calculated by multiplying PIP by expected number of units to be sold. The grid operator had to consider real conditions the following day, at the end of which it had to reconcile the unconstrained schedule with actual operations, determine what each generator produced, how much it owed the generator for output, and how much it owed generators that made plants available but could not produce due to grid constraints (the price the Pool would have paid for their output less the cost of producing it).

Buyers did not just pay PIP. They had to pay for all the expenses of running the network, especially for ancillary services (generation required to keep the system running reliably) and for reserve capacity needed to meet emergencies, and payments to generators whose facilities had been designated for use by the network but, were not needed. All those extra charges went into a pot called Uplift.

Customers paid the Pool Output Price (POP) to get the electricity from the Pool, during Table A periods:

$$POP = SMP + (LOLP)(VOLL - SMP) + Uplift = PIP + Uplift$$

During Table B periods:

$$POP = PIP$$

Table A users paid any uplift costs incurred during Table B periods.

155. Competitive commodity markets work on a one price basis. Think about apples. Buyers want ten apples, local farmers bring nine to market and offer to sell them for 50 p per apple. What about unsatisfied demand? Should customers fight to determine who goes without an apple? A distant farmer arrives with one apple. He wants 60 p in order to cover travel costs. Consumers are willing to pay 60 p. At that point, local farmers say that an apple is an apple. If the distant farmer can get 60 p, they should, too. So price goes up. Customers are satisfied. Supply and demand determine price. At that price, some sellers make a bigger profit than others, but they all get the same price.

The Pool did not take into account consumer reaction to high prices, a serious flaw. For instance, the operator decides that consumers will want 100 kWh, which requires the Pool to pay 6p per kWh, because the bids stack up this way:

Generator	kWh Offered	Price (p)	Cumulative kWh offered
A	25	0	25
B	10	2	35
C	30	3	65
D	30	3	95
E	10	6	105
F	20	7	125
G	30	8	155

If customers reduce demand 5% because they consider 6p too high a price, and demand falls to 95 kWh, the Pool would pay only 3p. The Pool needed to put consumers into the picture.

The Pool was an ingenious mechanism to select power plants for next day's operation in a flawed market. Whether VOLL provided signals for long term investment in capacity was another matter, but not one of immediate consequence given all the excess capacity at the time.

Regulation

Back in 1982, Stephen C. Littlechild of the University of Birmingham devised a regulatory formula for British Telecom. He argued that regulators should:

- Protect customers from monopoly power

- Encourage efficiency and innovation

- Keep the regulatory burden light

- Promote competition

- Consider impact on the market value of companies, which "implicitly depends upon the perceived complexity or unfamiliarity of regulatory arrangements, and the predictability of regulation." [156]

Littlechild modified a plan conceived by the Monopolies and Mergers Commission (MMC) to control prices charged by a monopolistic contraceptives manufacturer. The

---

156. _____, *op. cit.*, p. 272.

MMC wanted something simple, so it decided to let prices rise at the rate of inflation less 1.5%. As Alex Henney put it, "Basically, the idea of price control is to create a pseudo competitive market price which provides the licensee with an incentive to improve efficiency and beat the control." [157]

The ideal UK regulator sets simple rules, and runs with a small staff to minimize regulatory expense and interferes little with management. Competition protects consumers and encourages innovation better than regulation. Regulatory uncertainty sabotages privatization, investors prefer simple, predictable regulation and RPI-X meets those goals better than alternatives. The government chose RPI-X for telecommunications, gas and water and – no surprise – for electricity, and appointed Professor Littlechild the first Director General of Electricity Supply.[158]

The RPI-X formula showcases price. Still, the RPI-X formula started with the price that gives the utility a fair return on investment over the period covered, assuming that the utility works to improve its efficiency. Investors did not worry about the complex mechanism that underlay the formula. (Everything set for five years. So why worry.)

The government, though, decided "that prices in general would not increase in real terms on vesting and immediately following privatisation." [159] Maybe the RECs could beat projections and excel despite the price lid, but no prudent civil servant could bet on the skills of managers with so little commercial experience. So, to make the numbers work, the government put a cap on the expected revenue, determined a large part of the expenses by the way it arranged contracts within the industry, and REC share owners ended up with the residual.

The REC offering prospectus provided enough information to analyze in depth prospects for the base year. The package of dull, low-risk utilities serving a mature market would generate hefty profits, the prospectus said, although not in those words. (Table 14-9.)

---

157. _____, *op. cit.*, p. 273.

158. The *Economist* described the industry as a "regulator's nightmare" due to lack of competition, complexity of the licenses and inadequate number of regulatory staffers. It ran a mock job ad for regulator that sought a "Dynamic, thrusting individual … Ability to perform miracles … useful." (*Economist*, February 25, 1989, p. 26.)

159. _____, *op. cit.*, p. 120. "Vesting" was the delivery of the assets of the industry to the new corporations, which took place on April 1, 1990. "Privatization" was the subsequent sale of the corporations to shareholders during late 1990 and early 1991.

Table 14-9. Projected Combined Financial Data for RECs and National Grid
(Fiscal Year 1990/1991) (£ millions and %)

	RECs	National Grid
	[1]	[2]
Invested capital	___ (£ millions) ___	
Debt	1,914	901
Equity	5,333	916
Total invested capital	7,247	1,817
Income for invested capital		
Interest	195	128
Net profit	646	241
Total income for invested capital	841	369
Dividends	227	105
Undistributed profits	614	136
Net plant	5,660	1,965
Market value of common equity	5,182	NA
Enterprise value	7,096	NA
Rates of return on	___ (%) ___	
Invested capital	11.6	20.3
Enterprise value	11.9	NA
Net plant	13.0	18.8
Equity	12.1	26.3
Equity excluding National Grid	11.6	NA
Capital excluding National Grid investment	11.6	NA
Equity including undistributed earnings of National Grid	14.7	NA

Notes:
1. Return on equity excluding National Grid assumes 74% of NG investment by RECs financed by equity.
2. Net plant includes tangible assets only.
3. All National Grid dividends paid to RECs.
4. Enterprise value equals debt plus market value of equity.

Source:
Placement Memorandum Dated Dec. 11, 1990...

National Grid, the star of the show, projected a historical cost return of about 19% on regulated assets and 26% on stockholder equity and the RECs a respectable 13% on regulated assets and 12% on regulated equity (excluding earnings from and investment in National Grid). If the RECs had reported their share of National Grid's earnings, they would have earned close to 15% on equity. Investors at the IPO purchased stock in the RECs at close to historical cost so returns on that basis approximated what investors could expect.

In 1989-1991, UK Government ten-year bonds yielded about 10.5%. Projected REC returns on stockholder equity produced a stingy 1.5 percentage point premium over bond yields but the transmission showed an astounding 15.5 percentage point premium, and the combined REC/transmission investment a 4.5 percentage points premium, which is what counted for REC investors, since they owned the package of both. In comparison, in the United States, where ten-year government bonds yielded 8.3%, American regulators allowed about 10.5% on rate base and 12.5-13.0% on historical cost equity. The UK's 4.5 percentage point risk premium, then, looked comparable to the 4.2-4.7 percentage point American equity premium set by more experienced regulators. But American utilities actually earned about 10% on equity, so the true equity risk premium was more like two percentage points. UK investors, with built in protection against inflation, looked better off, deservedly so because they did buy into an experiment.

The Electricity Act of 1989 required the Director General of Electricity Supply to:

- Assure that the industry satisfies reasonable demands for electricity.
- Assure that license holders can finance activities that promote competition.
- Protect consumers, in terms of price, continuity of supply and continuity of service.
- Promote efficiency of supply and consumption.
- Promote research and development.
- Protect the public from physical dangers.
- Assure the health and safety of electricity industry workers.

The transmission company had "to develop and maintain an efficient, coordinated and economical system ... and to facilitate competition..." and the RECs had to "develop and maintain an efficient, coordinated and economical system of electricity supply" [160] and to provide service when requested, on a nondiscriminatory basis. But the law imposed no duty on generators. They could sell or not sell as they pleased.

The Director General, in theory, was independent of the government. In contrast to American counterparts, he operated alone rather than as part of a panel and had no obligation to hold quasi-judicial hearings before making a decision. Disaffected parties could challenge his decisions at the Monopolies and Mergers Commission or in court. The Office of Electricity Regulation (OFFER) had headquarters in Birmingham, a Scottish office in Glasgow and regional offices for each REC and Scottish utility. The staff eventually numbered just over 200.

---

160. _____, *op. cit.*, p. 78.

In a lengthy letter published in the REC prospectus, the Director General wrote:

> I see my prime task as ensuring that consumers and licensees reap the benefits that will flow from the restructuring and privatisation ... However, if the benefits... are to be fully secured, competition is a vital ingredient. It will be important for new sources of generation and supply to emerge. When companies know that consumers have freedom of choice, they try even harder to deliver a service that meets consumers' wishes.
>
> I regard the evolution of a competitive market as wholly compatible with the financial health of the industry. I would expect the profitability and financing of members of the industry to reflect their efficiency and the competitive opportunities available. I believe that companies should be capable of financing their licensed activities and paying an adequate return to their shareholders. [161]

The regulator intended to bring the benefits of efficient service to the public not by regulatory prescription but rather by incentives and by competition.

Scotland

Scotland retained its two utilities.

The Secretary of State for Scotland's 1988 privatization white paper read as if he were delivering a sales pitch to the locals while walking on eggshells. Privatization would move control from London to Scotland, create major new Scottish companies, and give Scots an opportunity "to acquire a major stake in the ownership of Scottish industry," [162] while acknowledging that North of Scotland's "identity is strongly related to its unique statutory obligation to collaborate in the carrying out of any measures for the 'economic development and social improvement of the North of Scotland...'" [163]

---

161. *Placement Memorandum Dated December 11, 1990.*, p. 46.
162. Industry Department for Scotland, *Privatization of the Scottish Electricity Industry*, Presented to Parliament by the Secretary of State for Scotland by Command of Her Majesty (Edinburgh: Her Majesty's Stationery Office, March 1988), p. 1.
163. _____, *op. cit.*, p. 2.

The boards were "distinct in character," [164] differed in size, and the government sought "a fully Scottish solution," [165] meaning not much would change.

	North	South
Generating capacity (MW)	3,221	7,444
Territory (sq. mi.)	21,000	8,000
Population (million)	1.25	4.0

Scotland had excess generating capacity (11,741 MW installed vs. 6,089 MW peak demand predicted for fiscal 1989/1990). Competing generators were unlikely to enter that glutted market but customers could still get the benefits of competition by comparing the two companies ("yardstick competition") and shareholders could "draw appropriate conclusions."[166]

The Secretary of State lauded South of Scotland's "impressive record"[167] as a nuclear operator, favored joint ownership of nuclear plants, recommended each utility have a "balanced set of generating assets"[168] and concluded that "customers and shareholders"[169] would, somehow, force the utilities to cooperate to reduce costs. The companies would have to assure competitors access to their networks and "ring fence" regulated from unregulated activities in order to prevent cross subsidies or favoritism. North of Scotland's successor would maintain the uniform tariff plan that subsidized rural customers.

The two utilities would jointly operate nuclear facilities, exchange plant and find reason to cooperate. Pressure from customers and shareholders would make them do the right thing.

In the end, the government put nuclear stations into state-owned Scottish Nuclear and moved the Cruachan pumped storage plant from the northern to the southern utility. The companies contracted to buy electricity from each other. They changed their names, South of Scotland to Scottish Power and North of Scotland to Scottish Hydro-Electric. As for their businesses:

- Local distribution networks accounted for about 21% of the electric bill.

- Both owned and operated transmission. Scottish Power also owned and operated the line to England and Wales. Transmission made up about 6% of the electric bill.

- The regulator set price sold to retail customers. Cost of generation and purchased power made up about 68% of the electric bill.

164. _____, *op. cit.*, p. 6.
165. _____, *op. cit.*, p. 5.
166. _____, *op. cit.*, p. 6.
167. _____, *op. cit.*, p. 8.
168. _____, *op. cit.*, p. 7.
169. *Ibid.*

- They could supply within their franchise areas and elsewhere. The regulator set prices for franchised customers.

- Appliance retailing, electric contracting and other activities accounted for about 5% of revenues and less than 2% of operating income before privatization.

The Scottish companies depended heavily on coal-fired and nuclear power (Table 14-10):

Table 14-10. Scottish Capacity and Generation (31 March 1991)

	Generating Capacity (MW)	Generation and Purchases (GWh)
	[1]	[2]
Owned		
Coal	3,888	12,719
Dual oil/gas	1,284	3,870
Conventional hydro	1,189	3,436
Pumped storage	699	440
Other	233	279
Total owned	7,293	20,744
Purchased and Pooled		
Scottish Nuclear	2,400	12,176
Other	250	3,094
Total purchased and pooled	2,650	15,270
Total owned and purchased and pooled	9,943	36,014

Note:
Generating capacity as of 31 March 1991. Generation and purchases for fiscal year ended 31 March 1991.

Source:
Prospectus, The two Scottish electricity companies share offers, 31 May 1991, p. 38.

With fuel and purchased power costs tied to the cost of living, the regulator could set the RPI-X framework.[170] Also, the "revenues ..." of each utility's "generation business are

---

170. British Coal would supply Scottish Power's huge Longannet station with a minimum of 2.5 million metric tons of coal per annum at an above market price for three years and 2.0-2.5 million metric tons per annum in the next two years. Price would rise if the RPI increase exceeded 5.5% in the first year and 5.0% in the second and then follow the cost of living. Scottish Power could purchase additional coal at lower prices. Hydro-Electric participated in contracts to the extent that it purchased coal-generated electricity from Scottish Power. The two signed up for Scottish Nuclear's output at a price adjusted annually by RPI for four years, and in the subsequent four years by a combination of fixed price plus inflation and base load generation prices in England and Wales. Beginning April 1, 1998, prices in England and Wales set Scottish prices. The Nuclear Energy Agreement would expire in 2005. Scottish utilities did not pay the fossil fuel levy except for second tier supply activities in England and Wales as a result of having contracted to buy power from Scottish Nuclear.

not directly subject to price control, but are constrained indirectly…," [171] another way of saying that each utility's generating division could charge its supply affiliate whatever it wanted for power, but the supplier could not charge customers more than a predetermined price set by the regulator.

The regulator set an initial price and X factor for each bill component. Price increased at a rate slightly below that of inflation. (See Table 14-11.)

Table 14-11. Initial Tariffs for Scottish Regulated Customers (Fiscal Year 1990/1991)

	Scottish Power		Hydro-Electric	
	p/ kWh	X	p/ kWh	X
Transmission	0.382	-1.0	0.388	-0.5
Distribution	1.251	- 0.5	1.265	- 0.3
Supply				
Generation/purchase	4.086	0.0	3.981	- 0.0
Supply expenses	0.298	- 0.5	0.226	- 0.3

Source:
Prospectus, pp. 116, 205.

The Director General could review prices, transmission in 1993 and distribution and supply in 1995. From fiscal 1994/1995 through fiscal 1997/1998, "the generation/purchase component of the supply price moves progressively to a basis determined by reference to the electricity purchase costs … in England and Wales." [172] Customers with "an average maximum monthly demand of 1 MW or less" fall under the regulated tariff. [173]

Projected returns were closer to those of generators than RECs. Leveraging boosted returns on equity, thereby producing an appearance of exceptional profitability for a combination regulated business and commodity producer operating in a market with excess capacity. (See Table 14-12.) The government also extracted a higher price from investors for Scottish shares.

---

171. *Prospectus, The two Scottish electricity companies share offers*, 31 May 1991, p. 116.
172. *Loc. cit.*, p. 18.
173. *Loc. cit.*, p. 117.

Table 14-12. Projected Combined Financial Data for Scottish Utilities (Fiscal Year 1990/ 1991) (£ millions and %)

	Pro-forma	Pro-forma excluding extraordinary items
	[1]	[2]
	___ £ millions ___	
Invested capital (end of period)		
Debt	626	626
Equity	1,080	1,129
Total invested capital	1,706	1,755
Net plant (end of period)	1,821	1,821
Market value of equity (offering price)	2,876	2,876
Enterprise value (offering price)	3,502	3,502
Income for invested capital		
Interest	73	73
Net profit	177	226
Total income for invested capital	250	299
	___ % ___	
Rates of return		
Return on invested capital	14.7	17.0
Return on enterprise value	7.1	8.5
Return on equity	16.4	20.0
Return on net plant	13.7	16.4

Notes:
See Table 14-4.

Source:
Prospectus.

The Scottish utilities, then, faced minimum competition and operated under an essentially regulated framework.

Selling the Industry

The challenge: sell 16 unfamiliar companies in less than a year, during a tumultuous period in which Labor threatened to renationalize if it returned to power, Tories defenestrated Mrs. Thatcher and the UK fought the first Gulf War. John Major's government had to sell the shares high enough to preclude charges that it was giving away valuable assets ("the family silver") but low enough for buyers (voters in the next election) to make a profit. It had to assure that all companies sold equally well. It sought to attract small shareholders but needed large investors for the offerings to succeed. It wanted to create competitive conditions in the market but not competitive enough to drive away investors. The Tories had years of experience selling off assets. The electric privatization, though, involved more moving parts and more restructuring.

The government whipped up competition between bankers hoping to sell the shares. Brokers published weighty analyses months before the offerings. A monster, Frank N. Stein, pitched the REC shares and the Star Trek crew sung the praises of the generators in television commercials. Company managements went on road shows, making scripted presentations before slick backdrops while black-attired theatrical technicians hovered on the sidelines. The government pulled out all the stops, with show business flair.

The REC offering came first. Its 800 page prospectus measured one and five-eighth inches in thickness, weighed five pounds and looked like a telephone book. Marketers printed up 230,000 copies of that tome and 12.5 million copies of a mini-prospectus offered up to those with less time for reading. The average REC stock would yield 8.4% and sell at eight times earnings. The government sold the smaller and northern RECs at higher yields and lower price-earnings ratios (that is, at lower valuations), in order to overcome investor prejudice that London bankers feared would inhibit their sales.

According to the placement document the RECs would maintain "a progressive dividend policy" [174] or a "progressive dividend policy with dividends ... in line with underlying profit ... to produce dividend increases in real terms." [175] Inflation had averaged 6.5% per year in the previous five years and hit 9% before the offering. Adding the 8% dividend yield to a 6.5% annual share price increase (assuming that price would rise at the same rate as dividends) produced a total return (dividend plus increase in share price) of 14-15% per year. UK government bonds yielded just under 10%. REC shares could earn an equity risk premium over bonds of 4-5% per year and a real total return of about 8%, projections in line with historical market experience. And so much of the return would come from the predictable dividend.

The offering priced each REC's shares at 240p on November 18, 1990. The purchaser paid 100p on application, 70p on October 22, 1991 and 70p on September 15, 1992. The installment plan made purchase attractive to buyers without ready cash to buy the minimum 100 shares all at once. The partially paid share, moreover, received the first dividend in full in October 1991, thereby earning a 10% dividend return on the initial investment in less than a year.

---

174. *Placement Memorandum Dated December 11, 1990...*, p. 350.
175. *Loc. cit.*, p. 193.

The government wanted individuals to buy and set aside 54.6% of the shares for them. Those who registered before November 14 to buy shares in their local REC could get vouchers against their electric bills or an additional share for every ten still held after three years, up to a bonus of 300 shares. Non- customers could collect a bonus of one share for every 20 held for three years, up to a total of 150 bonus shares. By midnight of November 14, the Electricity Share Information Office had received 7.3 million applications.

Institutional and foreign purchasers bought the rest of the shares. Many had no intention of holding for three years, or even long enough to collect the first dividend. They expected a big price pop on the first day of trading and intended to sell quickly. They had to buy a package of all RECs. Market conditions were uncertain. Stock brokers worried. They need not have, because 12.75 million purchase applications flooded in. Thanks to the outpouring of retail interest, managers of the offering did no have enough shares to meet the institutional demand.

On the first day of trading, the fully paid share package's price shot up 49p to 289p. But purchasers actually put down 100p for the partially paid share which rose to 149p, a 49% profit in one day. That was the profit that the traders sought. The government converted ownership in the RECs into cash receipts of £5.2 billion (before flotation expenses) from sale of stock plus £2.5 billion of debt that the RECs and National Grid agreed to pay it, a total of £7.7 billion, close to the government's investment (including any debt forgiven) in the RECs and National Grid at the time of the offering. The government retained one "£1 Special Share" in each REC, which it could vote only in special circumstances, primarily to prevent change in control of the company. The government did not give away the family silver, as some opponents feared. Nor did it maximize return either, as demonstrated by how the stock shot up stock on the first day of trading. A 20% price rise for an initial public offering is not extraordinary. A 49% price increase is another matter, though. But underwriting is an art, not a science.

Floating the generators was less straightforward. In mid 1990, the conglomerate Hanson plc almost bid to buy PowerGen for £1.4 billion. The government backed away from a deal when PowerGen finally agreed to something close to the government's demands for debt that PowerGen would pay to it. A sale to Hanson, anyway, would have given Labor an opportunity to slam the deal as a tax giveaway to the conglomerate. On February 1, 1991, the bankers issued the prospectus to sell 60% of the shares of both generating companies, right in the middle of the Gulf War. The government sold the companies in two tranches in the hope that the share prices would rise after the first offering, and the two step sale would maximize the take.

The offering priced shares of both generating companies at 175p each (100p down and 75p one year later). Those who held shares for three years received a one share bonus for every ten held, with a maximum of 400 bonus shares. The flotation priced the shares at about 7.7 times earnings, with a dividend yield of 6.3%. The proposed dividend would provide a yield of about 7.2% on the partially paid shares in the first year. Brokers extolled the benefits of duopoly in enhancing profits and the possibility of 10% per year growth. The companies promised "progressive" dividend policies. Assuming that earnings and dividends moved up as projected and stock price moved in tandem, investors might expect total annual returns of 16%, better than the RECs might provide. Higher returns, though, should accompany higher risk.

Investors sent in 1.9 million purchase applications. On opening day, price shot up 37p, so speculators in for a one day profit made a 37% return. The government collected £2.2 billion for the 60% of shares sold, plus £0.8 billion of debt that the generators agreed to pay it, less flotation costs. The book capitalization of the two firms added up to about £3.7 billion. If the government could sell its remaining holdings at the offering price, it would collect another £1.4 billion, for a total sale and debt value of £4.4 billion. (The government sold the balance of the shares in early 1995 for about £4.0 billion.) The government also retained a special share in each company. And, in the end, it came out ahead by rejecting a deal with Hanson.

By the Scottish flotation, interest rates had declined, the stock market rose, the Gulf War ended, and the Tories must have had little desire to enrich Scots who always voted for Labor. The Scottish prospectus, dated May 31, 1991, priced shares for the combined offering at 240p, with 100p payable on application, 70p in May 1992 and the 70p balance by April 1993. Customers that held on for three years would receive a bonus of one share for every ten held, up to a total of 300 shares, or a discount on the electric bill. Non customers who held shares for three years would receive a bonus of one share for every 20 held, up to 150 shares. The offering priced the shares at 12.7 times earnings, with dividend yield of 5.1%. The companies did not specify the dividend payment for the partially paid shares in the first year, but the offering document suggested a partial dividend in March 1992, with a yield of about 4.0% on the 100p first payment. On the first day of trading, the share prices rose an average of 18p over the offering price, that is an 18% return on the 100p first payment.

The government collected £2.9 billion from the Scottish offering plus £0.6 billion from debt that the utilities would have to pay to it, less flotation expenses. It retained special shares in both companies. The government's book investment (including debt forgiven) before the offering was about £1.7 billion. The government did not sell cheaply, and made a handsome profit.

Investors buying the entire package paid about nine times earnings (vs 15 times for the average UK stock) and collected a dividend yield of over 6% (vs the UK average of 5%). The risk-free alternative, a government bond, yielded 10%. The opening day price pop was not extraordinary measured against the fully paid price of the shares, so in that sense the government did not underprice the shares, but the partial payment plan for the shares created a bonanza for institutional investors who managed to get big positions from friendly investment bankers. It was found money for the well connected, courtesy of the Treasury. (See Table 14-13.)

Table 14-13. Stock Offering Summary (1990-1991)

Offering	Date	Prospective P/E Ratio	Dividend Yield	Offering + Debt as % of Book Capital	First Day Premium as % of First Payment	First Day Premium as % of Offering Price
RECs	Dec. 1990	8.0x	8.4%	99%	49%	20%
Generators	Mar. 1991	7.7	6.3	119	37	21
Scotland	June 1991	12.7	5.1	205	18	8

Note:
Generator sale assumes sale of 100% of holdings in 1991.

Sources:
Offering documents for RECs, generators and Scottish utilities.

The selling team learned to set prices better over time, as evidenced by the decreasing price pop on offering day. Improving markets helped, too. From December 1990 to March 1991, interest rates fell and stock prices rose by about 10%. From March to June 1991, interest rates held steady, while stock prices increased marginally. Within a half year, the government extracted £10 billion from investors for shares in 16 untried entities, had those firms on the hook for another £4 billion in debt, and would later collect another £4 billion for the still unsold shares in the generators. Not a bad haul from a complicated offering made in a big hurry.

Conclusion

Working on a tight schedule, the privatization team restructured a major industry, sold it to the public, got it operating without serious snafus, while limiting risks to consumers and investors. They seem to have worked out every angle. The British model became the standard for industry restructuring throughout the world. The problems emerged afterwards.

Chapter 15

Debugging the System (1991-1994)

If you think this is complicated, believe me you are just beginning to appreciate the utility industry's love of complexity. [176]

– I.M.H. Preston

Eighteen major industry players, British Coal, British Gas and assorted independent producers played in the new market. The regulator had to placate special interest groups, fend off interference, encourage competition, satisfy investors and give customers a fair deal. The government had rushed the process in order to keep it on schedule. The regulator could make corrections along the way as well as prepare a the 1995 regulatory review.

RECs Make Out

Distribution accounted for over 90% of REC invested capital and operating profit. Supply was a sideline. Wires made the money.

RECs operated under the RPI-X or RPI+X formula. Although distribution costs were largely fixed, volume sold determined revenue, which removed any incentive to promote energy conservation because RECs would collect less if they helped customers use less.

REC-owned suppliers sold mainly to franchised customers. RECs used those guaranteed sales as a basis to sign contracts to buy power from the generators, a deal that did not cover all costs because they were based on Pool Purchase Price, not Pool Selling Price. When Pool Uplift charges doubled from the first half of fiscal 1990/1991 to the following fiscal year, RECs supply arms lost money. They started with a low margin, had to maintain working capital, took credit risks, and could not fully protect themselves from Pool price swings.

Initially, the REC investment thesis looked like this:

7%	dividend yield on the initial price
+ 4	real growth in share price in line with growth of earnings and dividends
+ 5	growth in share price due to inflationary impact on earnings and dividends
= 16 %	per year total return in nominal terms

---

176. I.M.H. Preston, "Managing Change," in Leonard S. Hyman, ed., *The Privatization of Public Utilities* (Vienna, VA: Public Utilities Reports, 1995), p. 284.

The brokerage houses that touted the industry pushed that prognosis. [177]

On offering, the shares yielded over 8%. RECs had to generate 8% annual growth to meet the target, so they had to earn 16% on equity (paying 8% as dividends and reinvesting 8% to finance growth). The prospectus (see Table 14-10) showed a 12% return. The extra profit had to come from National Grid, which earned more than the RECs reported. Including NG, RECs could produce a 15% return on equity. The predictions looked plausible.

The RECs beat the projections. Operating income in the first four fiscal years rose 16% per year in real and 22% per year in nominal terms. Returns on shareholder equity reached the high teens to low twenties and National Grid consistently earned over 20%, exceptional numbers for regulated monopolies. From privatization to mid-June 1994 (before the regulatory decision on the earnings in the next five years), REC shares outperformed the market (Table 15-1):

Table 15-1. Average Annual Returns on REC Shares (12/90-6/94) (%)

	Total Return	Stock Price Only
Nominal returns:		
RECs	37	30
UK stock market	17	12
Real returns:		
RECs	34	27
UK stock market	14	9

Source:
Andrew Adonis, "Diverging views on UK utilities' identities," *Financial Times*, June 17, 1996, p. 19.

Note:
1. Data adjusted to include first day trading profit on REC shares.
2. Total return includes dividend plus stock price appreciation. Dividends estimated for total return.
3. Stock market defined as FT-SE-A All-share.
4. RPI used as deflator.

That stock performance, in the period before takeovers of RECs began in 1995, was 10-23% per year higher than justified by the level of risk involved, according to one study. [178]

Did that amazing performance occur because:

---

177. Chris Rowlands, "Electricity: Issues for Investors," Barclays de Zoete Wedd, Sept. 1990, p. 2.
Dr. John Wilson, "The Electricity Industry," UBS Phillips & Drew, 28 August 1990, pp. 14, 102.
178. Roger Buckland and Patricia Fraser, "The scale and patterns of abnormal returns to equity investment in UK electricity distribution," *Global Finance Journal*, 13 (2002) 39-62.

- The government sold the shares too cheaply? – Eliminating the opening day price pop (the best measure of underpricing) reduced nominal annual return from 37% to a still exceptional 32%.

- RECs, the government or both, honestly underestimated potential cost savings? – They had to make conservative estimates, in front of a public offering, but they knew that other privatized industries achieved significant operating savings. Why not this one?

- REC managements misrepresented potential savings? – Perhaps they took advantage of their superior knowledge of the businesses and negotiated hard, knowing that a fight about regulation would delay the offering, and the government was in a hurry.

Dieter Helm speculated that inefficiency "was much greater than realized at the time."[179] Looking at the situation differently, imagine a group of engineers running the least glamorous end of the industry realizing that they knew more than the Oxbridge crowd at the Treasury, that most prestigious of government departments. By undershooting profitability targets the year before they may have lowered governmental expectations. Soon they would run their own ventures, with a mandate to make money, not to serve the public interest beyond what regulations called for. Soon, their shareholders would pay them to outwit the regulator if possible. That was the job as designed by the very officials with whom they would negotiate. Why not start immediately?

Even after the extraordinary profitability became evident, the regulator did not step in. That inaction fueled a political backlash and prompted outside firms to take over the RECs.

Transmission Thrives

National Grid's network accounted for 80% of profits and 90% of capital investment. NG owned interconnectors to Scotland and France and pumped storage plants, too, bought and sold ancillary services for the network at no profit and ran the settlement service for Pool transactions at a nominal profit.

RECs and suppliers paid about 76% of transmission charges and generators the balance. Of course, in truth, consumers, indirectly, paid the entire bill.

The charges did not cover full costs of long distance transmission. New generators could site plants near northern gas fields and transmit electricity to southern customers or locate plants near consumers and transport gas to by pipeline. Generators chose the trans-

---

179. Dieter Helm, *Energy, the State and the Market: British Energy Policy Since 1979* (Oxford: Oxford University Press, 2003), p. 206.

portation service that underpriced its value, which meant NG. Power plants went up in the north.

The grid management framework put nobody in charge of Uplift (costs of constraints and ancillary services). The government anticipated that Uplift would equal about 2% of Pool trading value but the number hit 10% in 1993. NG, as network operator, passed Uplift costs on to buyers of electricity. It had no reason to control the costs.

In June 1992, NG tried to modify tariffs to better reflect inter-regional transmission costs. Users in the far reaches of the network complained. NG's new regulatory framework, which took effect on April 1, 1993, incorporated geographic pricing. It had a RPI-3 formula for total system charges (X = -1.9%, with an assumed 1.1% annual growth in demand, netting out to X = -3%). In effect, the new formula controlled revenues. It did not depend on growth of demand.

Then, in 1994, OFFER forced NG to take responsibility for Uplift. The system operator incentive scheme set an annual target for costs. National Grid earned a bonus or paid a penalty depending on deviation from the target. The new demands hardly dented the company's profitability. Within three years, though, Uplift fell by two-thirds. In America, by contrast, non-profit system operators took no active role to reduce costs. British consumers gained by forcing National Grid to do something other than just observing the damage.

National Grid's story raises questions. Did the nationalized firm operate so badly that the privatized NG had to look good? Was regulation too lax? Did privatization release unheralded creativity? Or did the government know from the beginning that NG started with extraordinary advantages and only poor management or punitive regulation would have prevented a good performance? They must have known. Look at the projections.

Generation, Fuel and Supply Complications

Until March 31, 1994, only large consumers (demand of 1 MW or more) could buy directly from the Pool or from second tier (competitive) suppliers. Those 5,000 customers accounted for around 30% of sales in England and Wales. Initially, host RECs snared about 58% of second tier business in their own territories, the Big Two another 34%, and others 8%. In the fiscal year ended March 30, 1994, the host REC's percentage fell to about 42%, the Big Two to 25%, and others took a 33% share. Host RECs tended to serve smaller customers. The supply businesses started out with thin margin. Its profits made up an insignificant portion of Big Two profits. But having customers lined up would play a significant role in their strategy.

For the next 45,000 customers (demand of 0.1-1.0 MW range) to join the competitive market, the industry had to install meters, arrange the payment system and protect customer information from unauthorized use. But, as Prof. Helm noted, supply profit margin was so slim that "there was little scope for significant reductions in customers' bills." [180] Maybe suppliers with more buying power could extract better deals from generators, the regulator intimated, and that would help customers.

Before privatization, the electric industry ran on coal and nuclear power. The European Community lifted restrictions on gas as a generation fuel in October 1990, just in time for privatization. North Sea fields supplied the gas. To burn it, RECs and independent power plant developers would build modern combined cycle gas turbine (CCGT) units that ran more flexibly, efficiently and cleanly than existing power stations. (Table 15-2.)

Table 15-2. Estimates for Coal vs Gas (CCGT) Generation

	CCGT	New Coal	Existing Coal
	[1]	[2]	[3]
Cost per kWh (p)			
Fuel	1.30	–	–
Operation and depreciation	0.50	–	–
Interest and profit	0.75	–	–
Total	2.55	3.6	2.6-3.0
Operations and construction			
Thermal efficiency (%)	53	–	37
Years to construct	3	6	–
Construction cost (£/kW)	400	800	–

Notes:
1. New coal – with pollution control equipment, burning imported coal.
2. Thermal efficiency – average coal plant in operation.

Sources:
1. Cost per kWh from Henney, op. cit., pp. 229-232.
2. "Presentation by John Rennocks, Executive Finance Director, PowerGen," 7 February 1992.

Independent developers had to move fast to secure gas supply before National Power and PowerGen began to compete for it. In February 1989, Enron, the now infamous American energy firm, launched Teeside Power, the world's largest cogenerator, to operate at ICI's enormous chemical works in northeastern England. ICI would buy steam and electricity from it. An existing connection to the grid would permit Teeside to sell excess power in

180. _____, op. cit., p. 263.

the Pool. Suppliers could bypass British Gas and pipe gas directly to the generating station. That 1,875 MW plant would account for 5% of British gas consumption and produce 3% of Britain's electricity once it opened.

Enron planned for a winter 1993 service date. It lined up gas and ordered turbines before receiving letters of intent from the RECs to buy power from the plant. Enron ran a nervy operation. In September 1990, Teeside wrapped up contracts with fuel suppliers, and power purchase agreements with ICI and four RECs that were its shareholders. Construction commenced in November 1990, before Enron secured permanent financing for the project in June 1991. Teeside achieved full commercial operation in April 1993.

By late 1993, nine independent power projects (5,500 MW of capacity) with REC contracts and investment were in operation or under construction. They could supply about 15% of REC needs, close to the regulatory cap on REC ownership of generation. The project owners could raise money because they had contracts to sell to RECs. In effect, the RECs indirectly financed the plants with their credit because the bankers could rely on them to pay the power bills.

PowerGen, in 1989, bought a gas field, announced a large gas-fired power project and established a gas pipeline operation. Then National Power announced its first major gas-fired power station. By late 1993, National Power and PowerGen, between them, had 4,700 MW of CCGT units completed or under construction and about 1,300 MW more in planning.[181] The Big Two acted as nervily as Enron when they ordered those power stations without securing contracts from customers.

Coal-fired generators had to deal with British Coal, a government owned monopoly. Gas-fired generators bought from British Gas (BG), a privatized monopoly. In November 1990, BG offered them gas at 16.1 p per therm.[182] BG first underestimated demand, then worried about running short and in late February 1991 raised prices 35%. Buyers could sign up at the old price if they acted within one week and furnished a contract guarantee from their parent companies. Independent power producers tend to establish generating units as separate businesses whose creditors have no recourse to the parent, so BG signed only three contracts, one guaranteed by a government agency, one by a REC, and one by National Power.

Two non-signers protested to the gas regulator who voiced doubts about the shortage and ordered British Gas to sign contracts with the aggrieved parties. BG ignored the order, which led to legal action. Finally, gas regulator James McKinnon, a Scottish accountant who did not hesitate to tussle with his charge, told BG that he would lower the price,

---

181. Gas service to a large Scottish station had already commenced in 1992.
182. The therm is a measure of quantity of heat, equivalent to 100,000 British thermal units, or the energy found in roughly 100 cubic feet of natural gas.

so BG did it instead and in September 1991 gave generators six months to sign for fiscal 1995/1996 delivery at 19.6 p per therm plus escalation for inflation. Four generators did. Then in October, BG raised price to 21.5 p per therm for 1996 delivery, making gas uncompetitive as a fuel.

According to projections, gas-fired capacity as a percentage of total generating capacity in England and Wales would rise from 0% in 1990 to more than 5% in 1993 and 12% in 1995. New gas units, being base loaded, would run most of the time. In a slow growing market that already had too much generating capacity, the "dash for gas" could only mean that efficient gas fired generators would push old coal burning units out of business.

The 1984-1985 strikes fired up Tory determination to reduce reliance on coal and get the government out of mining. The CEGB considered importing fuel. A 1988 cabinet paper predicted that coal consumption would decline sharply. British Coal neither moved quickly to adjust to its new prospects nor to reduce prices to slow the decline. BC's management misjudged the determination of their generation counterparts. BC's chairman wanted "to call their bluff, they need us." [183] Late in 1989, realizing that the Big Two would build CCGTs, British Coal signed a smaller contract at lower prices. The government, though, wanted to privatize BC and for that project called in N.M. Rothschild, the legendary banking house.

Rothschild predicted that BC's sales to generators would drop from 65 million metric tons in fiscal 1992/1993 (last year of the coal contract) to 30 million metric tons per year thereafter, and its work force plunge from 60,000 to 25,000 miners. To sell them, the government had to assure that the mines had contracts for their output. Parliament considered steps to boost British Coal: defer competition for customers with less than 1 MW of demand, subsidize coal mines, force the Big Two to buy more coal for inventory, limit licenses for CCGTs and restrict power imports from France – all to exit coal mining gracefully.

This rush to make the mines more valuable at somebody else's expense caused John Baker, chief executive officer of National Power to write to Michael Heseltine, President of the Board of Trade, in terms as blunt as an ex-civil servant could muster:

> ... my board is quite clear it goes beyond their duty ... to ...volunteer the shareholders to assume responsibilities which are not theirs ... The prospectus... made it quite explicit that we would be pursuing policies in relation to diversification of fuel supplies ... [184]

---

183. Henney, *op. cit.*, p. 288.
184. _____, *op. cit.*, p. 295. *The Financial Times* quotes Baker as telling Heseltine, that he had asked the generator to dig its "own grave... to solve your coal problem." (Michael Cassell, "Dark descent into a political minefield," *Financial Times*, March 25, 1993, p. 14.)

Once the competitive market expanded from 30% to 50% of sales, contracted coal volume could fall one third. RECs would no longer buy high priced power because they were about to lose customers who bought that power, and generators would not buy high priced coal unless they had contracts to sell power produced by that coal. But RECs could not violate their regulatory obligation to purchase economically. The rescue plan that allowed the government to unload BC hinged on the approval of Stephen Littlechild, who had to certify contracts as economical. The regulator's charge, though, did not include subsidizing British Coal.

Prof. Littlechild advanced "The Review of Economic Purchasing" to December 1992. In it he approved REC contracts to buy power from independent producers (many affiliated with RECs so not that independent). Then, in February 1993, he issued "Review of Economic Purchasing: Further Statement." In it he agreed that a REC-Big Two power purchase deal that served to back up another coal contract would not violate the economic purchasing requirement. Before issuing his December 1992 review, Littlechild told Parliament that he did not make energy policy. The decision may have averted an attempt to rescue BC by delaying the expansion of retail competition, but consumers had to pay for high priced coal in return for the uncertain benefit of keeping retail competition on schedule.

Electric generators purchased 65 million metric tons of coal in fiscal 1992/1993. They contracted with British Coal for 40 million metric tons in the next year and 30 million metric tons in each of the following four years and with private mines for three million metric tons in 1993/1994, and five million metric tons by the end of the contract. The starting price for coal, 151 p per gigajoule (GJ) [185] exceeded the likely delivered price for foreign coal. Price would decline to 133p/GJ in 1997/1998 (plus adjustment for inflation).

British Coal closed mines. In 1993, the government split it into five regions. The contracts created value for buyers: they would collect a stream of income for a fixed period of time. In 1994, three firms bought most of the mines, paying less than £1 billion on the installment plan.

Nuclear Electric and subsidiary Scottish Nuclear produced one fifth of all electricity. Nuclear Electric, chaired by nuclear engineer John Collier, planned to cut staff, improve AGR operating performance, extend Magnox station lives, reduce decommissioning costs, negotiate waste processing fees paid to British Nuclear Fuels (another government-owned entity) and double operating profits within eight years. Sizewell B, the last big nuclear unit under construction, a pressurized water reactor (PWR), went into service in 1995. The Scots focused on keeping their plants running well. And the government wanted to privatize Nuclear Electric. (See Table 15-3.)

---

185. The gigajoule is a unit of heat content equivalent to about 950,000 British thermal units. A metric ton (tonne) of coal has a heat content of roughly 27 GJ.

Table 15-3. Nuclear Performance (Fiscal Years 1989/1990 - 1994/1995)

	1989 / 1990	1994 / 1995
Output (TWh)	36.9	54.1
Operating costs / kWh (p)	5.2	2.7
Operating income (loss) (£ million)		
Nuclear Electric	(1,101)	(35)
Scottish Nuclear	(32)	150

Note:
Operating income calculated before interest costs, extraordinary items and the nuclear levy.

Source:
Helm, *op. cit.*, pp. 194-195.

In 1993, a Nuclear Electric official said the company could earn a profit in 1995 without the nuclear levy. Management had turned the firm around.

The government completed a review of nuclear power in 1995. Nuclear Electric, said its managers, as a privatized firm could not build a nuclear station without government guarantees and long-term contracts to sell its output. Scottish Nuclear officials wanted to replace an aging unit with a new one in 2011. Greenpeace, the environmental crusader, argued that power from a new nuclear station would cost several times more than from a CCGT. Nuclear power did not look like a competitive market player.

National Power and PowerGen, unlike Scottish generators had no customer base and un-like independent power producers, no long-term sales contracts with RECs. In the year before privatization, they had an 80% share of the England and Wales market, reported revenues of about 3.3p per kWh, and a pre-tax profit (excluding extraordinary items) of 0.4p per kWh generated. Four years later, in 1993/1994, they generated less and their market share fell to 69%, but they collected about 3.6p per kWh generated and made a pre-tax profit of about 0.7p per kWh. They earned returns on equity in the high teens, close to those of the RECs and marginally below those of the Scottish utilities.

The Scottish generating sectors showed strikingly similar results. Revenue per kWh gen-erated rose from 4.9 p to 5.7 p, and pre-tax net income per kWh from 0.7 p to over 1.2 p in the same period. The new model did wonders for electric company profitability.

The government hesitated to rely on markets when doing so interfered with cherished energy policies or political necessities. Policy makers, though, did push for greater ef-ficiency. Improved margins showed that the industry could cash in on the benefits of competition.

Pool Flaws Surface

The Pool's computer programs had to calculate prices accurately and assure that funds flowed to proper parties. The Pool passed those tests, after start-up glitches. Beyond that, Pool rules should not distort the market and should prevent players from unfairly disadvantaging consumers. The Pool flunked those tests. The regulator noticed.

Despite the prevalence of contracts, daily Pool prices mattered and generators had reason to manipulate those prices to their advantage.[186]

The other generators left pricing decisions to the Big Two. Gas-fired power stations partially owned by RECs had contracts that protected their revenues as long as they operated, so they bid low to assure selection. Nuclear generators had to keep plants in service around the clock so they put in low or zero bids to insure that the Pool selected them.

---

186. Generators can play it safe or gamble on prices.

Example 1: Generator A produces 1,000 kWh. It signs contract for differences with customer B who wants to buy 1,000 kWh at 5 p per kWh. A's production costs are 2 p per kWh.

Pool price is 4p. A sells 1,000 kWh into Pool, collects 4,000p from Pool, and B pays generator another 1p per kWh to cover difference between Pool and contract price, so A collects 5,000p, altogether. B pays a total of 5,000p for electricity as required by the contract (4,000p to the Pool and 1,000p to A). After expenses of 2p per kWh produced (2,000p), A nets a 3,000p profit.

Pool price rises to 6p. A sells 1,000 kWh, collects 6,000p from Pool, and sends 1,000p to B (reflecting difference between Pool and contract price for volume contracted). A keeps 5,000p and B pays 5,000 p (6,000p to Pool less 1,000p received from A). After paying 2,000p of production costs, A nets 3,000p profit.

The fully-contracted generator is indifferent to the level or direction of the Pool price, because it ends up with the same profit no matter what the Pool price.

Example 2: Generator A produces 1,000 kWh. It signs contract for differences with Contract Buyer B for only 500 kWh at a price of 5p. Production costs are 2p per kWh.

Pool price falls to 4p. A sells 1,000 kWh into Pool. It collects 4,000p from Pool and another 500p from B (difference between Pool and contract price times volume contracted), for a total of 4,500p. After paying 2,000p of production costs, A nets 2,500p.

Pool price rises to 6p. A sells 1,000 kWh into Pool. It receives 6,000p from Pool, but must give B 500p (difference between Pool and contract price times volume contracted), A ends up with 5,500p in revenue. After paying out 2,000p in production costs, A earns a 3,500p profit.

The under-contracted generator A gains from higher Pool price and loses when Pool price falls.

Example 3: Generator A produces 1,000 kWh. It contracts to sell 2,000 kWh at 5p per kWh to Customer B. It has to buy power that it cannot produce from the Pool. Its production costs are 2p per kWh. Pool price drops to 4p. A collects 10,000p (2,000 kWh times Pool price of 4p plus another 2,000p from B to compensate A for Pool price below the contract price). A has 2,000p in operating expenses for the 1,000 kWh that it generated and another 4,000p that it had to lay out to buy from Pool the 1,000 kWh that it did not generate at 4p per kWh. A nets 4,000p in profit.

Pool price rises to 6p. A still collects 10,000 p (2,000 kWh times Pool price less payment to B to compensate for Pool price over the contract price). A pays 2,000p for operating expenses for own generation of 1,000 kWh and another 6,000p to buy in Pool the 1,000 kWh that it did not generate on its own. A nets 2,000 p in profit.

The over contracted generator gains from a Pool price drop and loses when Pool price increases.

The Big Two generators owned "mid merit" coal fired power stations. They ran those plants only when Pool price made operation profitable. National Power and PowerGen could control power supply by bidding or not bidding those facilities, and they could control price through their bids. The system needed the output of some of those plants every day. It had to choose them, no matter the price. High-cost facilities operated during emergency or peak demand periods came last in the pecking order. Those units set price only on the few days when the system needed them.

The bidding might work this way. The network requires 1,000 MW for a half hour. Nuclear companies have 300 MW of capacity. Gas fired units on contract have 200 MW. The Big Two own efficient, low cost coal fired stations with 300 MW of capacity. They also own another 400 MW of mid merit plants not under contract. Another generator owns 100 MW of expensive-to-run peak plant that rarely operates. A total of 1,300 MW is available to meet 1,000 MW of demand, which will leave 300 MW unchosen for service. Assume that generator owners bid their entire capacity to operate. The prices bids need not reflect costs:

	Operating Cost per kWh (p)	Bid per kWh (p)	MW Bid	MW Selected
Nuclear	2	0	300	300
Gas with REC contracts	3	2	200	200
Coal with REC contracts	3	3	300	300
Mid merit	4	5	400	200
Peak	10	30	100	0
Total			1,300	1,000

The Pool selects all nuclear units (300 MW), gas units with REC contracts (200 MW), coal units with REC contracts (300 MW), only 200 MW of mid merit plants and no peaking units. It pays every generator the highest price required to fulfill all its needs, 5p.

The nuclear operator bid 0p to furnish power costing 2p to produce, knowing that the Pool required more power and other generators with operating costs of at least 3p (mainly fuel costs) would not sell at a price below fuel cost. The nuclear operator could not turn its plants on and off in reaction to price, so anything it collected above nothing was better than nothing.

Generators with contracts could not collect on them unless they sold the electricity into the Pool. They knew enough about the operating costs of the uncontracted units to bid low enough to get selected for operation. They would not benefit from higher prices.

Mid-level generators bid carefully. If one of the Big Two underbid the other might retaliate by slashing bids and neither would earn a decent profit. They knew each other's costs from the days of common ownership, they could figure out how the other bid into the market and without much practice, learn how to raise bids together and share business.

The two price setting generators, though, might not want to raise price high enough to attract entry by new competitors or to annoy the government. Note, in the example, that if the mid-merit generators had charged 6p, the Pool would have had to select them and the Pool price would have risen to 6p. If half the mid-merit generators bid 4p and others 6p, the Pool would have selected the 4p generators, leaving out the higher priced bidders, and the Pool price would have been 4p. Thus, in this example, the mid-merit generators had to keep their bid prices close together in order to control prices and to make sure that each got a share of the business.

The Pool selected no peak generators, so their bids had no impact on Pool price. On a cold day, though, when the Pool needed 1,100 MW and 200 MW of contracted units were out of service for repairs, the Pool would have to call on all 100 MW of peak capacity. Owners of peak units could bid as high as they wanted, and the Pool would have to pay. If they removed from service other units that they owned, they could force the Pool to pay an even higher price for the peak power. Not too many of those shortage days occur, though, which is why owners of peak units have to charge so much in order to support the unit for the rest of the year.

Pool pricing did not start as predicted. The Big Two could not jack up prices because supply from nuclear plants and the interconnectors exceeded expectations. The two coal generators, moreover, had to burn coal that they had contracted to buy, so they had to keep plants running, which forced them to bid low in order to get selected. In July 1990, for three and a half hours, price hit zero. In addition, the government had required the generators to sign contracts to sell more power to the RECs than they could produce on their own, giving them incentive to depress Pool prices, in order to pay less for power needed to fulfill the contracts.

Once privatized, in February 1991, the generators had to raise Pool prices to increase profit on uncontracted units and to justify higher prices when existing contracts expired. They tried to burn more coal in the winter and raise prices in the summer (when demand was low) in order to make up for profits lost when burning up excess coal supplies. Some contracts expired after April 1991, and Pool prices rose despite seasonally lower demand. Customers buying directly from the Pool complained to OFFER. Uplift costs (to preserve system reliability and compensate selected generators that could not run due to grid constraints) almost doubled. The RECs' contracts hedged against Pool purchase price, not against Uplift, so that expense came out of their pockets. On September 9, 1991, the Pool purchase price jumped to 16.8p per kWh from a 3p level, prompting Stephen Little-child to institute a "Pool Pricing Inquiry" (the first of a series) whose conclusions OFFER published in December 1991.

The report said that Uplift in the first half of fiscal 1990/1991 reached £204 million vs. £102 million in the same period the previous year. Two items accounted for the difference. Operational Outturn – payments made to unselected generators that were required to operate due to network constraints or other problems – rose from £33 million to £93 million. Those generators could exert monopoly power because the system had to use them. Littlechild focused on National Power's bidding practices for a unit offered for service at 3.5p/kWh in April and 7p in September. The second item, Unscheduled Availability – extra payments the Pool had to make to get plants declared available during periods when capacity was short – rose from £1 million to £32 million. Between July and October, the report indicated that PowerGen withheld between 500 and 3,000 MW on a daily basis. With less capacity available, the chances of not meeting customer demand increased, causing an increase in the Pool's formula price for payment for capacity declared available (although not scheduled for production). Later in the day, after price rose, PowerGen would declare the plants available and collect the higher payments.[187] The generators knew how to play the game. It took a swift regulator to keep up with them, and it would have taken an even swifter one to stay ahead of them.

Prof. Littlechild wanted more competition, required new licensing terms to prohibit "monopolistic or anti-competitive behavior" and threatened price controls. The generators, however, were adept at finding ways to extract profits. Pool prices continued to rise in real terms although they did remain below the original government forecast (Table 15-4).

Table 15-4. Pool Price and Volume (Fiscal Years 1990/1991 - 1992/1993)

| | Fiscal Years | | |
	1990/1991	1991/1992	1992/1993
Pool volume:			
TWh	268	269	267
£ million	5,134	6,292	6,607
Prices (p/kWh)			
System marginal price (SMP)	1.74	1.95	2.26
Pool purchase price (PPP)	1.74	2.08	2.28
Pool selling price (PSP)	1.84	2.24	2.42
PPP vs government forecast (%)	-22	-15	-16

---

187. Independent power plant operators in California attempted to fleece the California Independent System Operator by withholding capacity and then declaring it available after the price went up. Did the California activity evolve independently, as a response to a similar market condition, or did someone bring the idea across the ocean? Given the infatuation of California politicos for the UK Pool and their trips to check it out, why didn't they take more serious notice of the Pool Price Inquiry, undertaken five years before California organized its market?

Notes:
SMP-PPP = payment for capacity, PPP-PSP = payment for uplift.
Time weighted price per kWh.

Sources:
Henney, *op. cit.*, p. 242.
David M. Newbery, "Pool Reform and Competition in Electricity," IEA/LBS Lectures on Regulation,
Series VII 1997, November 1997, p. 33.

OFFER oversaw system reliability, too. National Grid required ancillary services from gas turbines owned by National Power and PowerGen. They owned more plant than needed for this purpose. PowerGen threatened to withdraw units from service if not paid properly. OFFER analyzed costs, concluded that some plants would make profits and others went unpaid for services rendered. OFFER's "Report on Gas Turbine Plant" (June 1992) argued that cost of that generation far exceeded the value of protection to consumers. It suggested that National Grid look into other options, including protective equipment attached to the transmission network and arrangements for customers to cut load at specific times.

In October 1992, OFFER's "Report on Constrained-on Plant" investigated payments to units whose operation was required due to grid limitations. OFFER let National Power off the hook, but criticized PowerGen for an affair involving closure of two stations required to support the local network near them. When informed of the potential closure, National Grid began to install new equipment to transmit replacement power to that local network. During the installation process NG had to deactivate the old equipment, which put additional constraints on the local network. PowerGen raised prices in the constrained areas, and collected £88 million from use of the power stations it would close, during the period when National Grid's facilities were out of service in preparation for the closure of the stations. (PowerGen had no obligation to do otherwise. It had no public service obligation. Profit maximization was the new creed.)

The study asked: should the Pool transfer responsibility for Uplift costs to National Grid and make NG control costs rather than simply pass them on to customers? NG offered to take charge of most Uplift expenses, using contracts to lower uncertainty. OFFER replied that NG had not addressed the magnitude of the costs. It should provide a specified level of service and pay a penalty when it fails to do so. OFFER subsequently instituted an incentive scheme that would persuade National Grid to make dramatic changes in its *modus operandi*.

Pool prices rose 20% from April to July 1992. In the "Review of Pool Prices" (December 1992), the regulator said that the two generators "wished to secure higher Pool prices. The extent to which each company was able to achieve this depended on the other company adopting a similar policy ... I do not suggest the two companies acted in collusion ... Nevertheless, I conclude that National Power and PowerGen together have market power, and exercised it ..."[188]

In April 1993, Pool prices rose 10%. In July, OFFER's "Pool Price Statement" said that prices rose because National Power and PowerGen wanted them to so they raised their bids. Prices spiked more frequently, the differential between day and night and week day and week end prices narrowed, prices increased when demand declined, and Uplift rose. Prof. Littlechild considered ending the requirement that all trading take place in the Pool. He questioned why the Pool paid units that had been constrained off, "which charges customers in order to compensate potential suppliers for not being able to deliver to the market." [189] For remedies, he mentioned price controls, sale of plant owned by the Big Two and a reference to the MMC.

To some extent, Dieter Helm noted, the regulator was in a bind. He had permitted RECs to purchase electricity from affiliated generators, not the sort of deal that normally leads to the best prices, in what appeared to be a strategy to expand competition. Forcing National Power and PowerGen to lower prices might make REC contracts look even worse and reduce the likelihood that new competitors would enter the generating business.

Littlechild made a deal with the Big Two in February 1994. Better a deal than a reference to the MMC. They agreed – voluntarily of course – to divest 6 GW of capacity (4 GW from National Power and 2 GW from PowerGen), and to limit Pool prices for the following two fiscal years.[190] Presumably, the price cap (set below the previous two years' peak prices) would protect the public until more competitors entered the market. Economists Richard Green and David M. Newbery wryly observed that, "as a measure of the ability of the two generators to control Pool prices, demand-weighted prices, were within 1 per cent of the capped level each year." [191] In the two year period, fuel prices declined, however, so the Big Two's profits rose. They finally divested the generating stations to Eastern Electricity Group in 1996.

Regulation Overworked

Stephen C. Littlechild, an economist, doubted that "... the purpose of utility regulation ... is to mimic the operation of the competitive market" because of "the difficulty of predicting ... a competitive market price ... in markets characterised by heavy capital

---

188. Henney, *op. cit.*, p. 248.

189. _____, *op. cit.*, p. 250.

190. The Big Two agreed to a price of 2.4p/kWh (weighted by time) or 2.55p/kWh (weighted by demand), measured in October 1993 prices, for fiscal years ended March 1995 and 1996.

191. Richard Green and David M. Newbery, "Competition in the Electricity Industry in England and Wales," *Oxford Review of Economic Policy*, Vol. 13, No. 1, (1997), p. 36.

investments that are location-specific and have long asset lives. One reason for regulation is ... because such markets do not operate in the same way as the more familiar commodity markets ..."[192]

Littlechild made three arguments for regulation – especially the style he advocated:

- Unregulated private monopoly providing an essential service is politically infeasible.
- Incentive regulation promotes efficiencies normally found in competitive markets.
- Regulation can promote new entry and competition in some sectors of the industry, thereby reducing the disadvantage of regulation.

Littlechild had to work with the cards dealt him: RPI-X formulas set in pre-privatization bargaining sessions, inadequate generating competition, fuel and supply contracts designed to put off the inevitable at British Coal, and a Pool susceptible to manipulation.

OFFER first pressed National Power and PowerGen to behave more competitively, then to sell stations in order to lessen their market power. It modified National Grid regulation to make the company responsible for bringing down its charges. The regulator was less forceful about cost of electricity purchased by RECs from affiliated generators. Perhaps just getting those generators into service would increase competition and eventually lower consumer prices.

Littlechild's treatment of RECs exemplified his approach. In their first year of private ownership, operating profits shot up 32%. He resisted calls to intervene and curtail profits that began in 1992. After all, profits not only beat expectations, but also reached levels beyond the normal range of a regulated natural monopoly. Prof. Littlechild countered that price controls "were set on the basis that there would be a reasonable period between reviews ... If a regulator is seen to interfere constantly ... there will be an adverse effect on the incentive ... to improve ... efficiency and reduce costs."[193] Interference would raise regulatory risk and cost of capital and in the long run elevate the cost of doing business and, therefore, costs to consumers.

In October 1993, OFFER began work on a REC pricing review set for spring 1994, without commissioning the plethora of detailed studies that other regulators required. Prof. Helm decried the absence of "the Treasury-like approach" that typified other regulatory work and said that "... the analysis ... was less sophisticated, and there were no consultation papers on the key building blocks, such as cost of capital and asset base ... For Littlechild, prices and costs were what mattered: how the industry was financed and valued was less

192 Stephen C. Littlechild, "Privatisation, Competition and Regulation," 29th Wincott Lecture, 1999, London: Institute of Economic Affairs, 2000, no pagination.

193. Helm, *op. cit.*, p. 210.

his concern." [194] (Prices were what mattered to customers, though, and if Littlechild got that right, what was the problem?) In spring 1994, a leaked letter from Littlechild to REC chairmen suggested that he would reduce prices 10-20% and set the same X factor for all RECs, and in August 1994 the final determination came out.

OFFER's process looked uncomfortably like rate of return regulation with bells and whistles attached. Determine rate of return, the asset base on which return is earned, and the next five years' capital spending and operating expenses in real terms. Then determine the revenues required to cover all expenses and the profit that provides the required return. OFFER's process differed from American rate of return regulation in three respects. First, prices were adjusted annually in line with the rate of inflation. Second, Offer added a kicker to the formula to incentive the utilities to run more efficiently. Third, Offer set prices for five years at a time. [195]

OFFER, however, skirted issues that typically clog regulatory proceedings, including efficiency of capital structure. The regulator acted as if dealing with an investor who made a cash outlay to purchase the regulated asset base in year one, expended money in each of the following five years to operate and expand the capital base, and then sold the asset base at the end of the final year for cost of investment (including money put in during the five years) plus a 7% return per year, plus adjustment for inflation. Investors could decide how to finance the venture.

---

194. _____, *op. cit.*, p. 211.
195. The REC case on the table looked like this:

OFFER determined a return that fairly compensate existing capital and attracted new capital as well. OFFER chose a 7.0% pre-tax real return apparently plucked from the midpoint of a range previously used by the MMC for a British Gas decision. (OFFER, used that return for the two Scottish utilities, then applied it to the RECs.)

Then OFFER set the regulated asset base on which the RECs would earn return as initial enterprise value (market value of REC stock plus debt) minus value of National Grid holdings plus 50 % of the sum calculated. The regulatory asset value exceeded value of the assets on the corporate books (historical costs) by about 25%. This calculation produced a regulated asset base that increased faster from the initial date of regulation than consumer prices but less than the market value of the RECs. OFFER projected regulated asset value growth over the next five years by adding net capital expenditures (spending on plant and equipment minus customer contributions and depreciation on new plant) minus depreciation of old asset base.

After that, OFFER projected five years' sales, capital expenditures and expenses and set prices to generate revenues needed to equal those costs plus a profit that would produce a 7% pre-tax real return on regulated assets. OFFER started the new period by resetting prices to a new level (P0). The 1994 price cuts ranged from 11% to 17%, to reflect the lower costs achieved in the previous five years as well as allow the RECs to earn the appropriate return.

Then OFFER set the X factor in the RPI-X formula. OFFER opted for RPI - 3 for all RECs but only RPI-2 after adjustments.

Table 15-5. Example: Calculating Revenue Requirement for Five Years (Real £)

	Year 1	Year 2	Year 3	Year 4	Year 5	Total
	[1]	[2]	[3]	[4]	[5]	[6]
Expenditures						
Operating expenses	61	60	58	57	56	
Capital expenditures	50	49	49	50	50	
Total	111	109	107	107	106	540
Present value of expenditures (7% discount rate)	107	98	91	84	78	458
Present value of opening and closing assets	563				-435	128
Present value of required revenues						586

Note:

All numbers rounded from those calculated by the MMC.

Source:

Monopolies and Mergers Commission, *Scottish Hydro-Electric Plc: A report on a reference under section 12 of the Electricity Act 1989* (London: HMSO, 1995), p. 18.

Regulator sets beginning value of assets at £563, projects spending for five years (all in real terms), then sets real cost of capital at 7% (pre-tax). That is, £100 invested should earn £7, from which utility might pay £2 income tax, leaving £5 of profit after tax to produce a 5% after tax return. With that information, regulator determines that utility would accept £458 now to cover outlay of £540 spread over five years, assuming that it could invest the money now at a 7% rate of return. (That is, present value of £540 received over five years is £458.)

Investors would pay £435 now for £563 paid at end of five years, at a 7% pre-tax return. Utility would immediately lose £128 of market value by dedicating that capital to the use of consumers. It requires a stream of revenue over a five-year period which has a present value of £458 to cover expenditures and £128 to provide a return on the investment, for a total of £586.

Assume that £138 per year over five years is the stream of revenue that would produce a flow of cash for which an investor expecting a 7% per year return would pay £586. If that utility sells 10,000 kWh per year, it will charge 1.38p per kWh. Assume, too, that in previous five year period, sales were 10,000 kWh per year, revenues £153, and consumers paid 1.53p per kWh. Regulator would begin a new five-year period by reducing price by 10%, to a new opening price (P_0 in regulatory parlance) of 1.38p per kWh.

In the price setting game, the utility produces projections to support higher prices and the regulator to support lower ones. The lower projections, though, if used in the decision, could become self-fulfilling as the utility reduces expenses down to the estimate. No point spending on items for which the regulator provides no compensation.

The stock market reacted jubilantly to the price order with its 7% real return. Why? Government bonds yielded 7%. Adding a 3% inflation rate and subtracting interest and corporate taxes translated that 7% real return on capital into a 9% nominal return on common equity, not high compared to bond returns. Did the market misread the decision? Or did the regulator miscalculate and investors saw that? [196]

After years of regulatory obfuscation, the United States Supreme Court declared, in the *Hope* decision written by William O. Douglas in 1944, that "… it is the result reached, not the method employed, which is controlling. It is not theory, but the impact of the rate order which counts."[197] In terms of methodology, the August REC finding may have failed the test of providing a clear explanation, but it had a well defined impact.

After the finding came out on August 11, REC share prices rose 10%. Brokerage houses celebrated: "Much better than expected," "licence to print money," "Running rings around Offer" and "Shareholders can only rejoice."[198] The *Financial Times* titled its editorial "Littlechild underpowered," complained that "The electricity companies have passed virtually all of the benefits of their improved efficiency through to shareholders …, " conceded that the new rules were "quite tough," but noted that City analysts calculated that the RECs could produce 6% real dividend increases in the future, which raises "questions about the effectiveness of Prof. Littlechild's new regime." [199]

Littlechild replied: "If companies are going to make significant profits they're going to have to work much harder … I have allowed these (current) price controls to run on because that's the period they were set for – five years. Now those terms are finished, it's time for me to crack down – and that's what I'm doing."[200] Investors thought not, and apparently neither did a conglomerate that decided to enter the electricity industry a few months later.

---

196. Errors in projecting expenditures can overwhelm like errors in calculating the correct rate of return. Using the example in Table 15-5, for instance, a one percentage point error in return (that is an error of 14% either way) would change the present value of revenue required (that is, the electric bill) by roughly £30 million, or 5%. A 14% miscalculation of expenditures would change the bill by £80 million or 14%.

197. *Federal Power Commission v Hope Natural Gas Co.*, 320 US 591 (1944).

198. Ian Cornwall and Alex Henney, *Regulation in England and Wales* (London: EEE Ltd. and Cornwall Consulting Ltd., no date), p. 27.

199. "Littlechild underpowered," *Financial Times*, August 12, 1994, p. 11.

200. Michael Smith, "Power price controls lift shares," *Financial Times*, August 12, 1994, p. 13.

Critics chided Littlechild for setting the a return on total investment without working up a weighted average cost of capital. Perhaps if he had followed standard regulatory procedures, he might have fixed a lower cost of capital.[201] While academics debated, others noticed.

The DGES had set return independent of components of capitalization. Thus the utility's owner could change capitalization (borrow money) without the regulator changing allowed return in response. The findings provided a high real return for a low risk business with steady cash flow. And the DGES' indicated a predilection for leaving things alone during the five years between price settings. All that added up to a business opportunity.

Corporate takeover artists extol the benefits of financially engineered takeover deals. In the real world, after the deal has closed and bankers have departed with their fees, the transaction may look less attractive, once the increased debt burden raises interest costs and depresses the bond rating, and the market adjusts the stock price down to reflect the increased risk levels and the projected savings do not materialize. Optimists do those deals.[202]

---

201. Cost of capital is the minimum return needed to attract capital to a particular investment. The investors may not earn cost of capital after they make the investment but they must believe that they will or they will not invest.

Modigliani and Miller (M&M) won a Nobel Prize for formulating possibly the most important insight in modern financial theory, that risk determines cost of capital, not how a firm raises funds. If a particular investment costs £100, and the risk of that investment calls for a 10% cost of capital, it has to earn £10 for its buyer to collect a satisfactory return. If cost of capital remains at 10%, and the investment earns £10, it maintains its value at £100. If the buyer of the asset manages to find a lender foolish enough to lend £100 to the buyer at 3% to make the purchase, that does not change the cost of capital from 10% to 3%. It just shows the value of a good sales pitch to a financial innocent. If the buyer is a firm with many investments that can raise the money at 8%, that does not change the cost of capital of the investment, either. (M&M and others modified the formula to take into account tax and other factors but that does not change the principle.) Littlechild's supposed lack of financial sophistication may have kept him closer to basics than his financial-engineering-oriented critics.

Regulators and bankers may produce a bogus cost of capital by incorrectly adding up costs of components of capital structure. For instance Utility borrows £50 at an interest cost of 5% (£2.50), and Regulator calculates cost for equity investment of £50 is 10% (£5.00) based on returns for similar investments, so cost of capital is £7.50, or 7.5%. Debt costs less than equity because it incurs less risk. Therefore, in order to reduce overall cost of capital, Regulator encourages use of debt, up to the point at which too much debt endangers the financial viability of the venture. (That is why regulators often calculate a cost of capital using what they consider an ideal capitalization.) If Utility borrows £60 at 5% interest (£3 interest) and raises the other £40 in equity money with a presumed cost of 10% (£4), then Regulator declares cost of capital as £7 or 7%. Note, in this simplified example, that the equity owner gets no benefit from borrowing more money that supposedly lowers cost of capital. The regulator lowers allowed return to reflect the new "cost of capital." The equity owner takes more risk but gets no additional reward.

202. Borrowing money (leverage) makes profits look better. Takeover artists love it.

Consider this example. Utility has to invest £100 in plant and equipment. Regulator decides that Utility should earn 10% (£10). Utility sells £20 of debt with a 5% interest cost (£1) and £80 of equity. Investment earns the allowed profit of £10. After paying £1 interest, £9 remains to produce a return of 11.3% (£9 / £80) on shareholders' money. A corporate raider decides to take advantage of the situation. Raider borrows £60 at 5% (£3 interest) and takes £20 out of its bank account to buy Utility's £80 of equity. Utility still earns £10, pays £1 interest on old debt, and £3 on new debt, leaving a profit of £6 for £20 of equity, which produces a 30% (£6 / £20) return on Raider's equity.

Consider another situation. Utility earns 10% on £100 invested in plant and equipment. It has £20 of debt outstanding at 5% interest (£1 of interest payments) and £80 of shareholder investment earns £9, or an 11.3%

Trafalgar House, a Hong Kong- controlled, distressed conglomerate, owner of the engineering arm of the old John Brown shipyard and the Cunard Line, made the first move on a REC – Northern Electric– in December 1994, desiring, according to Dieter Helm, "a secure cash flow and the ability to borrow against Northern's distribution assets."[203] RECs now had a new role: cash cow for a corporate raider.

Northern Ireland

Northern Ireland Electricity Service (NIES), established 1973, endured strikes during "The Troubles." The government almost sailed a nuclear submarine to Belfast to generate power there in 1974. Next year militants destroyed the transmission line between Northern Ireland and the Irish Republic. It remained out of service for 20 years. In 1991, NIES reincorporated as Northern Ireland Electricity, plc (NIE) and sold its four power stations in 1992. NIE's Power Procurement Business (PPB) contracted for each plant's output for the life of the station.

NIE owned and maintained transmission and distribution networks. The PPB dispatched power plants, ran the transmission grid and had a statutory monopoly of wholesale supply. PPB bought all power generated by the four stations, and sold it to retail suppliers who resold it to customers. The three RECs competing for retail supply business in Northern Ireland could not achieve a 1% market share. Customers with a demand of one MW or more could self-supply, or buy directly from the PPB, but none did.

The Director General of Electricity Supply for Northern Ireland regulated NIE. The price control formula utilized fixed and variable expenses categories, each of which had a different RPI-X.

---

return. Utility decides to pay the shareholders a special dividend of £60, which reduces shareholders' equity from £80 to £20. Where will £60 come from? Utility borrows it at 5% (£3 interest). After the transaction takes place, Utility still earns £10, now pays out £4 in interest, and has £6 left over for shareholders, who now earn a 30 % (£6/ £20) return. But how will the market value equity in a company with a huge amount of debt as opposed to one with less debt? If the stock market previously valued the shares at 12x, they were worth £108. If the market believes that the remaining stock is much riskier, it may value earnings at 6x the new earnings or £36. The shareholder then, collected a £60 dividend but lost £72 of market value.

Look at the situation another way. Corporation A has £1,000 of machinery and equipment, sells £2,000 of goods and shows an operating income (available to creditors and shareholders) of £200. The company has £100 of debt on which it pays 5% interest (£5), which means that it has £195 available to shareholders. The market values the debt at £100 and the stock at £1,950 (10x earnings for common stock). Therefore someone seeking to buy Corporation A would have to pay £2,050 to take over those assets and the business that produces £200 per year of profit. Corporation B, with identical assets, sales and operating income to Corporation A has borrowed £1,000 on which it pays 5% interest (£50), which leaves £150 for its shareholders. If the market valued Corporation B's debt at £1,000 and its stock at the same 10x earnings (£1,500) as Corporation A, it would have a market value of £2,500. Why would a buyer in the market pay more for the same business? Financial engineers and investment bankers say that they can convince the market to do so.

203. Helm, *op. cit.*, p. 215.

Northern Ireland's consumers paid high prices. They paid higher prices after privatization.

NIE ranked in size with the smallest RECs.

In 1993, the government floated shares of Northern Ireland Electric plc.

Adding Up Savings

In five years of privatization, generators cut staff by 50% and distributors by 20%. Privatization scotched plans for costly nuclear and coal units and pollution control equipment. Coal contracts reduced costs, although not as much as if generators had sought alternative suppliers. On the minus side, consumers forked out unnecessarily large sums to Électricité de France. Previously, France and the UK split savings from purchase of cheap French nuclear power. In the new market, the French not only got market prices for their power. But they also collected part of the nuclear levy. Alex Henney calculated net savings for England and Wales at roughly £1.9 billion per year, about 16% off the annual electric bill. (See Table 15-6.)

Table 15-6. Annual Savings to England and Wales Electric Industry from Privatization (£ millions)

Reduced employment	750
Cancellation of power plants and projects	950
Coal prices	500
French interconnection	-300
Net savings	1,900

Sources:

1. Alex Henney, "The Restructuring and Privatization of the Electricity Supply Industry," in Leonard S. Hyman, ed., *The Privatization of Public Utilities* (Vienna, VA: Public Utilities Reports, 1995), pp. 267. 269.
2. Prospectuses for RECs, generators and Scottish utilities.

Net operating savings of £950 million should have reduced bills 8% and prospective cancellations another of £950 million over what the bills would have been. But real price of electricity barely budged. The companies pocketed the savings. If they had earned five percentage points less on shareholders' equity, they would have collected about £800 - £900 million less per year, roughly the savings not passed on to customers.

Scottish electricity prices remained stable in nominal terms, too, so Scottish customers did not get the full benefit of cost reductions, either.

Privatization did more to stop projects that would have raised prices than to lower them. In the five years after privatization, though, electricity shares (including dividends) outperformed the market by about £800 million per year. From one pocket to another.

Conclusion

Allen Sykes characterized privatization as a: "... Pioneering Major, if Partial Success."[204] Nanki-Poo might have called that praise "Modified rapture!"[205]

Electric companies showed that they could cut costs, the Pool that markets can allocate power, generators that they knew how to game an inadequately competitive market, the government that even pro-market administrations do not hesitate to interfere for political reasons, and that a regulator sticking up for principles gets no respect. Consumers came out about even, a better outcome than Californians suffered a few years later, but no victory.

The government, before the sale, allowed electricity managers to act as if they ran independent firms not responsible to the government that owned them. Those managers had more knowledge of the business than their supposed masters at the Treasury and must have used that advantage when bargaining. The politicians seemed to have gotten their information on nuclear power late or just ignored it. Lord Rayner, onetime chairman of Marks & Spencer and former government official and adviser to Margaret Thatcher, attributed errors of water privatization to "haste and dogma."[206] That diagnosis applied as well to electricity.

Yes, Allen Sykes was right. Privatization, to date, was a partial success – not a bad outcome.

---

204. Allen Sykes, "The British Experience: The Pioneering, Major, if Partial Success," in Leonard S. Hyman, ed. *The Privatization of Public Utilities* (Vienna, VA: Public Utilities Reports, 1995), p. 293.

205. W.S. Gilbert, *The Complete Plays of Gilbert and Sullivan* (NY: The Modern Library, n.d.), p. 360.

206. Geoffrey Owen, "Twin-track talent at the top," *Financial Times*, February 4, 1991, p. 34.

Chapter 16

The Tories Finish the Job (1995-1997)

... my American friends kept asking me....: What's a nice free-market boy like you doing in a place like a regulator's office? [207]

– Stephen C. Littlechild

To complete the privatization, the government still had to sell nuclear generator and cede control over the other companies by letting the golden shares expire. Despite threats from Labor, outsiders bought electric companies. The regulator rectified previous missteps and the Tories tried to wrap up the loose ends.

Regulation and Politics

The 1994 price control formula overjoyed investors, encouraged Trafalgar House and provoked a storm of criticism. In March 1995, Prof. Littlechild, due to "widespread public concern," [208] began a review of the 1994 decision. REC shares fell. Then in May 1995, the MMC reviewed OFFER's Scottish Hydro-Electric price order. It adopted the 7% pre-tax return then popular in British regulatory circles, as had OFFER in the August 1994 decision, but used a different regulated asset base: market value at privatization adjusted for inflation (raising it 13% to 1994/1995 values) plus capital expenditures minus depreciation (all in 1994/1995 prices). No special adjustments. No arbitrarily write up assets. In July, OFFER issued new REC pricing rules, reduced the adjustment to the original regulatory base from 50% to 15%, ordered 10-13% price reductions for fiscal 1996/1997, and changed the X factor from -2 to -3.

The government wanted more. After prompting, the RECs agreed to give customers a share of the profit they would make selling shares of National Grid at the end of the year. So National Grid paid the RECs a special dividend of about £1.5 billion and they passed on £1.3 billion (about £50 per customer) as reductions to the electric bill, a refund big enough to reduce the RPI that year. In December 2005, RECs sold National Grid shares for about £3.0 billion, versus a book value of about £0.8 billion. They collected roughly £0.7 billion in ordinary dividends during their share ownership, too. After taxes on profits, they earned about 23% per year on that investment, a hefty return for a low risk business.

---

207. Stephen C. Littlechild, *op. cit.*, p. 1 of lecture text (Part 1)
208. Helm, *op. cit.*, p. 216.

The Labor Party wanted even more, calling in the fall of 1995, for a windfall profits tax. On September 11, 1996, it published "Vision for growth: A New Industrial Strategy for Britain," which included a windfall profits tax on utilities justified by "the excess profits arising from initial regulatory price caps." [209] Labor wanted regulatory reform, too. Regulators would have to "justify their proposals" to a "non-executive board," and insure profit sharing when, during the five year price period, profits exceeded a "pre-determined level." [210]

RECs or Sitting Ducks?

Golden shares expired on March 31, 1995. After that the government could no longer arbitrarily prevent changes of control at the RECs, which had marketable assets and inordinately high cash positions, making them juicy targets for predators. Cash and equivalents at the eight RECs rated by Standard & Poor's rose from £504 million at the end of fiscal 1992 to £1,579 million at the end of fiscal 1994, up from 8% to 24% of capitalization. The average American utility, in contrast, held only 2% of capitalization in cash and equivalents. The RECs also owned National Grid shares then valued by the press at £3 billion. A potential buyer could have analyzed the average REC early in 1995 this way (Table 16-1):

Table 16-1. Takeover Analysis: REC Finances Simplified

	£	%
Initial capitalization (book) and earnings		
Debt	150	15
Equity	850	85
Total	1,000	100
Net earnings	175	
Dividends from National Grid	-15	
Earnings from REC business	160	
Interest on cash (after tax)	-10	
Earnings from distribution and supply	150	
Calculation of transaction:		
Purchase REC stock at 2x book value	1,700	
Strip out excess cash	-200	
Strip out proceeds of National Grid sale	-400	
Net cash outlay	1,100	

Earning £150 on £1,100 outlay = 13.6% return on net cash outlay.

---

209. Andrew Wright and Kevin Lapwood, "UK Utility Countdown," Merrill Lynch, 24 October 1996, p. 13.
210. _____, *op. cit.*, pp. 12-14.

REC sells National Grid shares with cost of £150 for £500, pays £100 tax on £350 profit, leaving after tax profit of £250 to add to common equity. Cash proceeds from sale of £500 less £100 tax on profits leaves £400 for special dividend payment to REC owners.

The common equity of the REC, after all transactions, becomes:

Beginning of period equity	850
After tax profit on National Grid sale	250
Dividend to deplete cash balance	-200
Dividend from National Grid sale	-400
Common equity after transactions	500

The new capitalization of the REC becomes:

		(%)
Debt (unchanged from before)	150	23
Common equity	500	77
Total	650	100

Even stripped of extraneous assets, the average REC retained an unimpaired ability to raise funds. The buyer of a REC could improve return by borrowing part of the purchase price at an interest rate below the return earned on the asset. The purchaser, for example, borrows £600 of the required £1,100 net cash outlay at 7% interest (interest cost of £42). The purchaser, thus, puts up only £500 of its own money. The return calculation (omitting the tax benefits of the borrowing) looks like this:

Earnings	£150
Interest on loan	-42
Net income to buyer	108

The £108 profit on a £500 investment produces a 21.6% return. The buyer taking advantage of tax deductions could earn close to 24%. High returns for safe investments.

Buying and Selling RECs

Trafalgar House's run at Northern Electric, even before the golden share expired, anticipated a frenzy of takeovers. Economists characterized this activity as "competition in the capital market."[211] Threat of takeover might scare "lazy managers"[212] into action in fear that they would lose their jobs when their firms fell prey to those who knew better how to reduce costs. More realistically, the RECs looked cheap to outside firms seeking ways to boost earnings and give the appearance of dynamism via a muddled acquisition strategy,

---

211. Helm, *op. cit.*, p. 221.
212. Green and Newbery, *op. cit.*, p. 38.

helped by the fact that British managements did not often resist a deal when presented with the right price.

Nothing indicated that Trafalgar House could manage the REC better. The *Financial Times* charitably described Trafalgar House's record as "indifferent."[213] Trafalgar House, though, could get more from the deal than operating savings, excess cash and strong cash flow – it could recover £223 million pounds of taxes (£2 per Northern share) and save on future taxes. Trafalgar House bid £11 per share, about 24% above market price, for a total of £1.2 billion. It could partially pay for the bid by stripping out the REC's excess cash, selling the National Grid shares and collecting an immediate tax refund:

	(£ millions)
Bid price	1,200
Excess cash	-70
National Grid (after tax)	-215
Tax savings	-223
Net cost	692

Northern's purchaser could expect (excluding adjustments for interest or tax benefits):

Net income	103
Loss of National Grid dividend	-7
Loss of interest on cash	-3
Net income after transactions	93

Trafalgar House offered to buy Northern shares at about 7.4x earnings, equal to a 13.5% return on its investment not counting tax benefits from interest or other savings. If Trafalgar House borrowed half the net cash needed, after tax return might exceed 22%.

To fight the bid, Northern proposed to pay a £5 per share special dividend, which reduced its value as an acquisition by draining off excess cash before Trafalgar House could lay its hands on the money and by creating debt as well. Northern appealed to government and regulator to stop the offer but neither would intervene. Stephen Littlechild's announcement, on March 7, 1995, that he intended to review his 1994 ruling on REC prices, however, did kill the deal. Northern's share price plunged. Trafalgar House backed out. But that did not end the story. In August, Northern's shareholders approved the special dividend of £5 per share (about £600 million), payment of which required Northern to borrow money and clear out extraneous assets. A year and a half later, Northern sold out to an American firm.

---

213. "High wire act," *Financial Times*, Dec. 15, 1994, p. 21.

The Trafalgar House/Northern merger attempt looked like a purely financial transaction with no readily discernible operating or strategic merits. The 14 acquisition proposals between September 1995 and March 1997 with one exception fitted into three categories:

- English generator desiring to integrate vertically buys distribution and supply – The Big Two tried to do so before the 1998 date when all customers could choose suppliers. The government vetoed the deals as anti-competitive.

- British utility buys British utility – Water companies bought RECs to achieve operating savings (they served overlapping territories) and create larger markets for multiple services. Mergers did not reduce competition and might lower costs. In the one electric- with- electric merger, Scottish Power bought Manweb, a REC with no generation, which enabled the buyer to enter the larger England and Wales market. The government approved all these deals.

- American utility buys UK utility – American invaders swallowed seven RECs after the Energy Policy Act of 1992 freed them from restrictions on foreign investments. They had strong cash flows but uncertain domestic growth prospects. Investments abroad might accelerate growth, diversify sources of income and provide insights into how competitive markets worked. American utility executives who had little or no experience abroad felt comfortable in the UK. After all, the British spoke the same language, queued up politely, adhered to the rule of law, and UK electric stocks looked cheap.

Hanson's purchase of Eastern, the largest REC, which had already committed to acquire generating assets from National Power and PowerGen, was the exception. Hanson, which controlled Peabody, American's biggest coal miner, later packaged all energy assets as The Energy Group, which it spun off as a separate company in February 1997.

American buyers ignored Labor's windfall profits tax warning. Perhaps they believed that no fair-minded government would impose a tax on latecomers who had no responsibility for and gained nothing from the windfall profits. Or they thought that Tories would remain in power forever. More likely, they succumbed to blandishments of persuasive investment bankers. The post August 1994 change in regulation did not deter them, either. Eventually they realized that owning RECs offered no strategic or financial advantages. They disposed of most of their holdings when they liquidated diversification efforts undertaken in an epidemic of Enron envy.

Buyers paid 12.7x earnings for targeted company shares (11.7x, excluding generators). The average UK stock traded for 16x earnings, and electric and water utilities at about 8x earnings. In the USA, the average stock sold for 20x earnings and utilities for 12x. British utility stocks were bargains! The purchaser could borrow to buy the company, pay a lower interest rate (after tax) on the loan than the return earned on the purchase and tell share-holders that the deal increased net income.[214] An accretive purchase, as they like to say.

Buyers thought that they could trim costs at RECs, sell additional products and servic-es to customers, establish a beachhead in Europe and learn skills that could serve them elsewhere. Some of those reasons made sense for British buyers. Americans, however, achieved few advantages, which is why most sold the RECs within a few years of purchase.

Generation and Supply

Late in January 1995, Stephen Littlechild warned National Power and PowerGen "that he was monitoring their actions following record electricity prices,"[215] just ahead of the government's sale of its remaining shareholdings in the companies. The second offering, however, took in £3.6 billion. The two offerings together produced a total of £5.6 billion, plus £0.8 billion of debt from the generators and £0.3 billion of dividends, or £6.7 billion altogether, not bad considering that the government's equity position in the two com-panies, in September 1990, before financial restructuring, added up to only £3.7 billion.

The Big Two planned for the day when all customers could choose suppliers. They want-ed their own stables of customers. In September 1995, PowerGen offered to buy Mid-lands. In October, National Power bid for Southern Electric, another REC. Both mergers would have raised the buyer's share of generation and supply markets and increased their net income and capitalization by about 30%. Each targeted REC's supply sales equaled about 30% of each generator's output. OFFER opposed the deals, the MMC okayed them, and Ian Lang, Secretary of State for Trade and Industry, vetoed both after Atlanta-based Southern Co. muddied the waters.

Southern already controlled the South Western Electricity REC and coveted National Power. If the MMC let National Power purchase the REC and Southern Co. then acquired National Power, the Atlanta firm could pick up a generator and REC at the same time, and achieve a dominant position in the England and Wales market. On April 17, Southern Co. said that "it might try to acquire the United Kingdom's largest generator,"[216] but would

---

214. How to buy a REC. The buyer pays 12x after tax earnings for the REC shares. The buyer earns £1 for every £12 spent, an 8.5% return (£1/ £12 = 8.5%). The buyer borrows the £12 at an 8% interest rate (the going rate at the time), but the additional £0.96 of interest expense reduces income tax paid by £0.26, so after-tax interest cost is only £0.70. The buyer of the REC, then, adds £0.30 to its income as a result of the transaction (£1.00 after-tax income of REC less £0.70 after tax cost of borrowed money equals incremental profit from deal of £0.30). This analysis, of course, ignores the risk of taking on £12 of debt just to make an extra £0.30, but managers often ignore that sort of risk, in order to goose up reported earnings.

215. Peggy Hollinger, "Different pulls of power," *Financial Times*, Jan. 30, 1995, p. 18.

216. Emory Thomas Jr. and Kimberley A. Strassel, "U.S. Utility Weighs a Major U.K. Bid," *Wall Street Journal*, April 18, 1996, p. A15.

hold off making an offer until the government decided on the National Power-Southern Electric deal. Next day, National Power rejected a meeting with Southern Co. without a real offer, and then said it was not for sale, anyway. A few days later, Ian Lang turned down both generator-REC mergers and extended the golden share duration just in case anyone else made unwanted advances on generators. In May, National Power declared a £1.1 billion special dividend, making it less attractive to predators by reducing its cash horde. In August, it purchased an Australian power station, thereby reducing the company's liquidity even more.

The generation market remained uncompetitive. On March 30, 1991, the Big Two owned 78% of generating capacity in England and Wales and controlled plants that set price most of the time. As of March 30, 1996, they still owned 62% of available capacity and the Competition Commission (formerly MMC) in 1996 estimated that they set prices 85% of the time.[217]

The Big Two's share of output fell under 50% in fiscal 1996/1997 after they divested capacity to Eastern – but not by a clean sale. They leased the plants with a payment mechanism that discouraged price cutting. Eastern paid via an "earn-out" of £6 per megawatt hour (MWh) generated. Fuel and other operating expenses amounted to at least £10 per MWh. Roughly speaking, then, Eastern could not clear a cash profit unless it collected over £16 per MWh (vs average monthly system marginal price of between £15 and £25 at time of the deal). Higher prices enabled Eastern to pay for plants faster. The Brattle Group, a consulting firm, wrote that "Eastern's plants submitted higher bids and set SMP at higher levels than when the same capacity was under the control of National Power and PowerGen. ... The persistence of high prices could be explained by neither changes in fuel costs, demand nor unit availability."[218] Bringing in a third competitor helped maintain high prices! After the lease deal the Big Two set prices maybe 67% of the time. (See Table 16-2.)

Table 16-2. Generation and Supply Market Shares (England and Wales) (Fiscal Years 1990/1991 - 1996/1997) (%)

	1990-1991	1994-1995	1996-1997
	[1]	[2]	[3]
Generating capacity			
National Power and PowerGen	78	65	49
Nuclear	14	17	17
Others	8	18	34
Total	100	100	100

---

217. Competition Commission, *National Power Plc and Southern Electric Plc: A report on the proposed merger*, April 2006, p. 25.
218. The Brattle Group, "Electricity Markets: The Price of Power, *ENERGY*, 2000, no. 1, p. 3.

	1990-1991	1994-1995	1996-1997
Generating output			
National Power and PowerGen	74	60	46
Nuclear	17	22	24
Others	9	18	30
Total	100	100	100
Supply sales (over 1 MW)			
National Power and PowerGen	36	40	48
First tier REC	57	37	27
Second tier REC	4	15	19
Others	3	8	6
Total	100	100	100
Supply sales (1 kW to 1 MW)			
National Power and PowerGen	–	8	8
First tier supplier	–	68	51
Second tier supplier	–	23	35
Others	–	1	6
Total	–	100	100

Sources:

1. Offer, *Review of Electricity Trading Arrangements,* Background Paper 1, February 1998, pp. 50,51.
_____, *Review of Energy Sources for Power Stations*, Submission by the Director General of Energy Supply, April 1998.
2. Standard & Poor's, *Global Utilities Credit Review*, Oct. 1997.
3. Littlechild, *op. cit.*
4. Lester C. Hunt, module coordinator, Dept. of Economics, U. of Surrey, "EC610: Energy Economics and Technology, Autumn Semester 2005."

The supply business looked different. First tier REC suppliers dominated home markets. The Big Two specialized in large customers (over 1 MW) and became national market leaders, though generally second to the local supplier in each REC market. REC suppliers did best with small customers. The supply business had low margins. Building a small customers base was a costly a proposition. National Grid and PowerGen would have had to start from scratch so they either avoided small customers or tried to buy them *en masse* by acquiring RECs.

The Herfindahl Hirschman Index (HHI) measures market concentration. Markets dominated by a few firms tend to function less competitively. American antitrust authorities, at the time, categorized markets with HHI ratings under 1,000 as competitive, and over 2,500 as highly concentrated. Only the supply market for large customers came close to qualifying for a competitive rating in the England and Wales markets.(See Table 16-3.)

Table 16-3. Estimated HHI in England and Wales Markets
(Fiscal Years 1990/1991 - 1996/1997)

	1990/1991	1994/1995	1996/1997
Generating capacity	3,400	2,400	1,500
Generating output	3,200	2,400	1,500
Large customer supply sales	800	1,000	1,200

Notes:
1. HHI calculated by adding together the square of the percentage market share of the large market participants.
2. For supply calculation, national market assumed,
3. All calculations are approximate, for purposes of determining direction and magnitude of concentration.

Plant divestitures, improved nuclear plant performance and commissioning of new power stations reduced concentration in the generating market, but not to competitive levels.

The Lerner Index assesses market power by profitability.[219] The higher the profitability the less competitive the market. Andrew Sweeting of Northwestern University concluded from a Lerner Index analysis that after the two-year agreement to control prices ended, National Power and PowerGen's "behavior was consistent with them ... tacitly colluding to raise Pool prices ..." and "the divestiture of plants negotiated by OFFER ... did not prevent generators exercising significant market power..."[220] They got away with it because fuel prices had fallen, so the generators could keep Pool prices unchanged, make a bigger margin, and nobody would notice.[221] "This made it harder for the regulator to intervene."[222]

The top generating companies continued to dominate the market despite regulatory admonitions, new entries and divestiture of plants. No doubt, consumers paid up, as a result.

Nuclear Privatized

Tories wanted the government out of nuclear generating but could not sell the Magnox power stations, with their short expected lives and no cash reserves set aside to cover de-

---

219. Technically the index measures the difference between price and marginal cost as a percentage of price.
220. Andrew Sweeting, "Market Power in the England and Wales Wholesale Electricity Market 1995-2000," Massachusetts Institute of Technology Center for Energy and Environmental Policy Research, August 2004, p. 3.
221. That is, if previously fuel cost 3p per kWh and the generator sold electricity at 5p per kWh, it earned a profit of 2p per kWh. If fuel price falls to 2p per kWh and the generator continues to sell electricity at 5p per kWh, it makes a profit of 3p per kWh, but customers do not notice because they still pay 5p per kWh.
222. _____, op. cit., p. 21.

commissioning expenses. Those plants produced about 7% of electricity sold in England and Wales. In April 1996, the government transferred the rest of Nuclear Electric (eight advanced gas reactors in England and Scotland and the pressurized water reactor, Sizewell B) to a new corporate entity, British Energy, scheduled for privatization in July. The government ended the nuclear levy in England and Wales and eliminated the premium price paid to the Scottish nuclear generators. British Energy would have to sink or swim as a commercial entity. The operating turnaround of the transferred assets must have lulled policymakers and executives into thinking that they had created a commercial nuclear enterprise (Table 16-4).

Table 16-4. British Energy Before Privatization
(Fiscal Years 1990/1991 - 1995/1996)

Fiscal year ended March 31	1991	1992	1993	1994	1995	1996
	[1]	[2]	[3]	[4]	[5]	[6]
Advanced gas reactors						
Output (TWh)	35.4	41.0	49.8	54.4	54.9	53.3
Capacity (GW)	7.6	7.8	8.1	8.3	8.4	8.4
Load factor (%)	53.1	59.9	70.6	74.6	75.0	72.4
TWh per employee	6.9	7.9	10.2	12.2	13.0	12.7
Pressured water reactor						
Output (TWh)	–	–	–	–	0.2	7.9
Capacity (GW)	–	–	–	–	1.2	1.2
Load factor (%)	–	–	–	–	1.5	75.2
TWh per employee	–	–	–	–	–	11.1
Financial record (p/kWh)						
Revenue	–	2.54	2.70	2.82	2.65	2.54
Variable cost	–	2.64	2.23	2.03	1.91	1.95
Operating profit before valorization	–	-0.70	-0.05	0.29	0.28	0.21

Source:
Nick Pink, "British Energy," SBC Warburg, July 1996, pp. 34-37.

Notes:
1. Variable costs include fuel cycle, materials and services and staff.
2. Valorization is the annual revaluation of nuclear liabilities, a non-cash charge.
3. Load factor, average load as a percentage of peak load, measures utilization of plant.

British Energy raised rated capacity at existing facilities, opened a new plant, reduced production costs and raised load factor – ingredients for a successful company launch. But it had no plans to build additional power stations in the UK and expected to retire 61% of its capacity by the end of 2009. The company was liable for fuel cycle and plant decommissioning expenses, too. The Tories could neither sell British Energy as a growth stock because it would shrink over time nor as a blue chip because of its unfunded obligations. It was, instead, an income stock whose fat dividend exceeded its earnings.

British Energy reported an operating profit before a provision for "back end costs" (nuclear waste disposal and plant decommissioning). It revalued those costs annually, a process called "valorization." The revaluation exceeded income from operations in the four years preceding privatization. But it required no current outlay of cash. It represented a future payment.

Valorization added to the liability for future costs, but the company funded only a small part of it (for decommissioning of power plants). British Energy could invest cash flow supposedly set aside for future liabilities in any way, which effectively meant that it was not set aside. If the investments paid off, the company would have money to meet nuclear obligations. If not, the government, would have to pay, because it could not just leave nuclear materials out there. If the company could extend power plants' lives (which investors assumed at time of the offering), then it could delay expenditures for nuclear liabilities. Putting off expenses reduced their discounted value and the valorization expenses that the company had to report.

Regulated US nuclear generators charged a price that covered costs, profits and a set-aside for future nuclear cycle expenses that was really set aside in a separate fund invested under strict rules. British Energy, in contrast, took whatever price the market offered, which might not cover costs plus a set-aside. Essentially, it collected a cash flow of uncertain size, did what it wanted with the cash and acted as if it had made provisions for the future.

Brokers gushed: "Powerful cash generation" and "Scope for output to rise and costs to fall"[223] and "substantial cash flow," "Market share is protected" and "Nuclear risks ... are minimal."[224] Ian Graham and Adam Forsyth of NatWest Securities, on the other hand, recommended that prospective investors read page 18 of the prospectus which warned that "... in the longer term the Directors expect only to recommend dividends that are covered by profits..."[225]

"Both maintenance and growth of dividends"[226] said the directors, depend on these conditions:

223. Daniel Martin, Simon Taylor and Piers Coombs, "British Energy, The Final Analysis" Barclays de Zoete Wedd, 22 May 1996, p. 1.
224. Michael Sayers and James Hutton-Mills, "British Energy: A Unique Investment Opportunity," Morgan Stanley, 20 April 1996, p. 2.
225. Ian Graham and Adam Forsyth, "British Energy," NatWest Securities, 24 June 1996, p. 4.
226. _____, op. cit., p. 4.

1. The "Pool Purchase Price will not fall more than 5 per cent in real terms..." [227]

2. The company will not face "major unanticipated plant outages..." [228]

3. The company can extend plant lives.

4. It can implement "significant cost savings." [229]

5. It can achieve "satisfactory returns ... from reinvestment." [230]

6. Regulation does not change in a way that adversely affect earnings.

7. Inflation does not increase significantly.

The case for investing rested on optimistic assumptions. Analysts forecast declining fuel costs, more generation competitors and flat to lower electricity prices so a five percent electricity price fall was not implausible. Nuclear plants worldwide suffered from unplanned outages so why not British ones? Unexpected capital expenditures could eat into free cash flow. The company had no demonstrated investment skills and no publicly stated investment plans, so how could the market evaluate its ability to invest in a way that would insure commercial success and meet future nuclear obligations? British Energy also bore a heavy debt burden: either 82% or 91% of capitalization, depending on which bond rating agency (S&P or Moody's) made the calculation. On the positive side, the company could extend power station lives, giving it more time to accumulate funds to deal with nuclear liabilities.

The NatWest analysts, after reviewing caveats on page 18 of the prospectus, wrote that "the chances of one or more of the above coinciding look to us to be high, and the prospect of offsetting positive items does not look great. However, the size of problems which would force a dividend cut ... may not occur early after flotation ... but the picture painted in the Prospectus is not one which encourages [us] to look for anything less than the highest yield in the sector." [231] They believed that British Energy's shares required a dividend yield of 8%. The flotation priced British Energy to pay even more.

The British Energy offering took place in July 1996, with proceeds "barely half of what [the government] expected when it announced the sale in March." [232] "The deal prices the operator of eight nuclear power plants below its book value of £1.54 billion." [233] As a result of poor demand, the government retained 11.5% of the stock for sale later in the year.

---

227. *Ibid.*
228. Graham and Forsyth, *op. cit.*, p. 5.
229. *Ibid.*
230. *Ibid.*
231. Graham and Forsyth, *op. cit.*, p. 6.
232. "Low Price in British Energy Privatization," *New York Times*, July 15, 1996, p. D4.
233. "U.K. Nuclear Sell-Off Disappoints," *Wall Street Journal*, July 15, 1996.

Two years later, the National Audit Office (NAO) reviewed the sale. The government collected £1.44 billion cash for shares plus £600 million of company debt but retained £3.7 billion of nuclear liabilities. It had aimed to "maximise net proceeds and to launch a robust and viable British Energy" because it "sought to ensure that the privatised business would be able to meet its nuclear liabilities, so that these liabilities would not in the future fall to the government by default."[234] The NAO concluded that the government had created a strong company, and an independent trust fund (£228 million upon privatization) for nuclear decommissioning would reduce future risk to the government which had set up "the company's financial structure in a way ... likely to maximise proceeds."[235] Yet, the NAO warned that putting nuclear liabilities onto British Energy "cannot remove all residual risk that future governments may have to meet some of those costs" and argued that it should "consider the case for retaining the power over the disposal of nuclear plant" beyond the 2006 expiration date of its golden share.[236]

The government privatized British Energy thinking to rid itself of nuclear liabilities through a combination of free market prices and clever investment strategies. Nuclear hope springs eternal.

Transmission Unchained

The RECs owned a valuable asset, National Grid,. They wanted to unlock that value. National Grid, at the same time, wanted its own market listing.

RECs sold all NG shares in a December 1995 public offering that priced them at a three times book value, a record for UK utilities. Aside from earning a high return on shareholders' equity (analysts predicted close to 30%), National Grid also owned a telecommunications venture at a time when telecoms were hot items in the market.

National Grid, itself, without making major capital expenditures, had reduced controllable costs, despite a growing volume of business, showing that a focused incentive regulatory framework could not only reduce costs for grid users but also produce high profits for the regulated company. (See Table 16-5.)

---

234. National Audit Office, National Audit Office Press Release, "The Sale of British Energy," 8 May 1998, p. 1.
235. *Ibid.*
236. National Audit Office, *op. cit.*, p. 2.

Table 16-5. National Grid Operations (Fiscal Years 1993/1994 - 1996/1997)

Fiscal year March 30	1994	1995	1996	1997	% change
	[1]	[2]	[3]	[4]	[5]
Units transmitted (TWh)	267	269	279	289	8.2
Peak demand (GW)	47.7	45.9	47.9	50.1	5.0
Circuit kms (thousands)	13.3	13.4	13.5	13.5	1.5
Real cost of congestion and ancillary services (1997 £)					
Total (£ millions)	533	425	232	297	-44.3
Per TWh	2.00	1.58	90.83	1.03	-48.5
System unavailability (%)	6.0	5.1	4.1	4.0	–

Sources:
1. S&P Credit Review, *Global Utilities*, Oct. 1997.
2. EEE Ltd., Cameron McKenna LLP and Steven Stoft, "A Proposal for a Study of the Independent Transmission Company Plus (ITC-PLUS)," Jan. 2003.
3. Offer, *Report on Distribution and Transmission Performance*, 1997/1998.

Pool, Prices and Customers

Retail electricity prices in real terms fell about 15% in the three fiscal years ending March 1997, largely because falling coal costs were passed through the contracts with the RECs .

Pool prices acted oddly from April 1, 1994 to March 31, 1996, during which time the Big Two had pledged to keep Pool prices flat. Extra charges for Uplift and capacity, blamed on power plant outages, spiked in the winter of 1994. OFFER told the Big Two not to do it again. Extras shot up in the winter of 1995. Next year, they fell in winter but rose other times. During those two fiscal years Uplift and capacity charges per kWh rose two and a half times, adding about 15% to the basic price. Regulator says keep price flat, okay, but let's charge extra for extras.

Could the Big Two manipulate Pool bidding procedures and plant availability, depress prices most of the year (deterring entry of potential competitors that might operate base load generation) and raise prices when only they could supply needed capacity and services? It looked that way. The Pool needed more price setting competitors and adding Eastern to the roster did not help.

Coal interests charged that the Pool discriminated against them. Coal miners still played a role in British politics. No government could cavalierly disregard their complaints.

The Tories considered breaking up the generators, eliminating the Pool, and allowing all players to make their own arrangements to buy and sell power. The choice, however, would fall to the Labor Party ("New Labor") which won the May 1997 general election, ending 18 years of Conservative rule.

The New Labor Platform

When Hugh Gaitskell tried to ditch Clause IV ("common ownership") of Labor's constitution in 1959, the party reacted by printing Clause IV, Part 4 on membership cards. Tony Blair had better luck. In 1993, he wrote a pamphlet criticizing Clause IV. Nationalization was a means, not an end. Labor had to modernize the means employed to reach its goals. After becoming party leader, Blair sponsored a new version of the clause in 1995. It spoke of "common endeavour ... means to realise our true potential... wealth and opportunity ... in the hands of the many ... where we live together, in a spirit of solidarity..." [237] "Common ownership" morphed into "common endeavour." Blair replaced socialist dogma with new age piffle.

Labor had already, proposed regulatory changes and a windfall profits tax. The party platform for 1997, entitled "New Labour Because Britain Deserves Better" said:

> New Labour is a party of ideas and ideals, but not outdated ideology.... The old left would have sought state control of industry. The Conservative right is content to leave all to the market. We reject both approaches.

> ... We will put concern for the **environment** at the heart of policy-making.

> ... we will reform Britain's competition law. We will adopt a tough ... approach...

> In the utility industries we will promote competition wherever possible. Where competition is not ... effective ... we will pursue tough, efficient regulation in the interest of customers ... We recognize the need for open and predictable regulation which is fair both to consumers and to shareholders and at the same time provides incentives to managers to innovate and improve efficiency. [238]

In short, Labor promised more of the same, but in a more environmentally friendly, more efficient, more consumer friendly fashion. No more socialism.

---

237. Labour Party, *Labour Party Rule Book 2013* (London: Labour Party, 2013), p. 3.
238. Labour election manifesto 1997- Labour Because Britain Deserves Better, (*http:// www.psr.keele.ac.uk/man/lab97.html*).

Conclusion

The privatized industry piled up high profits and attracted the attention of acquisition-minded outsiders. The regulator had to tackle deficiencies in the generating market and trim his policies to fend off accusations of too much generosity. As for consumers, service measures improved and prices fell, largely due to reduced fuel costs. The industry enjoyed seven fat years under the Tories. Would Tony Blair's Labor government bring in leaner years?

Chapter 17

Privatization's Seven Fat Years (1990-1997)

There are two ways of keeping profits down, competition and regulation.[239]

– Richard Green

Tories privatized electric companies, closed or sold coal mines, implemented a market and freed consumers from thralldom to monopolists. Electric companies worked assets harder and replaced old generating plants with clean, modern units. UK restructuring became a model for the world. Ambrose Bierce, though, translated *cui bono* as "What good would that do me?"[240] After seven years of restructuring, the British public might have asked that question.

Prices Fall and Costs Fall More

Real price of electricity declined 1.6% per year from 1989 through 1996. Operating costs fell more. Therefore operating income (profit) per kWh expanded (Table 17-1).

Real coal prices fell 45%, accounting for over two fifths of the drop in the electric price, because the contracts passed on those savings to consumers.

The industry squeezed more output from its nuclear plants: 6,700 hours of operation per unit of capacity in 1996 vs 5,700 in 1989. Consumers did not benefit directly. Reduced costs increased profits rather than lowering prices, because nuclear operators charged market prices. Lower costs, though, enabled the government to drop the fossil fuel levy, which did lower prices.

---

239. Richard Green, "England and Wales – A Competitive Electricity Market?" Program on Workable Energy Regulation, University of California, PWP-060. September 1998, p. 11.
240. Ambrose Bierce, *The Devil's Dictionary* (NY: Dover, 1958), p. 26.

Table 17-1. Real Price of Electricity, Costs and Profit in United Kingdom (1989-1996) (2005/2006 p per kWh)

	1989 [1]	1990 [2]	1991 [3]	1992 [4]	1993 [5]	1994 [6]	1995 [7]	1996 [8]
Revenue	7.42	7.31	7.36	7.55	7.45	7.22	6.96	6.63
Expenses								
Fossil fuel costs	1.53	1.33	1.26	1.20	1.04	0.97	0.94	0.93
Other expenses	4.63	4.60	4.57	4.72	4.66	4.54	4.26	3.95
Total expenses	6.16	5.93	5.83	5.92	5.70	5.51	5.20	4.88
Operating income	1.26	1.38	1.53	1.63	1.75	1.71	1.76	1.75

Notes:
1. Revenues of all final consumers of electricity divided by electricity made available.
2. Fossil fuel cost estimates based on government data.
3. Operating income (pre-tax) for 12 RECs, 2 Scottish companies, 2 major generators and Northern Ireland Electric. For 1996, 1.72p after adjustment for non-electric acquisition in year.
4. Other costs include power purchases from outside UK, all other operating expenses, and profit or deficit of nuclear entities.
5. GDP deflator used.
6. Fiscal and calendar year data used. Author's estimates to adjust data to calendar year.

Sources:
1. Department of Trade and Industry and National Statistics, for revenues and electricity available.
2. Company information from company reports, S&P, Moody's and brokerage reports.

Margin (pre-tax operating income) which provides the return on invested capital rose 37% per kWh while capital invested per kWh rose only 15%. The privatized companies, then, did what they were supposed to do: squeeze more margin out of each kWh sold and raise return on capital invested. They did it exceptionally well.

Spending on fossil fuel declined as coal prices dropped and nuclear power (generated in both France and the UK) replaced coal. The industry reduced costs across the board, but the nuclear subsidy (the fossil fuel levy) and additional taxes partially offset the savings.

Fossil fuel expenses per kWh in 1989	1.53p
Coal price reductions	- 0.34
Shift to nuclear generation	- 0.13
Other fuel expense reductions	- 0.13
Fossil fuel expense per kWh in 1996	0.93

Other expenses per kWh in 1989	4.63p
Nuclear subsidy (fossil fuel levy)	+ 0.45
Fuel shift fossil to nuclear	+ 0.10
Additional taxes	+ 0.23
Other savings	-1.46
Other expenses per kWh in 1996	3.95

Despite higher taxes, companies squeezed 1.28p per kWh off costs while reducing prices only 0.78p. The difference went into profits (operating income). Nothing wrong with that, all legal, but how likely in either a truly competitive or closely regulated business?

Fuel mix changed (Table 17-2). Fuel costs fell, but Pool price dropped less (Table 17-3) and profits rose as a percentage of revenue (Table 17-4). In competitive markets, price declines toward marginal cost but this market was scarcely competitive. Market share of the three top generators and their successors never fell below 76% (Table 17-5). In concentrated markets, firms observe each other and retaliate swiftly to price cuts and attempts to grab market share. They make sure that price cutting does not pay without a single illegal meeting.

Overall, consumers realized half the savings made, and half of their share came from coal price reductions passed on to them through the contracts. Still, half is better than nothing.

Table 17-2. Production by Fuel and Interconnector Import of Electricity into England and Wales
(Fiscal Years 1989/1990-1996/1997) (TWh and %)

Fiscal Year Ended 31 March	1990	1991	1992	1993	1994	1995	1996	1997
	[1]	[2]	[3]	[4]	[5]	[6]	[7]	[8]
TWh								
Coal	184.7	185.5	187.6	168.9	144.3	140.8	129.5	111.7
Nuclear	42.5	47.6	51.9	57.8	63.7	61.9	65.0	71.8
Gas	0.0	0.3	0.8	6.2	29.6	38.9	56.1	74.4
Oil	14.4	11.4	7.7	7.7	7.4	5.1	3.9	3.6
Interconnector	12.6	20.3	23.1	23.1	23.1	25.2	26.2	27.2
Other	1.8	1.8	1.3	1.6	1.4	2.0	2.3	2.6
Total	256	266.9	272.4	265.2	269.5	273.9	283.0	291.3
%								
Coal	72.1	69.5	68.9	63.7	53.5	51.4	45.8	38.4
Nuclear	16.6	17.8	19.1	21.8	23.6	22.6	23.0	24.7
Gas	0.0	0.1	0.3	2.3	11.0	14.2	19.8	25.5
Oil	5.6	4.3	2.8	2.9	2.7	1.9	1.4	1.2
Interconnector	4.9	7.6	8.5	8.7	8.6	9.2	9.3	9.3
Other	0.8	0.7	0.4	0.6	0.6	0.7	0.7	0.9
Total	100.0	100.0	100.0	100.0	100.0	100.0	100.0	100.0

Source:

Office of Electricity Regulation (OFFER), "Review of Energy Sources for Power Stations, Submission by the Director General of Electricity Supply," April 1998, p. 7.

Table 17-3. Estimated Fuel Costs vs Pool Purchase Prices
(England and Wales)
(Fiscal Years 1990/1991-1996/1997) (p/kWh)

Fiscal Year Ended 31 March	1991	1992	1993	1994	1995	1996	1997	Average
	[1]	[2]	[3]	[4]	[5]	[6]	[7]	[8]
Nominal (p)								
Pool purchase price (PPP)	1.74	2.08	2.28	2.42	2.40	2.39	2.37	2.24
Fuel cost per kWh								
All	1.26	1.27	1.27	1.14	1.16	1.03	1.09	1.17
Fossil only	1.15	1.20	1.24	1.09	1.24	1.12	1.20	1.128
PPP margin over fuel costs								
All	0.48	0.81	1.01	1.28	1.24	1.36	1.28	1.07
Fossil only	0.59	0.88	1.04	1.33	1.16	1.27	1.17	1.06
Real (2006 p)								
Pool purchase price (PPP)	2.62	2.96	3.14	3.25	3.17	3.07	2.94	3.02
Fuel cost per kWh								
All	1.90	1.81	1.75	1.53	1.54	1.32	1.35	1.6
Fossil only	1.74	1.8	1.71	1.46	1.65	1.41	1.49	1.62
PPP margin over fuel costs								
All	0.72	1.15	1.39	1.72	1.63	1.75	1.59	1.42
Fossil only	0.88	1.06	1.33	1.79	1.52	1.66	1.45	1.38

Notes:
Average time weighted pool purchase price (PPP).
Fuel costs for England and Wales derived from data for UK, adjusted for fuel mix in England and Wales.
Interconnector fuel costs based on fuel mix of Scotland and estimate for France.
Fossil fuel costs for generation within England and Wales.
GDP deflator (fiscal year 2005/2006 = 100).

Sources:
PPP from OFFER and Merrill Lynch publications.
Fuel costs from Department of Trade and Industry (and predecessor) publications, based on tonnes of oil equivalent costs estimated from British Nuclear and British Energy reports.

Table 17-4. Disposition of Real Revenue in England and Wales
(Fiscal Years 1989/1990-1996/1997)
(2005/2006 p/kWh and %)

Fiscal Year Ended 31 March	1990	1991	1992	1993	1994	1995	1996	1997	Average
	[1]	[2]	[3]	[4]	[5]	[6]	[7]	[8]	[9]
Real p/kWh									
Fuel	2.17	1.90	1.81	1.75	1.53	1.54	1.32	1.35	1.67
E&W profit	1.29	1.43	1.61	1.71	1.82	1.88	1.74	1.80	1.66
Other	4.13	3.85	3.83	4.20	4.11	3.70	3.73	3.29	3.86
Total	7.59	7.18	7.25	7.66	7.46	7.12	6.79	6.44	7.19
% of revenue									
Fuel	28.6	26.5	25.0	22.8	20.5	21.6	19.4	21.0	23.2
E&W profit	17.0	19.9	22.2	22.3	24.4	26.4	25.6	28.0	23.2
Other	54.4	53.6	52.8	54.9	55.1	52.0	55.0	51.0	53.6
Total	100.0	100.0	100.0	100.0	100.0	100.0	100.0	100.0	100.0

Notes:
Derived from total UK revenues, adjusted to England and Wales (E&W).
E&W profit is pretax operating income of 12 RECs, National Grid, National Power and PowerGen.
Other includes other expenses, operating taxes and profits of other market participants.

Source:
Company reports, *Digest of U.K. Energy Statistics* (various issues) and government fuel statistics.

Table 17-5. Electricity Production and Imports in England and Wales Market by Producer (Fiscal Years 1989/1990-1996/1997) (TWh and %)

Fiscal Year Ended 31 March	1990	1991	1992	1993	1994	1995	1996	1997
	[1]	[2]	[3]	[4]	[5]	[6]	[7]	[8]
Traditional								
National Power	123.8	121.3	117	108.6	94.6	92.3	90.8	70.2
PowerGen	76.5	75.8	75.2	73.5	70.2	70.9	65.3	62.7
Eastern	—	—	—	—	—	—	—	19.3
Total	200.3	197.1	192.2	182.1	164.8	163.2	156.1	152.2
Nuclear								
Nuclear Electric	42.5	46.5	50.7	56.8	62.5	60.8	63.7	—
British Energy	—	—	—	—	—	—	—	50.3
Magnox	—	—	—	—	—	—	—	20.2
Total	42.5	46.5	50.7	56.8	62.5	60.8	63.7	70.5
Interconnectors								
Scotland	4.2	4.5	6.3	6.6	6.4	8.2	10.1	10.2
France	8.4	15.8	16.8	16.5	16.7	17.0	16.1	17.0
Total	12.6	20.3	23.1	23.1	23.1	25.2	26.2	27.2
Other	0.6	2.9	6.4	3.2	19.1	24.7	37.0	41.4
Grand total	256.0	266.8	272.4	265.3	269.5	273.9	283.0	291.1
Traditional and nuclear share of market (%)	94.8	91.3	89.2	90.0	84.3	81.8	77.7	76.5

Sources:

1. Lester C. Hunt, module coordinator, Dept. of Economics, University of Surrey, EC 610: Energy Economics and Technology, Autumn Semester 2005/2006, Topic 3 from Helm & Jenkinson (1988).
2. Offer, "Review of Energy Sources for Power Stations, Submission of Director General of Electricity Supply," April 1998, p. 11.
3. Offer, "Background Paper 1, Electricity Trading Arrangements in England and Wales," Feb. 1998, p. 44.

Notes:

May not add due to rounding.

Nuclear production by other than by Nuclear Electric, British Energy or Magnox are included in Other.

Service and Reliability

Consumers want reliable service. Regulated firms focus on meeting the regulator's standards for reliability. After teething problems, the industry improved its grades on the regulator's check list (Table 17-6), satisfying the one customer that counts, the regulator.

Table 17-6. Service Quality in Great Britain (Fiscal Years 1989/1990-1996/1997)

Fiscal Year Ended 31 March	1990	1991	1992	1993	1994	1995	1996	1997
	[1]	[2]	[3]	[4]	[5]	[6]	[7]	[8]
Transmission								
Incidents per year	10	11	9	9	8	8	5	14
Unsupplied energy per incident (MWh)	120	290	20	80	60	10	30	10
Annual unavailability (%)	—	—	9.0	7.1	6.0	5.1	4.1	4.0
Distribution								
Minutes lost per connected customer	—	227	103	106	96	97	87	88
Supply interruptions per 100 customers	—	112	88	95	85	88	91	89
Failure to meet guaranteed standards per 100,000 customers	—	—	51	49	33	21	15	9

Notes:
1. Transmission data for National Grid. Approximated from charts in source.
2. Distribution data for all Public Electricity Suppliers (14 companies) in Great Britain.

Sources:
OFFER, "Report on Distribution and Transmission System Performance 1997/ 98." Department of Trade and Industry, "The Social Effects of Energy Liberalisation: The UK Experience," June 2000, p. 9.

Could the market somehow determine the right reserve margin? The British, fortunately, did not have to answer that question because the UK had so much extra capacity. As OFFER put it in an 1998 report:

The total of generation capacity in the England and Wales system has remained around 60 GW over the past eight years. 16 GW of new plant has been built ... but this has been offset by 13 GW of closures and the mothballing of a further 5 GW ... [241]

In six fiscal years ended March 31, 1997, sales and peak demand grew 1.0% and 0.6% per year, respectively. Reserve margin fell from 33% to 24%, both high numbers, but as OF-FER noted:

> ... the number of generating sets has fallen by 35%, whilst the total capacity on the system has remained essentially constant. This has occurred because ... smaller units have been closed and they have been replaced by larger ones. Since the size of the largest unit ... has also increased, reserve requirements have grown because the system must be able to withstand the loss of its largest in-feed. [242]

In England and Wales, the 20 largest fossil and five largest nuclear plants accounted for over two-thirds of capacity at privatization. Old plants made up the bulk of the remainder. Closing down facilities over 30 years old would have reduced plant count 34% (86 to 57) and capacity (including interconnectors) 10%. Eliminating small, aged facilities would hardly affect reliability. (Numbers for the UK as a whole told a similar story.)

Reliability requires a diverse fuel mix, something the British learned during the coal strikes. Before privatization, British coal accounted for 70% of generating output, nuclear 20% and other sources 10%. Seven years later, the mix was roughly 45% coal, 30% nuclear, 20% gas, and 5% everything else – a cleaner, more efficient and diverse generating mix.

When surveyed about reliability in 1997, 15% of customers said they were "Totally satisfied" and 4% were "Neither satisfied nor dissatisfied" or "Totally/Very/Fairly dissatisfied." Not surprisingly, given the disinterest in the topic, 71% wanted to pay "Nothing" to "help ensure that there were no power cuts in the future" [243] People took reliability for granted.

Financial Analysis

In a successful privatization, the seller receives value for assets sold and the buyer a return commensurate with risk. Competitive forces and robust regulation prevent firms from earning abnormal profits.

---

241. OFFER, "Review of Electricity Trading Arrangements, Background Paper 1, Electricity Trading Arrangements in England and Wales," Feb. 1998, p. 22.
242. _____, op. cit., p. 23.
243. _____, "Reviews of Public Electricity Suppliers 1998 to 2000, Price Controls and Competition Consultation Paper," July 1998, p. 58.

The accounting framework defines profitability. British nationalized industries used current cost accounting (CCA) to adjust assets and depreciation expense to present day values. Most businesses use historical cost accounting (HCA) in which cost – with certain exceptions – determines valuation. The accounting framework affects reported profitability but not cash flow. In truth, though, only when the corporation winds up its affairs can its accountants furnish an accurate measure of its profitability. Until then, investors have to rely on estimates, educated guesses, subjective judgments and occasionally fudged books.

In a way, the market provides the most objective profitability measure. Investor buys stock on January 1 for £100, collects a £5 dividend during the year, sells stock on December 31 for £110, makes a £15 profit (£5 dividend plus £10 capital gain) on £100 investment, or a 15% total return. Total return varies by year, but a multi-year average tells a meaningful story. Market data also help to determine cost of capital.

Return should rise as risk increases. Therefore, "risk free" Treasury bonds should earn the lowest returns, stocks more, and riskier stocks more than safer ones. (On average in the 1990s UK stocks earned roughly 3-5% per year more than bonds.) [244] Results diverged from theory. Utility stocks beat the market and their owners certainly got more than their money's worth. (See Table 17-7.)

Table 17-7. Estimated Annual Total Returns on Shares of British Electric Companies Publicly Traded from Privatization to 31 March 1997 (%)

Sample (#)	Initial Date	Electric Companies			Stock Market		
		Capital Gains	+ Dividend Yield	= Total Return	Capital Gains	+ Dividend Yield	= Total Return
Generators (2)	2/91	23	7	30	10	4	14
Scottish (2)	5/91	7	6	13	9	4	13
RECs (2)	12/90	23	8	31	12	4	16
National Grid (1)	12/90	33	8	41	12	4	16
Privatized Electric	2/91	16	7	23	10	4	14

Notes:
1. During the period covered, UK government (10 year gilt) bonds yielded about 8%. The bonds earned a total return of about 9.0-9.5 %.
2. Generators: unweighted average of National Power and PowerGen. Special dividend included in capital gains.

---

244. Looking at 10 year trailing averages that compare return on the FTSE-100 stock index to that of long term government bonds, Alistair Byrne wrote, "A downward trend is evident with the 4%-to-6% range for most periods ending in the first half of the 1990s giving way to a 2%-to-4% range in the second half of the decade." Alistair Byrne, "How have we done?," *Professional Investor*, May 2002, p. 27.

3. Scottish: unweighted average of ScottishPower and Scottish Hydro.
4. RECs: unweighted average of Southern and Yorkshire (terminal price as of 2/97, when acquired). Special dividends included in capital gains. Southern and Yorkshire were only RECs still independent at or near 3/97.
5. National Grid: calculation of return based on assumption that public could have purchased NG shares on 12/90, at same valuation as used when shares were contributed to RECs, and would have collected same dividends as RECs did during their holding of NG. Special dividends included in capital gains.
6. Privatized electric: reported in Alistair Buchanan, "UK Electricity Sector," Donaldson, Lufkin & Jenrette, July 14, 1999, p.11. Probably understates performance by missing initial privatizations. Special dividend estimates included in capital gains.
7. Stock market returns calculated from FTSE 100 index.
8. All calculations by the author, derived from available data in various financial publications. Due to initial partial payment for shares, partial payment of dividends, bonus share payments for holding shares and irregular timing of special dividends, the numbers shown should be viewed as approximations for purposes of comparison, rather than as precise calculations. For those reasons, calculations round to nearest whole number.
9. All returns before taxes.

An investor who purchased a market-weighted portfolio of electric shares at privatization would have earned about 28% per year (pre-tax total return) versus 15% from a portfolio of large British stocks and 9% from government bonds. What accounted for that amazing profitability?

- The government underpriced the shares – True, but only in retrospect. The opening day price surge added three percentage points to annual total return. Yet the government could not sell voters stocks that would decline. Brokers projected 8-10% growth. Add a dividend yield of 7-8% and investors could expect a 15-18% annual total return. Government bonds then yielded about 10% and stocks required a 4-6% equity premium, so any successful share sale would have to offer the potential for a 14-16% annual return. That is what the share pricing seemed to offer. Based on market conditions at the time, the offering prices may have been on the low side but not excessively so.

- Initially, nobody realized the profit potential of the companies – Possibly true. If company executives knew, they did not tell. The 8-10% per year estimated growth in nominal earnings built in a 5% inflation rate. Inflation ran at 3%. Sales volume grew 1% per year. Industry earnings, however, advanced 11% annually. The industry reduced costs, managed generating prices and returned excess capital to shareholders, a surprising display of commercial skills from the ex-government bureaucrats running the industry.

- The companies earned returns unjustified by risk levels – Definitely true. Buckland and Fraser showed that REC shares earned an excess ("abnormal") return of 15%

per year.[245] That excess, due to extraordinary profitability, if reflected in billings, would have added about 7% to the average bill.

- The stock market revalued the shares to reflect a new view of the companies – Not really. To simplify, if a stock sells at 10x earnings of £1 in year one (£10) and 12x earnings of £1 in year five (£12), that 20% revaluation adds 3.7% per year to the stock's total return. The price/earnings ratio of electric stocks rose from 9x on opening to 10x at the end of fiscal 1997, which added 2% per year to total return. In the same period the market (FTSE-100 Index) rose from 12x to 16x earnings. The increased electric stock valuation, then, was less a vote of confidence than a case of rising tide lifts all boats.

Still, even after normalizing for miscalculations and market revaluation, returns remained impressive. (Table 17-8):

Table 17-8. Normalized Total Returns from Stock Investments

	Electric	FTSE 100
Actual average annual total return	28%	15%
Adjustments for:		
Initial mispricing	-3	–
Underestimate of growth	-3	–
Rise in price/earnings ratio	-2	-4
Normalized nominal total return	20	11
Rate of inflation	-3	-3
Normalized real return	17	8

Sources and Notes:
See Table 17-7.

Twentieth century UK stock investors earned average real returns of 5.2% per year: 13.5% in the most profitable twenty-year period and -1.7% in the least profitable.[246] Electric industry investors struck gold by comparison. The companies took limited financial risks, earned high returns in their businesses and paid shareholders the funds that they could not employ profitably. Only later did they resort to financial engineering to make the numbers look better and to fend off predators by taking on debt.[247]

---

245. Buckland and Fraser, *op. cit.*, p. 46 The paper, suggests that "The cost of capital used" in the price determinations "may be overly and significantly generous..." (p. 56), a kind way of saying that the regulator let the utilities make too much money.

246. Elroy Dimson, Paul Marsh and Mike Staunton, "Irrational Optimism," *Financial Analysts Journal*, Jan./Feb. 2004, pp. 19-20.

247. The analyses exclude data from British Electric, which are incomplete for the period, and generally undecipherable, as evidenced by the fact that Moody's, S&P and various brokerage houses could not agree on figures for rev-

The RECs reported steady margins, high returns on capital and equity, despite conservative financial policies maintained until the first takeover attempt scared them and Eastern took on debt to buy assets. They earned over 20% on a stockholder equity (24% with National Grid consolidated) which constituted about 70% of capital. They earned almost double the return on equity of American utilities and they did it with a lower risk capitalization. They had their cake and ate it, too. What other developed country offered such returns on a regulated business?

National Power and PowerGen, despite the supposedly competitive market in which they operated, improved margins while maintaining high equity ratios, too, until they disbursed special dividends and sold debt to expand abroad. They earned under 20% on book equity, less than the safer RECs, and they kept equity at a conservative 75% of capitalization. Assuming that market value at end of first day of trading better represents the value of competitive assets, they earned about 15% on equity, at least in line with market expectations for the business.

National Grid was the undisputed industry star, earning an astounding 28% on its shareholder equity. (Assuming that NG had gone public at the same time as the RECs and sold at an equivalent premium at the end of the first trading day the return would have fallen to 24%, a number that surely would have attracted buyers.) Admittedly, NG shareholders took more financial risk that the others, having a lower (but still high by utility standards) 61% equity ratio. Yet it is hard to imagine a better deal for investors: the low risk, monopolistic owner of a vital link in an indispensable business making so much money.

The Scots got the short end of the stick. On the basis of book returns, the Scottish companies earned 24% on equity, not bad considering that they also took more financial risk, with a 66% equity ratio (still a high number by utility standards). But the government cleverly marketed the companies at steep prices, and appropriated the value of those high profits for itself, and left investors in Scottish utilities with an unexceptional 11% return earned on market value (after the first day price pop).

Overall, fiscal years 1991-1997 were truly fat years financially for the regulated utilities measured by return on assets (using book or market value for assets). The generating companies, the ones that had to compete, did less well. Imperfect competition seemed to work less well for shareholders than regulation. (See Tables 17-9 and 17-10.)

---

enue, net income and capitalization. The numbers have been adjusted, wherever possible, to exclude non-recurring or extraordinary items, in order to present a picture of normal operations. For RECs, numbers include estimates for data missing due to mergers and acquisitions. Analyses are based on regulated and unregulated operations as reported on the books.

Table 17-9. Financial Statements for UK Electric Companies by Sector
for Fiscal Years 1990-1991 through 1996-1997
(Annual Averages) (£ billions)

	RECs	National Grid	Scotland	Generators
	[1]	[2]	[3]	[4]
Revenues	15.06	1.38	2.57	6.99
Operating income				
Pretax	1.79	0.60	0.55	1.11
Aftertax	1.51	0.45	0.45	0.78
Net income				
Pretax	1.65	0.55	0.49	1.13
Aftertax	1.38	0.39	0.39	0.80
Capitalization (book)				
Equity	6.71	1.38	1.62	4.11
Debt	2.27	0.87	0.82	1.37
Total	9.57	2.25	2.45	5.48
Capitalization (offering price)				
Equity	6.56	1.35	3.42	4.64
Debt	2.87	0.87	0.82	1.37
Total	9.42	2.22	4.24	6.01
Capitalization (first day premium)				
Equity	7.63	1.62	3.66	5.40
Debt	2.87	0.87	0.82	1.37
Total	10.50	2.49	4.48	6.77

Notes and Sources:

See Appendix Tables Q, S, U and W.

Table 17-10. Financial Analysis of UK Electric Companies by Sector for Fiscal Years 1990-1991 through 1996-1997 (%)

	RECs	National Grid	Scotland	Generators
	[1]	[2]	[3]	[4]
Return on Capital (pre-tax operating income)				
Book	18.7	26.7	22.4	20.3
Offering price	19.0	27.0	13.0	18.5
First day premium	17.0	24.1	12.3	16.4
Return on capital (after-tax operating income)				
Book	15.8	20.0	18.4	14.4
Offering price	16.0	20.3	10.6	13.0
First day premium	14.4	18.1	10.0	11.5
Return on equity (after-tax)				
Book	20.6	28.3	24.1	19.5
Offering price	21.0	28.9	11.4	17.2
First day premium	18.1	24.1	10.7	14.8
Debt ratio				
Book	30.0	38.7	33.5	25.0
Offering price	30.5	39.2	19.3	22.8
First day premium	27.3	34.9	18.3	20.2
Margin (pretax operating income as % of revenues)	11.9	43.5	21.4	15.9

Notes and Sources:
See Appendix Tables R, T, V and X.

The more a company borrows, the greater the risk to its owners and creditors who require a higher return as compensation for the added risk. British electric companies avoided heavy borrowing but still earned higher returns than utilities elsewhere.

How much did excessive returns add to electric bills? Using 10% after tax on invested capital as a benchmark for normal return, consumers in England and Wales paid the RECs and National Grid an extra £0.9 billion and £0.3 billion per year, respectively, or about 8% of the average electric bill during the seven year period. Using a less stringent benchmark of 15% on equity, customers in England and Wales paid 5% per year too much for electricity. To put it another way, in retrospect the government could have sold the shares for an additional £3.8 billion and the buyers of the shares would have still earned a fair return on investment. [248]

Regulation and Industry Structure

According to the prospectus, the DGES had to assure that regulated companies operated efficiently and reliably, while also protecting the "interests of electricity consumers ... in respect of charges and terms of continuity of supply and quality of ... services." [249] In the unregulated sector, the DGES had to "promote competition in the generation and supply of electricity." [250]

The regulatory regime promoted operating efficiency, quality of service remained stable and the companies could attract capital, but did the regulator protect the interests of consumers when setting the price? More to the point, should the regulator have lowered the price when he saw how much money the electric companies made in their low risk businesses?

To be fair, Prof. Litttlechild did not make the deal. Should he have broken it by reviewing prices sooner than expected? Prof. Littlechild wrote, in 1992, that prices:

> were set on the basis that there would be a reasonable period between reviews ... If a regulator is seen to intervene constantly ... there will be an adverse effect on the incentive for that company to improve its efficiency and reduce costs. [251]

He argued, as well, that reviews of price ahead of schedule would raise regulatory risk, thereby increasing cost of capital, and consumers would pay higher prices to cover higher capital costs.

Dieter Helm said "there is nothing natural or sacrosanct about five-year periods," [252] that not confronting the pricing error sooner was a mistake. Economists Michael Crew and Paul Kleindorfer, in contrast, warned about "the importance of ... commitment issues in regulation – the notion that the regulator will not renege on the terms of the price cap. The behavior of ... Littlechild in March 1995 in resetting price caps and X

---

248. The calculation of a higher price is the discounted value, at a 12% rate, of the excess revenue less income taxes.
249. *Placement Memorandum Dated December 11, 1990* .., p. 33.
250. *Ibid.*
251. *Ibid.*
252. 213 Helm, *op. cit.*, p. 210.

factors in response to public pressure raised the question of whether companies would pursue efficiency as strongly." [253]

Which raises these questions:

- Did the government make a five-year deal?
- Should regulators protect firms from the consequences of their actions?
- Should management just follow the rules or consider the consequences of doing so?

The government did not commit the regulator to a fixed period. Other British regulators, however, set prices for five year periods. Ernest Liu, an extraordinarily thorough and careful analyst working for Goldman Sachs, the banking house in charge of the international distribution of shares, and as close to the government as a firm could get, wrote that:

The distribution X factors for the RECs are expected to last for a five-year period. [254]

The Goldman Sachs report noted, too, that:

The UK system has no profit cap. [255]

and, if the RECs were able to control costs, reduce debt and cut interest costs:

The RECs would have no obligation to share benefits… with customers under the current distribution RPI + X framework, at least given the individual X factors that have been assigned for the first five years. [256]

The British investment houses agreed. Anthony White, formerly in the electric industry, wrote for the broker James Capel & Co.:

The terms of the licences may be modified with the licensee's agreement or if the Monopolies and Mergers Commission confirms that modification would be in the public interest. [257]

---

253. Peter Lowe, "The Reform of Utility Regulation in Britain: Some Current Issues in Historical Perspective," *Journal of Economic Issues*, March 1998, p. 171.

254. Ernest S. Liu, Liz Christie, James P. McFadden and Ashar Khan, "Electricity Privatisation in England and Wales: An International Perspective," Goldman Sachs, London, September 1990, p. 12.

255. _____, *op. cit.*, p. 4.

256. _____, *op. cit.*, p. 5.

257. Anthony White, "Reshaping the Electricity Supply Industry in England and Wales," James Capel & Son, Ltd., London, February 1990, p. 3.

In addition:

> The DGES ... is likely to review the operation of the price conditions in the Public Electricity Supply Licence and the Transmission Licence after five and three years respectively. [258]

According to John Wilson, formerly a consultant, writing for UBS Phillips & Drew:

> The X-factor is set for a period of three, four or five years depending on the particular business. [259]

> If any of the company's [sic] achieve a return greater than expected this can be passed on to shareholders. However, if any of the companies over achieves by a significant margin, the regulator at the review state, will almost certainly impose an X-factor that would reduce the return to an adequate level. [260]

The prospectus said:

> Assuming it is not modified under the procedure for modifying licence conditions.. distribution price controls will remain in force for the duration of each ... licence.. Save where DGES otherwise agrees, the earliest date on which such a request may become effective is 31st March 1995. [261]

The overall message to investors, then, read like this:

> The regulator sets prices for five years (three for National Grid). Nobody will bother the companies during that period, although, legally, speaking, some condition might call for modification of the price setting, but making changes requires the company's agreement or imposition by a higher power, so do not worry about it.

The government, controlled the pre-offering message. Investors believed that they had a five year deal. Prof. Littlechild played fair with investors. A deal is a deal, even a bad deal.

Prof. Helm wrote that the DGES "neglected the political dimension of the industry." [262] Politicians react to blatant profitability. A politically savvy regulator might have proposed "voluntary" modifications to avoid a furor but preserve the structure. But company ex-

---

258. _____, *op. cit.*, p. 37.
259. John Wilson, "The Electricity Industry," UBS Phillips & Drew, London, August 1990, p. 6.
260. _____, *op. cit.*, p. 120.
261. *Placement Memorandum Dated December 11, 1990..*, p. 39.
262. Helm, *op. cit.*, p. 211.

ecutives share responsibility. They seem to have cherished their roles of unreconstructed capitalists extracting maximum profit from consumers over the short term, while myopically assuming that nothing would change as a result. Smart managers know when to share spoils. Consumers probably paid 5-8% more than necessary to provide those high returns. Giving back some of that money might have averted Labor's actions after winning the next election.

In the generating sector in England and Wales, concentration declined. National Power and PowerGen closed down old facilities and transferred power stations to Eastern. New entrants (in part owned by the RECs) took an increased share of the market. The revived nuclear sector, split into two companies, increased its market share. (See Table 17-11.)

Table 17-11. Breakdown of Electricity Output Sold in England and Wales by Generator (Fiscal Years 1990/1991-1996/1997) (%)

	Fiscal 1990/1991	Fiscal 1996/1997
National Power	45.5	24.1
PowerGen	28.4	21.5
Eastern	–	6.6
Nuclear Electric	17.4	–
British Energy	–	17.3
Magnox Electric	–	6.9
New entrants	–	13.0
First Hydro	0.6	0.8
Interconnectors	7.6	9.3
Others	0.5	0.5
Total	100.0	100.0

Source:
OFFER, "Review of Electricity Trading Arrangements, Background Paper 1, Electricity Trading Arrangements in England and Wales," February 1998, p. 44.

Yet, OFFER observed, National Power, PowerGen and Eastern, still "typically set prices for around 80% of the time. First Hydro," owner of pumped storage facilities used at peak times, set price "for the remaining 10-20% of the time."[263] Real prices in the Pool crept up despite a downward drift in fuel costs and an increased number of generating companies in the market, demonstrating that policies to make the market more competitive needed work.

---

263. OFFER, "Review of Electricity Trading Arrangements, February 1998," p. 45.

The Big Two's share of the large customer (over 1 MW) market increased. The first tier REC suppliers started off with two-thirds of small customers (in 1994/1995) and retained only half of that market segment two years later (see Table 17-3). In Scotland second tier (non incumbent) suppliers took roughly half the market share they achieved in England and Wales. By 1996/1997, second tier suppliers in Scotland other than Scottish Power and Hydro-Electric accounted for "less than 15% of all second tier sales."[264]

Supply earned a slim average pre-tax margin of 1.1% of sales in fiscal years 1991-1997. With so slim a margin, suppliers had little room to offer enticements to consumers. They all bought electricity from the same place, the Pool, so they had similar costs and had to charge similar prices. Perhaps consumers profited by having so many supply competitors taking financial risks on their behalf, to supply them with a product on an almost non profit basis. Other than to provide a stable market for generating output, why would any firm bother to stay in supply?

As for regulated utilities, OFFER seemed to sidestep the obvious question of what the numbers meant. The financial data produced by OFFER showed an industry with unusually healthy profitability and robust cash flows in excess of internal needs (Tables 17-12 and 17-13).

Table 17-12. Financial Statement Analysis in Real Terms for Public Electricity Suppliers in Great Britain (Fiscal Years 1990/1991-1996/1997) (Million 1996/1997 £)

Line	Annual Average	Calculation
1. Debt	£3,792	
2. Equity (flotation value base)	11,513	
3. Capitalization	15,305	
4. Debt	24.8%	(1/3) × 100
5. Equity	75.2	(2/3) × 100
6. Capitalization	100.0	
7. Earnings before interest and taxes	£2,447	
8. Taxes	599	
9. After tax earnings available for capital	1,848	
10. Interest charges	207	
11. Net income	1,641	

---

264. OFFER, "Review of Public Electricity Suppliers 1998 to 2000, July 1998," p. 68.

Line	Annual Average	Calculation
Return on capital	12.1%	(9/3) × 100
Return on equity	14.3	(11/2) × 100
Interest coverage ratio	11.8	(7/10) × 100

Source:
OFFER, "Review of Public Electric Suppliers 1998 to 2000, Price controls and Competition Consultation Paper, July 1998," pp. 21-22, 103-108.

Notes:
1. RPI used as deflator, as reported by OFFER.
2. Starting equity base is calculated as market value of companies at flotation, translated to 1996/1997 £.
3. Calculations of average equity and capital are based on the assumption that changes in balances are evenly spread over time.

Table 17-13. Analysis of Sources and Uses of Funds of Public Electricity Suppliers of Great Britain in Real Terms (Annual Averages) (Fiscal Years 1990/1991-1996/1997) (1996/1997 £ millions)

Sources of funds		Uses of funds	
Operating cash flow	2,740	Capital expenditures	1,625
Sale of assets	58	Other expenditures	309
Customer contributions	237	Total uses of funds	1,934
Dividends received	291		
Interest paid	-182		
Taxes paid	-545		
Dividends paid	-948		
Total internal sources	1,651		
Funds from cash balance and external financing	283		
Total sources of funds	1,934		

Source:
See Table 17-19.

The companies earned real returns of 12.1% on capital and 14.3% on equity (Table 17-18) while ten year UK government bonds yielded 8.5% in nominal terms (5.6% real). Thus the industry earned a healthy premium over the risk free return. Interest coverage of 11.8x exceeded utility norms several times over, showing how little financial risk the companies took. They made good returns and spent large sums on new plant. Assume that market capitalization after the offering (about £18 billion) represented value of assets serving the public. The firms made annual capital expenditures (net of depreciation and customer contributions) of about £1 billion, expanding plant about 6% per year, while kWh sales and customer count rose less than 2% and 0.5% per year respectively. Regulation did not discourage the capital spending that increased regulated asset value ("rate base" in American terms). More likely, the high returns encouraged spending. [265]

Northern Ireland

Electricity users in Northern Ireland started off paying more than the British. By the end of fiscal 1997, they paid still more.

Northern Ireland Electricity Service was incorporated as a public limited company (plc) in 1991 and floated in June 1993. By then, the government knew how to price the stock, which rose to a 26.5% premium over the first payment (100p) but only a 12.1% premium over the fully paid price of 220 p. The initial price, at 120% of book value, was not cheap. Only the Scots did worse. Investors holding shares through the end of fiscal 1997 earned a pretax total return of about 20% per year plus bonus shares for customers vs market returns (represented by the FTSE 100) of about 16% per year. The privatized Northern Ireland Electric earned a high return on equity, over 26%. Consumers fared less well. Domestic users paid more than in Great Britain due to high fuel costs (generation depended more on oil), subsidies paid to industrial customers and, said the utility, because the network had "physical characteristics" that made it "relatively more costly to operate." [266]

Real price in Northern Ireland rose absolutely and relatively compared to in Great Britain (Table 17-14). Only the timing of a subsidy payment and a revenue delay kept the comparison from looking worse. Industrial and commercial customers, also, paid more than British counterparts.

---

265. American regulators worried that rate of return regulation encouraged needless investment ("padding the rate base") – called the "Averch-Johnson effect" after two proponents of the idea. Harvey Averch and Leland L. Johnson, "The Behavior of the Firm Under Regulatory Constraint," *American Economic Review*, 52 (5): 1052-1069.
266. Mergers and Monopolies Commission, *Northern Ireland Electricity Plc: A Report on reference under Article 15 of the Electricity (Northern Ireland) Order 1992* (London: HMSO, April 1997), p. 69.

Table 17-14. Real Electricity Prices for Northern Ireland Electricity (NIE) Customers (Fiscal Years 1990/1991-1996/1997) (1996/1997 p)

| | Fiscal Years ended 31 March | | | |
	1991	1993	1997	% change 1991-1997
NIE rank among 15 public electricity suppliers (#1 is highest price)	#4	#3	#1	–
Real price for domestic customer (p/kWh)				
NIE	10.09	10.29	10.29	2.0
Comparable RECs	10.25	10.46	9.18	-10.5
Great Britain	9.58	9.76	8.48	-11.6
Transmission and distribution component of domestic real price				
NIE	3.18	3.30	4.27	34.3
Comparable RECs	3.85	4.16	3.58	-7.0
Great Britain	3.22	3.43	2.97	-7.8
Energy and other supply component of domestic real price				
NIE	6.91	6.99	6.02	-12.9
Comparable RECs	6.40	6.30	5.60	-12.5
Great Britain	6.40	6.33	5.51	-13.4

Source:
MMC, *Northern Ireland Electricity Plc: A report on reference under Article 15 of the Electricity (Northern Ireland) Order 1992* (London: HMSO, April 1997), pp. 70-74.

Note:
Comparable RECs: South Wales, Manweb, and South Western.

In sum, neither consumers nor investors did as well in Northern Ireland as in Great Britain.

Conclusion

Social and economic programs often produce unintended consequences. Electricity privatization produced no disasters, although the regulatory and competitive framework gave companies room to take advantage and they did. Electricity prices fell but consumers still paid too much.

Electric companies made extraordinary profits in light of the risks taken. Predatory firms took a sudden interest in buying electric companies, another sign that the business was too profitable.

Cost of capital constitutes a large part of the electric bill. Regulation should prevent utilities from employing more capital than necessary and assure that return on capital reflects risk. Competitive producers should earn no more than a competitive return. The UK's electricity market seemed to have strayed off that ideal path into a swamp of profitability.

Flaws in the regulatory and competitive framework emerged:

- The one dimensional RPI-X formula focused on price, with other aspects of product and service more like side issues. Management, as a result, concentrated on cost cutting, not on product and service innovation.

- Creating a competitive market required more than additional competitors. Bidding arrangements, psychology, network imperfections and financial engineering all played a role, and the regulator had a hard time keeping track of all of them.

- Information mattered, and companies had more of it than regulators or politicians and they knew how to use that advantage. British regulators had less information than American ones and more faith that the market would solve the problems.

- The efficient one person regulator failed to gain public confidence for decisions made in private and lacked the in-house check on bad decisions that comes when a panel rather than one person thinks through the problem.

A new government, was on the way, though, with its another set of solutions.

Chapter 18

Labor Restructures (1997-2001)

The guiding principles must be transparency, consistency and predictability of regulation. [267]

– Margaret Beckett

New Labor wanted to tax utilities, protect the environment and coal miners, put energy regulation under one roof and customers first. The regulator wanted to revamp the market and the companies to vertically reintegrate. Customers first?

Windfall Profits Taxed

Gordon Brown, future Chancellor of the Exchequer, announced his windfall profits tax plan in September 1996: Labor would not raise taxes but needed funds for new social programs. A one-time tax on windfall profits somehow did not count as a tax.

Once elected, in May 1997, Tony Blair's Labor government moved speedily, announcing in July a "windfall tax on the excess profits of the privatised utilities." [268] Electric companies would pay £2.0 billion of the expected £5.2 billion take.

Parliament imposed a 23% tax on the difference between nine times annual after tax profit earned in the four years after privatization and market value at flotation (adjusted for government-owned shares unsold at flotation). The tax collector, Inland Revenue, explained that "The price/earnings ratio of 9 approximates the lowest average price/earnings ratio of the ... companies during the relevant periods..." [269]

Those who collected the windfall did not pay the tax. The utilities did, years later, when owned by different investors. It was easier to tax a handful of companies that could not move their assets than to find the real beneficiaries of the windfall.

The government explained the tax to Parliament on July 10, 1997:

---

267. HC Deb 30 June 1997, cc 20-21w, quoted in Donna Gore and Grahame Danby, Grahame Allen and Patsy Roberts, "Utilities Bill," House of Commons Library Research Paper 00/7, 26 January 2000, p. 9.
268. Inland Revenue, press release, 2 July 1997, p. 1.
269. _____, *op. cit.*, p. 2.

The windfall tax will fund the New Deal... It is a one-off tax which puts right the bad deal which customers and taxpayers got from the privatisation of the utilities. [270]

As for the possible impact of the tax on the regulatory process:

The Government consulted with the regulators fully on the design of the windfall tax ...we shall not be publishing their advice... [271]

As for the size of the tax, the government explained:

Excluding the temporary effects of the windfall tax ... Public Sector Borrowing Requirement is forecast to be £13 ¼ billion in 1997-98 and £5 ½ billion in 1998-99. [272]

That is, the government needed the money to fill a hole in the budget.

The tax had these virtues:

- Foreign firms owned much of the industry. – They could complain, but Labor had warned them. Five American groups bid to buy UK electric companies between September 1996 (when Gordon Brown spelled out his plans) and July 1997 (when Labor acted on them). Unless their bids discounted for the tax, what were they thinking, that Labor could not win or that Labor would not keep its promise?

- Utilities could afford to pay it. – At the end of fiscal 1996, they had about £2 billion of cash (7% of capitalization). They could dip into their overstuffed bank accounts or they could borrow the money, reducing their high 56% ratio of equity to capital to a still healthy 49%. Most companies had high bond ratings (AA/A) and maintained those ratings afterwards. The tax would barely dent their finances.

- The tax would not affect electric bills. – The companies could not pass it on to consumers (voters) without regulatory approval or covert price manipulation. And any price increase would be insignificant, anyway. [273] A clever tax. Nobody would notice it.

The tax should have warned investors that government can recover profits legitimately earned in the past if deemed too high in the future. The tax negated the first principle of price cap regulation, that the regulator sets price not profit. It created an asymmetric risk

---

270. Commons Hansard – Written Answers, 10 July 1997, column 603.
271. _____, *op. cit.*, column 604.
272. *Ibid.*
273. If the companies borrowed the money to pay the tax and repaid the loan over 20 years, the interest and principal repayment would raise prices by less than 2%.

for investors. Government could seize "excessive" profits retroactively but was unlikely to reimburse firms for inadequate earnings. Those risks increased the cost of capital.

"Don't worry, it won't happen again," the government explained. A Merrill Lynch report written in March 1998 declared that "… investors' confidence in utility shares returned quickly" upon hearing the government assurances of the "one-off" nature of the levy.[274]

The tax fleeced investors who had nothing to do with the supposedly bad deal inflicted on consumers, minimally diluted the financial strength of the industry, told managers that unexpectedly profitability could lead to critical scrutiny, but most importantly, laid bare the risk of retroactive regulation via taxation – not an auspicious start for a government promising fair, predictable regulation.

Restructuring Ahead

Coal contracts expired on March 31, 1998. RJB Mining, the major coal company still in business, threatened to fire 5,000 miners before Christmas.

Labor had close ties to the miners. Late in 1997, to protect the coal industry, Blair's government imposed a moratorium on licenses for new gas-fired power stations, a futile measure because the Tories had already okayed construction of enough gas-fired stations to meet demand growth for the next ten years. In four subsequent years, major power producers opened one 393 MW coal-fired power station and closed 1,353 MW of old ones. Then the *Review of Energy Sources for Power Generation*, launched in December 1997, examined competitive, environmental and security issues. The Merrill Lynch analytical team wrote: "The … review … is essentially all about coal, and finding ways to retain a market for the remnants of the domestic mining industry."[275]

The March 1998 Green Paper on regulation, *A Fair Deal for Consumers: Modernising the Framework for Utility Regulation*, concluded that:

> The government should set a clear framework for regulation … regulators should operate at arm's length from Ministers.

> The regulators should be given a new primary duty to protect …consumers.

> RPI-X should be retained … if …this best serves consumers' interests… RPI-X could be supplemented with … mechanisms to share unearned benefits promptly

---

274. Simon Flowers and Ian Graham, "UK Utility Regulation," Merrill Lynch, 23 March 1998, p. 5.
275. Ian Graham and Simon Flowers, "UK Electricity," Merrill Lynch, 9 June 1998, p.

... when companies benefit from specific factors outside their control, or when they have deliberately misled the regulator ... when the price cap was set.

... the regulators should provide reasons for key decisions...

Regulation of gas and electricity ... should be integrated... The licencing of electricity supply and distribution should be separated.

The energy regulators ... should ... ensure that disadvantaged customers benefit...[276]

President of the Board of Trade Margaret Beckett explained:

The privatised utilities were sold too cheaply ... price controls ... were too lax. The result was a bad deal for consumers and taxpayers. The Windfall Tax put right the mistakes of the past. It is time to ... to ensure that consumers are better served in the future.[277]

The more consumer-friendly and socially oriented regulation, with its notion of "un-earned benefits" seemed to imply more retroactive ratemaking. The RPI-X formula looked less dependable. Would consumers have more protection and investors less?

Merger of energy regulators had to wait for legislation in 1999. To circumvent that problem, the government replaced the Director General of Gas Supply (OFGAS) in November 1998 with the candidate designated to head the new merged agency.

Moving on to the *Review of Energy Sources for Generation*, Labor faced a dilemma: how can a pro-environment and pro-free market (the "New Labour") government rationalize an attempt to save the coal industry? It argued that reliance on gas increased risks and that the market was prejudiced against coal. The outgoing regulator, Prof. Littlechild, replied in April that:

There are no reasons to expect ... catastrophic failure of gas supplies which could jeopardise ... electricity supplies; or that ... reserves of gas are likely to run out; or that there would be undue exposure to risks of gas price increases; or that gas-fired stations cannot operate ... flexibly, or that the gas transmission network could not be ... reinforced to accommodate ... gas-fired stations ...

---

276. U.K. Parliament, Select Committee on Trade and Industry, Sixth Special Report, 25 March 1998, Appendix 1, Annex A, P/98/240.
277. *Ibid.*

There have been concerns about interruptible supplies. Significant back-up supplies are available ...[278]

As for the claim that the market was rigged against coal:

> In fact ... the prospect of tightening emission constraints has ... dictated the replacement of much coal-fired plant by about the turn of the century, and the electricity market has ... been constrained to favour coal in the meantime.[279]

> ... the Vesting coal contracts and the restriction on competition in electricity supply have explicitly favoured coal ... at the expense of franchised customers.[280]

Peter Mandelson, legendary New Labor political spin doctor, became Secretary of State for Trade and Industry in July. His department produced a White Paper on fuel for power generation in October. It called for:

- Reform of the electricity trading arrangements in England and Wales;
- Seeking practical opportunities for divestiture by the major coal-fired generators;
- Pressing ahead with competition in electricity supply for all customers;
- Separating supply and distribution in electricity markets;
- Resolving the technical issues about the growth of gas, including the proper remuneration of flexible plant; and
- Continuing to press for open markets in Europe.[281]

According to the paper, the Big Two had jacked up prices, thereby luring gas-fired generators into the market. Forcing the Big Two to divest plants (to reduce their market power) and reforming the Pool (to reduce market manipulation) would lower prices and discourage new, gas-fired generators. Until then the government should not license gas units. The paper urged a range of greenhouse gas mitigation measures, rather than an assault on coal. The regulator, though, had already forced power plant divestitures and proposed Pool reform. Generators commissioned 13 gas fired and one coal fired station of more than 50 MW in 1998-2001. So much for the moratorium.

---

278. Office of Electricity Regulation, "*Review of Energy Sources for Power Stations,*" Submission by the Director General of Electricity Supply, April 1998, p. 3.
279. _____, *op. cit.*, p. 14.
280. _____, *op. cit.*, p. 15.
281. Donna Gore and Grahame Danby, Grahame Allen and Patsy Richards, "Utilities Bill," Research Paper 00/7, 26 January 2000, House of Commons Library, p. 14.

Malcolm Keay, an energy economist, described the 1998 White Paper, as "essentially a short-term fix for coal, which virtually ignored nuclear and said little about climate change …," a paper that was "soon overtaken by events." [282]

Energy Links, a consulting firm, in 1998 declared that the UK's mining industry had become "competitive to supply world priced coal … a stunning achievement … "but gas fired generators have driven down their costs so much that "gas at today's prices is able to produce electricity on a marginal cost basis cheaper than from coal from … anywhere…" [283] Furthermore, the UK could buy "competitive supplies of coal … from diverse world markets" if required. [284] And the UK had "an immensely robust and diverse gas supply network … " [285] In short, the UK could rely on gas and could get coal from elsewhere if needed.

In October, John Battle, Minister of State at the Department of Trade and Industry, launched a snappily entitled study: *A Fair Deal for Consumers: Modernising the Framework for Utility Regulation: Public Consultation Paper on the Future of Gas and Electricity Regulation.* The paper advocated reform of electricity trading arrangements (RETA) and separation of distribution from supply functions. Stephen Littlechild, in November, backed separation of supply and distribution and said that RETA would promote competition and flexibility.

Before leaving office, Peter Mandelson appointed Callum McCarthy Director General of Gas Supply (effective November 1, 1998) and Director General of Electric Supply (effective January 1, 1999). McCarthy had worked in the Department of Trade and Industry, as an investment banker and after regulating energy became chairman of the Financial Services Authority.

In October 1999, Helen Liddell, Energy Minister, introduced the new policies with exquisite political correctness, touching all bases with pitches for a "primary duty to put consumers first…," ensuring "a fair balance between … consumers … and the legitimate interests of electric and gas companies…," encouraging a "long-term stable framework…," getting utilities to innovate, determination to "replace the electricity pool with a new competitive market," and "…give the regulator greater discretion to respond quickly and appropriately to events…" [286]

---

282. Malcolm Keay, "Oxford Energy Comment," July 2006, p. 1.
283. Energy Links Consultancy, "Gas Moratorium, Battle for Power," Executive Summary, January 1998, p. 1.
284. _____, *op. cit.*, p. 2.
285. _____, *op. cit.*, p. 3.
286. Department of Trade and Industry (DTI) press release, P/99/ 812.

The Utilities Act 2000 would:

- Separate the supply function from distribution.
- Put gas and electric regulation into the same agency.
- Modify licenses to increase flexibility.
- Implement the New Electricity Trading Arrangement (NETA) to replace the Pool.
- Encourage renewable energy and energy efficiency.

The Act abolished the DGES and replaced it with a Gas and Electric Markets Authority (GEMA) that set policy and decided major issues. The Secretary of State for Trade and Industry appointed its chairman, who also served as chief executive of the Office of Gas and Electricity Markets (OFGEM), the regulator. "In fulfilling its obligations, Ofgem works closely with the Department of Trade and Industry (DTI) and energywatch"[287] (a consumer watchdog group). John Bower, an Oxford economist, described the Utilities Act as "a defining moment."[288] It gave the Competition Commission final say over license modifications and put OFGEM under the thumb of the Secretary of State, who could "issue guidance"[289] to GEMA and OFGEM.

Most of the new rules modified regulation or added new functions to the industry structure. Replacing the Pool with NETA fundamentally altered it.

RETA Becomes NETA

In October 1997, the government asked DGES to review electricity trading arrangements (the Pool). Its July 1998 report, *Review of Electricity Trading Arrangements: Proposals* (RETA) criticized the Pool for its complicated pricing and capacity payment mechanisms, for insufficiently valuing flexible generation, for market power issues, and for difficulty in changing governance. Dieter Helm categorized "three lines of attack:"[290]

- The Pool paid all generators the same System Marginal Price (SMP) – The Big Two owned price setting plants and could bid them to raise Pool prices for all. Other generators bid low just to get selected, letting the two price setters do all the work. Paying bid price would end price rigging and Big Two dominance. Consumers would seek the lowest bids. Generators setting high prices could not assume that competitors charged the same price as they did. Prices would fall.

287. Office of Gas and Electricity Markets (Ofgem), *Annual Report 2001-2002* (London: The Stationery Office, July 2002), p. 73.
288. John Bower, "Why Ofgem?," Oxford Institute for Energy Studies, August 2003, p. 2.
289. *Ibid.*
290. Helm, *op. cit.*, p. 308.

- Buyers and sellers had to do all business in the Pool and adhere to Pool rules – This compulsory process stifled innovation. In normal commodity markets, contracts and structures evolve to meet the market's needs. The Pool, in contrast, fixes rules and market structure in advance, forcing market participants to conform to them.

- The Pool did not recognize the value of demand side management (control of usage at the customer level) or incorporate it into the bidding – If the system operator can choose between paying an expensive power plant to generate or paying less to a customer to reduce usage, why not pay the consumer rather than the generator? (US utilities and the CEGB had demand side management programs.)

The government could have hired a multitude of consultants, empaneled committees of "stakeholders," and after much debate and more billable hours, modified Pool rules. Instead RETA abolished the Pool altogether: let buyers and sellers make deals, let markets develop for required financial instruments and let generators dispatch their own plants, rather than await orders from the system operator.

In the new market, buyers and sellers would report their plans for every half-hour to National Grid the system operator, at "gate closure," several hours ahead of actual operation. If the generator produced less than promised or the buyer took more than planned, the malefactor would pay a penalty to the "balancing market" (operated by an affiliate of National Grid), which would make up for the imbalance through purchase or sale in the market. Those who produced more or consumed less power than planned would receive payments from the balancing market, for their contribution to meeting the deficiency.

The balancing market's formula for deficiency charges and surplus payments was designed to encourage buyers and sellers to stick to their plans, not to provide a market to pick up or dispose of power needed to fulfill requirements. Generators that did not deliver the committed power and buyers who took more than originally requested would end up as losers.

Critics attacked: the government provided insufficient evidence to back up claims, showed ignorance of other electricity markets, and made dubious benefit calculations for the New Electricity Trading Arrangements (NETA, son of RETA). NETA, they said, would not achieve its goals, implying the need for some more research before acting.

David Newbery summed it up this way: the Pool suffered from market manipulation, design issues and Big Two dominance. Modify the market to improve it and break up the duopoly, don't get rid of it. As for the decline of coal usage, chalk that up to environmental restrictions. Changing the trading arrangements would not save the coal industry. [291]

---

291. David M. Newbery, "The Regulator's Review of the English Electricity Pool," Department of Applied Economics, Cambridge, 28 August 1998.

RETA's proponents said that contracts "will be driven by the normal balance of commercial considerations which guide price formation in other markets."[292] Market participants will bargain, and maybe get a better deal than they could in the Pool. But the same firms that owned power plants that dominated Pool pricing would continue to own them, so would the new arrangement really change the balance of power?

RETA would inject demand side bidding into a one sided market. In the Pool, buyers could not signal that they would reduce purchases if price rose above a certain point. Nor did the Pool reward consumers who voluntarily reduced demand and in doing so reduced prices for all other Pool customers because the Pool could avoid buying from high cost peak generators. Consumers (the demand side) who reject high priced power can depress prices as effectively as new generators (the supply side) whose entry undercuts prices. Adding in the demand side is like bringing more competitors to the market. Newbery argued that the Pool already had the means to take into account customer response to price. He concluded that RETA could not back its claims, and the government should experiment before making wholesale changes because:

> ... the industry has a demonstrated tendency to undertake reforms whose costs are extremely large for relatively modest improvements ... and the potential costs of mistakes are relatively high.[293]

One year after he left office, Stephen Littlechild commented that:

> I am pleased that the regulator is taking forward reform ... to replace the Pool by ... bilateral contracting processes ... There may be scope for more precise ... signals than hitherto proposed ... But, the fundamental aim, to replace half a market by a two-sided one, and to replace a mechanical price setting mechanism by prices negotiated in the market ...seems ... a sound one. Whether or not it will also deliver the ... price reductions that some have associated with it, it will reduce the scope for collusion, bring greater flexibility and efficiency, and hence be conducive to lower prices over the long term.[294]

Newbery calculated gross benefits from CEGB privatization and restructuring (before costs) at about 4% of the electric bill but a good part of the saving had nothing to do with the Pool's formation, just as savings in the United States often were unrelated to the market model.[295] Other economists calculated that productivity improvements in the distri-

---

292. _____, *op. cit.*, p. 11.
293. _____, *op. cit.*, p. 19.
294. Littlechild, *op. cit.*, no pagination.
295. Leonard S. Hyman, "Restructuring Electricity Policy and Financial Models," *Energy Economics*, Vol. 32, Issue 4, pp. 751-757.

bution sector attributable to the regulatory regime equalled 4% of the bill.[296] Overall, Prof. Newbery found that "Almost all the gains were reaped by shareholders, with the public exchequer making a … small net return … while consumers as a whole lost."[297]

The Pool modified the CEGB's merit order that dispatched lowest cost plants first to one that selected plants that made the lowest bid. Although most buyers and sellers set their prices using CFDs, Pool prices mattered because the Pool set the tone for private arrangements, and because those without CFDs had to buy or sell at Pool prices. By most accounts, though, whatever benefits the Pool brought went directly into the pockets of the electric companies.

Implementing NETA would take over three years. Meantime, the new regulator had to fashion policies and introduce competition to all consumers, and electricity suppliers had to develop strategies to deal with a tougher, more hands-on regulator and a new industry structure.

Regulator in Action

The regulator had to bring competition to small customers, review prices, discipline generators, adhere to new rules imposed by the Utilities Act 2000, and implement NETA. A full plate.

Roughly 4,000 large customers with demand of one MW or more already purchased about 30% of all electricity in the Pool or from competitive suppliers. From April 1, 1994, another 50,000 customers with demand of 100 kW or more could do the same, thus bringing competition to over half the market. Metering, communications and billing problems slowed the second handover. The industry also missed the April 1, 1998 starting date to supply all 26 million customers competitively. That project stretched from September 1998 to May 1999. RECs had no reason to hurry. They would lose customers, were uncertain how to recover costs, and then OFFER decided to regulate supply prices into 2002, so the RECs had even less reason to rush.

David Newbery characterized retail competition for small customers as one of those "reforms whose costs are extremely large for relatively modest improvements…"[298] Dieter Helm questioned benefits claimed for the 1994 opening because the suppliers all bought electricity in the same place, the Pool, and the average margin on sales was 2% "so there was little scope here for significant reductions in customers' bills" from the

---

296. Preetum Domah and Michael G. Pollitt, "The Restructuring and Privatisation of the Electric Distribution and Supply Businesses in England and Wales: A Social Cost Benefit Analysis," July 2000, p. 33.

297. Newbery, *op. cit.*, p. 5.

298. _____, *op. cit.*, p. 19.

cost side.[299] The suppliers could pressure generators to lower prices, but generators had market power, and could resist.

Fully opening the supply market would hurt the coal industry. RECs would not renew power contracts that backstopped the generators' coal contracts so the generators would not renew the coal contracts, and that precipitated the Labor government's frantic review of fuel policy.

Generators reexamined business strategy. Should they remain commodity producers or get deeper into retail supply? If RETA eliminated the Pool, they would have to deal directly with suppliers, so why not own them because owning a supplier provided a hedge against price swings. Suppliers tended to raise or lower retail prices slowly, so rising generation prices reduce supplier profits while falling generating prices raised them. Owning both businesses reduced the impact of price volatility on the company and lowered its risk level.

RECs feared, though, that full competition would lead to a reallocation of costs between distribution and unregulated supply operations and endanger recovery of costs currently charged to the regulated distribution business. So why hurry? Still, fully opening the retail market to competition offered a chance to sell services to more customers. (The biggest supplier, Centrica, bought the Automobile Association in 1999, adding road side repairs to its portfolio of offerings.) Eventually, suppliers discovered that selling a package of low margin, or money-losing services did not add up to a profitable business.

The regulator had price reviews on the agenda, too. Under Stephen Littlechild, OFFER appeared doubtful about its ability to elicit hard information from the utilities and it paid relatively little attention to cost of capital findings. Under Callum McCarthy, OFFER (which merged into the Office of Gas and Electricity Markets or OFGEM, in June 1999) took a less relaxed view. McCarthy did what other regulators did, hired consultants to pore over utility plans and to produce recommendations of cost of capital based on the latest methodology.

The new OFGEM, tightening the screws:

- Switched the regulatory asset value (RAV) formula to that used by other regulators.
- Adopted the capital asset pricing model (CAPM) formula for cost of capital used by other regulators.
- Made substantial reductions in the beginning of period price level (P_0) that brings the return on RAV to the level set for the upcoming price period.
- Increased the size of X factor reductions in price.

---

299. Helm, *op. cit.*, p. 263.

The new regime skeptically viewed utilities' projections and proposed prices that allowed only the most efficient to earn cost of capital, a goal seemingly out of step with the Electricity Act 1989, which, in deathless prose, required the government "to secure that licence holders are able to finance the carrying on of the activities to which they are authorised to carry on,"[300] as well as out of step with New Labour's fairness and balance doctrine. McCarthy backed off but proposed big price reductions, anyway, which RECs accepted without protest, possibly an indication of how bloated were their cost and capital expenditure estimates.

In April 2000, OFGEM tackled the noncompetitive propensity of generators by adding a market abuse condition to their licenses. Two appealed to the Competition Commission (CC), which found in their favor in December. OFGEM argued that "the special characteristics of electricity markets made them especially vulnerable to the exercise of market power close to real time..."[301] OFGEM asked for more stringent standards than the CC normally applied. The CC, however, chose to focus on market concentration measures and concluded that:

> ... the structure of the generation sector has changed in the last 18 months... We expect that this will significantly reduce the scope for the harmful exercise of market power. Moreover there are reasons to expect the structure to become even less concentrated and hence more competitive.[302]

OFGEM would have to find another way to rein in the oligopoly that replaced the duopoly.

Applying CAPM probably affected findings less than the decline in overall capital costs evidenced by falling Treasury bond yields. The X factor turned negative but the real damage came from the big first year (Po) price reductions, which slashed earning power. (Table 18-1.)

---

300. *Electricity Act 1989*, chapter 29 (London: Her Majesty's Stationery Office, 1989), p. 2.
301. Competition Commission, *AES and British Energy: A report on references made under section 12 of the Electricity Act 1989* (2001), p. 200.
302. _____, *op. cit.*, p. 37.

Table 18-1. Regulatory Decisions. (1990-2001) (%)

Year	Companies	Po Price Reduction	X	Pretax Real Cost of capital	Indexed Bond Yield	Risk Premium
[1]		[2]	[3]	[4]	[5]	[6]
1990	RECs	–	+1.3	–	4.2	–
1990	National Grid	–	0.0	–	4.2	–
1993	National Grid	–	-3.0	–	3.3	–
1995	RECs	14.0	-2.0	7.0	3.7	3.3
1996	RECs	11.5	-3.0	7.0	3.7	3.3
1997	National Grid	20.0	-4.0	7.0	3.4	3.6
2000	RECs	24.5	-3.0	6.5	2.3	4.2
2001	National Grid	–	-1.5	6.5	2.5	4.0

Notes (by column):
2. One off price reduction in first year. Sample average for RECs.
3. X in RPI-X formula. Positive sign indicates price rise above rate of inflation. Average for RECs.
4. Pretax cost of capital in real terms.
5. Pretax simple average of annual averages for UK Treasury inflation indexed bonds (2.5% Index Linked Treasury Stock 2016 and 2% Index Linked Treasury Stock 2006).
6. Difference between regulatory finding of cost of capital and average yield on indexed bonds.

Sources:
EEE/ Cornwall, *op. cit.*, pp. 19, 26,37.
Bank of England
Domah and Pollitt, *op. cit.*, p. 32
Steve Thomas, "The Impact of Privatisation on Electricity Prices in Britain," Presentation to the IDEC Seminar on Public Utilities, Sao Paulo, August 6-8, 2002, p. 9.
Tim Jenkinson, "Regulation and Cost of Capital," forthcoming in the *International Handbook on Economic Regulation*, edited by Michael Crew and David Parker, Edward Elgar, 2006, p. 17.

OFGEM, by changing the regulated asset base to market value after privatization plus net new plant investment, all in real terms, eliminated arbitrary adjustments from the process.

Regulatory consultants defined cost of capital as:

... return that needs to be earned ... in order that the financial markets will ... invest in that project or firm's securities ... As such, it is immediately clear that cost of capital is determined by the financial markets (and not by its regulator).[303]

---

303. *The cost of capital in the water industry*, volume 2, Main report, A response by the Water Services Association and the Water Companies Association to the OFWAT Consultation Paper, Nov. 1991, p. 1.

The Capital Asset Pricing Model (CAPM), already used by the other UK regulators to determine capital costs was a favorite in academic circles and often an ingredient of American rate findings. [304] No more fudging. OFGEM had a formula, now.

 In projecting expenditures, OFGEM made greater use of expert witnesses – hired guns brought in to duel over numbers, one side producing evidence of need to make large expenditures which consumers would pay for through a higher price, the other side showing that the utility could operate with less funding and lower prices. Clients knew consultants' opinions and methodologies from previous work, so they could pick their consultant accordingly. For example, financial analyst (and future regulator) Alistair Buchanan, wrote in a 1999 report:

> A further worry will be ...that Ofgem appointed [PKF] to handle the ... review. In 1997 PKF proposed a 10% annual cut in ... costs... Indeed PKF has already made its mark by arguing that 66% and not 90% of corporate costs can ... get allocated to distribution. [305]

Consultants' reports could bound the scope of the regulator's decision, affect public perception of its fairness and give the regulator ammunition for whatever decision he intended to make.

Buchanan summed up his expectations for the new regime in this way:

> The concept of permanent regulation has been introduced, which when combined with a yardstick approach potentially creates a garrote on the industry...

> Significant costs are being removed from the monopoly distribution business...

> The hitherto unregulated large ... consumers ...may well fall under ... formal price regulation ...

> Limited efficiencies left for companies...

> Rate of return for English REC's simply too high: so big price cuts...

> Changes to price setting mechanism may be delayed beyond 1999... [306]

Did the new regulator threaten to remove the "incentive" from "incentive regulation?" Previously utilities either outsmarted the regulator or rose to the bait of incentives by finding new ways to cut costs more than they cut prices. This time around, it might not be as easy.

---

304. See Appendix: Technics and Metrics for a discussion of CAPM.
305. Alistair Buchanan, "UK Electricity Sector," July 14, 1999, Donaldson, Lufkin & Jenrette, pp. 9-10.
306. *Ibid.*

Buying RECs

The RECs' small size made them easy takeover targets. Through 1997, two water companies bought local RECs, hoping to reduce costs by consolidating services and to profit by selling products to a larger customer base. Scottish Power bought a REC, thus acquiring customers south of the border. Seven other acquisitions with no strategic rationale or operating synergies were propelled by a desire to goose up reported earnings via leveraged buyouts and to play in the charged up power game led by Enron. In the generation sector, an American utility took National Grid's cast-off pumped storage facilities. On a larger scale, Eastern, owned by Hanson, acquired generating assets from the Big Two. Less than a year later, Hanson spun off Eastern, after bundling it with coal mines in the United States and Australia as The Energy Group, which subsequently disappeared in another ill-fated merger.

After its election, Labor began to dismantle the Pool, institute stricter regulation and implement retail competition for all customers. Electricity managers began to rethink plans. Generators moved ahead with a strategy that Stephen Littlechild had thwarted: re-integration of the business. American owners decided to get out. One small REC bought another, possibly anticipating a need to reduce costs through consolidation once tougher regulation began. Two utilities already in the England and Wales market bought RECs. Scottish Hydro purchased a large one and later the supply division of another. Électricité de France (EDF), which owned a REC, bought the supply division of another and set up a joint venture with the supply division of a third.

In 1999, Ian Graham of Merrill Lynch wrote that as early as 1995 PowerGen planned to become a "fully integrated electric and gas company"[307] and that all big generators believed in a vertically integrated approach. He argued that the suppliers' "biggest risk is the acquisition cost of power."[308] Customers want a stable price, suppliers wants to avoid over paying for electricity, the new market (NETA) will create uncertainty, and the generators' biggest risk is the price they received. The regulator will not allow a "cosy relationship" between supplier and generator, he wrote, but owning generation provides a "virtual hedge."[309] Graham predicted that five or six vertically integrated players would dominate supply and generation: Centrica, PowerGen, Eastern, Scottish Power, Scottish and Southern (formerly Scottish Hydro), National Power and, perhaps, British Energy. Centrica, using the British Gas brand, lacked generation but had the biggest customer base, as a result of being the leading gas supplier.

PowerGen finally snared a REC in 1998: East Midlands. It had to sell 4,000 MW of generation to get the deal approved. PowerGen laid out its strategy in its fiscal 1998 annual report:

---

307. Ian Graham, "PowerGen," Merrill Lynch, 18 January 1999, p. 5.
308. *Ibid.*
309. Graham, *op. cit.*, p. 6.

The trend toward vertical integration in the UK is already well established, with a number of major players already active in both electricity and gas. The next step in the strategic development of PowerGen's UK business is to consolidate our position through the acquisition of a supply and distribution business.[310]

National Power followed, selling its largest power station, Drax, in order to get clearance to buy Midlands' supply business. The company's 1999 annual report boasted that "These transactions will transform our UK organisation into a broadly based energy business with better matched generation and retail operations."[311]

Both generators warned of increased competition. Both aimed to gather five million supply customers. National Power later simplified its structure by spinning off foreign operations into International Power, while retaining domestic operations as a newly renamed Innogy.

In early 2000, Merrill Lynch listed four ingredients for success in the NETA market: customers, integration, flexible generation and spare capacity. Four generators had them. The report concluded that "the companies least well placed to thrive ... are new entrants without existing capacity. In the absence of a customer base ... the risks and hence cost of capital on new stations must rise."[312] In 2000, just before NETA, eight generating firms produced 75% of electricity and controlled over 70% of the supply market. Would the number of competitors shrink to create an oligopoly that would control the electricity market?

Eastern, the only integrated utility in England and Wales, was part of The Energy Group. Hanson's ill-matched combination of US and Australian coal mining and English electricity attracted the notice of Pacificorp, a northwestern US utility already operating in Australia. American utility stocks, in 1997, sold at 15 times earnings. The Energy Group, at less than 10 times earnings, must have seemed a bargain. Pacificorp offered £3.7 billion for The Energy Group's stock, 20% above market price, on June 13, 1997. Then, another American utility in the race to look more like Enron, Texas Utilities (also owner of an Australian utility) outbid Pacificorp, paid £4.5 billion for The Energy Group's shares, closed the deal in September 1998, sold the mines, renamed the company TXU Europe and set up a joint supply venture with London Electricity (owned by EDF). After that, TXU purchased Norweb's supply division.

UK companies decided to employ their new found competitive prowess in the United States. British Energy partnered with Philadelphia-based PECO Energy, a major nuclear generator whose hard-nosed management did not need British nuclear expertise. Scottish Power purchased would-be predator Pacificorp, and, subsequently learned the intricacies of state regulation the hard way. National Grid bought New England Electric System (NEES)

---

310. PowerGen plc, *Report and Accounts 1998*, p. 2.
311. National Power, *Annual Review (1999)*, p. 3.
312. Ian Graham and Simon Flowers, "Electricity Generation," Merrill Lynch, 27 January 2000, p. 7.

and used it as a base to create an electric and gas distribution network in the northeastern United States, but failed to develop an independent transmission business there.

Between September 1995 and February 2001, 35 major transactions took place in an industry with only 17 large participants. From September 1995 through April 1997, buyers focused on RECs (Table 18-2A). After a period of digestion and contemplation, activity resumed in February 1998 as players that intended to remain shored up their supply businesses, and those who wanted to exit or to diversify outside the UK began to do so (Table 18-2B). Deals made from the beginning of 2000 up to the end of Pool trading in 2001 were largely attempts to bulk up for the new market (Table 18-2C). If nothing else, the deals showed a willingness to plunge into non-strategic ventures just because everyone else was doing it, a tribute to the persuasive skills of investment bankers who go unpaid if their deal does not goes through.

Table 18-2A. Major Non-Government Utility Transactions (1995-1997)

Date	Buyer	Asset Purchased or Merged
9/95	Southern Electric Intl, subsidiary of US utility Southern Co.,	South Western Electricity, a UK REC
9/95	Hanson, UK conglomerate	Eastern Group, a UK REC
10/95	ScottishPower, UK utility	Manweb, a UK REC
11/95	North West Water, a UK water utility. Merger forms United Utilities	NORWEB, a UK REC
12/95	Edison Mission, subsidiary of US utility Edison Intl	First Hydro (generation owned by National Grid)
1/96	Central & South West, a US utility	SEEBOARD, a UK REC
1/96	Welsh Water, UK water utility. Merger forms Hyder	South Wales Electricity, a UK REC
6/96	Avon Energy (US utilities GPU and Cinergy)	Midlands Electricity, a UK REC
6/96	Eastern, UK REC subsidiary of Hanson	Generating assets from National Power and PowerGen, UK generators
7/96	PPL, a US utility	25% share of South Western Electricity, a UK REC, from Southern Co.
8/96	ScottishPower, a UK utility	Southern Water, a UK water utility
12/96	CE Electric (US power producer Cal Energy and constructor Peter Kiewit)	Northern Electric, a UK REC
1/97	Dominion Resources, a US utility	East Midlands Electricity, a UK REC
2/97	Entergy, a US utility	London Electricity, a UK REC
2/97	Investing public	Hanson spins off Eastern as The Energy Group
4/97	Yorkshire Group (US utilities American Electric Power and Public Service of Colorado)	Yorkshire Electricity, a UK REC

Note:

Transaction date.

Source:

"Electricity Companies in the United Kingdom – a brief chronology," Policy Research, 21st October 2002, Electricity Association.

Table 18-2B. Major Non-Government Utility Transactions (1998-1999)

Date	Buyer	Asset Purchased or Merged
2/98	Cal Energy, US generator	Peter Kiewit's 50% of Northern Electric, UK REC
6/98	PPL, US utility	26% of UK REC South Western Electricity stock from Southern Co.
7/98	PowerGen, UK generator	UK REC East Midlands electricity from Dominion Resources
9/98	Texas Utilities, US utility, renamed TXU.	The Energy Group, a UK REC and generator, later renamed TXU Europe
12/98	Électricité de France (EDF)	UK REC London Electricity from Entergy
12/98	Scottish Hydro-Electric, UK utility. Merger forms Scottish and Southern Energy (later SSE).	Southern Electric, a UK REC
6/99	National Power	Supply division of Midlands Electricity
7/99	GPU	50% of Avon Energy (Midlands Electricity) owned by Cinergy
9/99	London Electricity, a UK REC owned by EDF	Supply division of South Western Electricity
11/99	ScottishPower, a UK utility	Pacificorp, a US utility

Note and Source:

See Table 18-1A.

Table 18-2C. Major Non-Government Utility Transactions (2000-2001)

Date	Buyer	Asset Purchased or Merged
2/00	British Energy, UK nuclear generator	Supply division of UK REC South Wales Electricity
3/00	National Grid, UK transmission utility	New England Electric System, US utility
4/00	London Electricity (UK REC owned by EDF) and TXU Europe (UK REC owned by TXU)	Joint venture of supply divisions of both firms
8/00	TXU Europe	Supply division of UK REC NORWEB
8/00	Scottish and Southern Energy, a UK utility	Former South Wales Electricity supply division, from British Energy
9/00	Western Power Distribution (formerly South Western Electricity), a UK REC jointly owned by PPL and Southern Co.	South Wales Electricity, a UK REC owned by Hyder
10/00	Public investors	National Power, UK generator, splits into Innogy (UK assets) and International Power (non-UK).
10/00	Southern Energy (renamed Mirant and spun off from Southern Co.), US generator.	19.7% of Western Power Distribution, UK REC, transferred from Southern Co.
2/01	Innogy, UK generator (formerly National Power)	95% of Yorkshire Electricity, UK REC, from US utility owners

Notes and Sources:
See Table 19-2A.

Giving Customers Choice

In competitive markets, buyers have a choice of sellers. That doctrine led the organizers of the UK's electricity sector to open competition to all customers. OFFER had great hopes for what retail competition could do for small customers. Its "back of the envelope" calculations provided to Parliament in 1997 showed huge benefits (Table 18-3):

Table 18-3. OFFER Calculates Benefits of Retail Competition (1997)

	£ million per year
More efficient purchasing	300
Lower generating costs	200-400
Other	100
Gross annual benefits	600-800
Operating costs	22.5
Net benefits	577.5-777.5
One time set-up costs	£150 million

Source:

Richard Green and Tanga McDaniel, "Competition in Electricity Supply: will '1998' be worth it?," University of California Energy Institute, May 1998, p. 9.

Those calculated benefits worked out to about £30-40 annually per customer less the one time £7.70 cost of setting up the competitive market. The actual project, completed five months late, cost £850 million. Did mass market retail competition in the almost two-year period before NETA began to operate measure up to preliminary billings?

The National Audit Office, in 2001, declared that "Competition has reduced electricity bills for many customers although some have experienced problems," and "The 6.5 million customers who had changed their electric supplier ... have seen their bills fall by £299 million since the start of competition" (£45 per year per customer), after which the NAO added, "About half of this reduction is attributable to competition ..."[313] The NAO's breakdown of savings, however, attributed only £83 million directly to the introduction of competition.

The other 19.6 million customers saw bills fall by £450 million (£23 per year per customer). The NAO report asserted that "The 19 million customers who have not switched suppliers could save up to £674 million, or 13% of their annual bills were they to switch."[314] That is, non-switchers would gain an additional £224 million, a number presumably arrived at by assuming that upon switching the previously unswitched customers would receive discounts similar to those who already switched, a dubious assumption because suppliers gave bigger discounts to bigger users, and high income customers (presumably the bigger users) switched more than low income customers. The non-switchers consisted of 15.4 million customers who received no benefit from retail competition and 4.2 million who collected some discounts from existing suppliers, which adds up to 19.6 million, rather than the 19 million cited in the text.

---

313. National Audit Office (NAO), *Giving Domestic Customers a Choice of Electricity Supplier*, Report by the Controller and Auditor General, HC 85 Session 2000-2001: 5 January 2001, p. 2.

314. _____, *op. cit.*, p. 5.

All customers, whether switchers or not, paid a £4 annual fee in fiscal years 1998/1999 to 2004/2005 to pay for the meters that made retail competition possible. Customers (usually low income) with prepaid meters paid up to £15 per year more to cover the extra costs of their meters. Overall, two-thirds of customers not only did not receive benefits from direct competition but probably paid more than they would have without it. (See Table 18-4.)

Table 18-4. NAO Calculates Reduction in Domestic (Residential) Electricity Bills Since Inception of Retail Competition by Customer Category

	Switched	No Switch-No Benefits	No Switch-Some Benefits	No Switch-All
	[1]	[2]	[3]	[4]
Number of customers (millions)	6.5	15.4	4.2	19.6
Reduction in annual electric bill since onset of retail competition				
£ millions total	299	346	104	450
£ per customer	45	22	25	23
% of electric bill	15	8	9	8
Reduction in annual electric bill directly attributable to retail competition				
£ millions total	83	0	8	8
£ per customer	12	0	2	0
% of electric bill	4	0	1	0

Notes by Column:
All customers who switched retail suppliers.
Customers who did not switch retail suppliers and received no benefits directly attributable to retail competition. All data except number of customers estimated by author.
Customers who did not switch but received some benefits attributable to retail competition. All data except number of customers estimated by author.
All customers who did not switch suppliers. Reduction in bill directly attributable to retail competition (total, per customer and % of bill) estimated by author.

Source:
NAO, Controller and Auditor General, Giving Domestic Customers a Choice of Electricity Supplier, 2001, pp. 1-5.

OFGEM, in June 2003, reported the results of a J.D. Powers survey of consumers that showed them both aware of and indifferent to retail competition (Table 18-5):

Table 18-5. Consumer Attitudes Toward Switching Suppliers in 2001 (%)

Percentage of customers aware of other suppliers	94
Percentage of customers very satisfied with supplier	70
– Switchers	65
– Non-switchers	73
Percentage of customers who found it very easy to compare prices	34
– Switchers	50
– Non-switchers	24
Percentage of customers who ever switched electricity suppliers	
– Summer 1999	11
– Summer 2000	19
– Summer 2001	38

Source:
Ofgem, "Domestic gas and electricity supply competition, Recent developments," June 2003, pp. 14, 18, 21, 37.

OFGEM argued that competition between suppliers would drive down generation prices. Gordon MacKerron, consultant at National Economic Research Associates (NERA) countered that "scope for cutting purchase costs is limited ... because suppliers are not always negotiating price with independent generators. Six major suppliers ... have interests in over 50% of ... generating capacity ... In other words, these suppliers are 'buying' ... from themselves..."[315]

Why would big suppliers fight to lower generating prices? They made profits in generation, not in low margin supply. Furthermore, the most aggressive supply entrant, Centrica, had no generation to speak of and was losing money, so why make life easier for it?

Most consumers knew that they could switch suppliers, one-third were dissatisfied with their supplier, one-third switched (with one-third of switchers dissatisfied with the new supplier) and only one-third thought they could easily compare prices. While customers knew that they had choice, only a minority thought the possible savings worth the trouble.

---

315. Gordon MacKerron, "Costs and Benefits of 100% Electricity Market Opening," NERA, April 2001, p. 3.

Two years after introduction of direct selling to all retail customers, net benefits remained slim, going mostly to wealthier customers, with the promise that something better was on the way.

Conclusion

Tories believed in the efficacy of the market, as a first principle, favored a light- handed regulation and instituted policies that interfered with commercially based decisions. Laborites aimed to impose a social and environmental template on the proceedings, gave lip service to light- handed regulation, and clamped down more on industry players. Both predicted big benefits from their policies and exaggerated those benefits afterwards.

Conservatives and Labor saw NETA as the next big thing in electricity reform. The idea emerged from the Conservative regulator and the Labor regulator took up the cudgels for it. Would the new system live up to its expectations?

Chapter 19
After the Pool (2001-2011)

Everything we do is designed to protect and advance the interests of consumers... [316]

– Callum McCarthy

British electric companies concentrated on acquiring customers and foreign utilities on acquiring British electric companies. The nuclear generator crashed. Renewable and carbon free energy took the spotlight. For a short time, NETA helped customers. But not for long. After a decade of NETA, the government moved on to newer issues, carbon constraints and energy security.

NETA become BETTA

Academics and consultants debated the virtues of "uniform price auction" (the Pool) vs "discriminatory" pricing (also known as "pay-bid"). Do buyers come out ahead if everyone pays the same price, or when they can shop around?

Economists ran simulations, did experiments and examined "the literature," possibly with more rigor than OFGEM. They concluded, more or less, that:

- Uniform price auctions are more vulnerable to collusive behavior because bidders can signal their intentions to each other. But they also lead to more efficient operations because the Pool dispatches the most efficient plants first.
- Prices are higher but less volatile in a pay-bid market.
- When demand is elastic (responds to price), buyers benefit from pay-bid. When demand is inelastic, there is no difference between the two markets.
- No conclusive evidence demonstrates the virtue of uniform price vs pay-bid in a real (as opposed to a simulated or experimental) market.

After the switch from Pool to NETA, what happened?

OFGEM's 2001-2002 annual report boasted that, in its first year:

---

316. Callum McCarthy, "Foreword by the Chairman and Chief Executive," *Ofgem Annual Report 2001-2002* (London: The Stationery Office, July 2002), p. 4.

> NETA has performed far better ... than ... expected. Wholesale electricity prices are now 40 per cent lower than they were when reform started and most electricity is now traded like any other commodity.
>
> ... As it was designed to do, the corporate governance of NETA ... has allowed change to be made effectively. [317]
>
> NETA ... achieved its main aim of creating more competitive trading arrangements and putting pressure on wholesale prices. [318]
>
> Greater competition and generous generator capacity margin saw prices fall by some 18 percent since Go-Live and by 40 per cent since reforms were proposed in 1998. [319]
>
> The impact of NETA on customers' prices also began to be seen. Industrial and commercial contracts are most responsive to wholesale changes so have seen the fastest fall in prices. Domestic prices also started to fall. [320]

OFGEM said that only part of the wholesale price drop ended up in customers' pockets because suppliers bought power on long-term contracts so they would not pass on price reductions until they received them, because some customers had package deals that offset lower electric with higher gas prices and because environmental charges had begun to kick in.

The industry quickly adjusted to NETA. OFGEM claimed it promptly approved market modifications and would shorten the lag between scheduling to gate closure from three-and-a-half hours to one hour, thereby aiding intermittent power producers unable to schedule too far ahead. Small generators, though, still had trouble marketing output. The upcoming British Electricity Transmission and Trading Arrangement (BETTA) would extend NETA to Scotland, put one operator in charge of all British transmission, and open a market for Scottish generation. (BETTA went live in April 2005.)

The Comptroller and Auditor General showed that from 1998 when NETA became government policy to October 2002, only part of the wholesale price decrease trickled down to retail customers because "... suppliers may be reluctant to pass on falls in price that they expect to be unsustainable..." [321] Why did retail prices drop in advance of wholesale

---

317. Office of Gas and Electricity Markets, *Annual Report 2001-2002* (London: The Stationery Office, 2002), p. 6.
318. _____, *op. cit.*, p. 25.
319. _____, *op. cit.*, p. 27.
320. _____, *op. cit.*, p. 28.
321. National Audit Office, *The New Electricity Trading Arrangement in England and Wales*, Report by the Comptroller and Auditor General, HC 624 Session 2002-2003 (London: The Stationery Office, 9 May 2003), p. 3.

for customers that switched? OFGEM suggested that happened "partly because suppliers anticipated the fall in wholesale prices up to two years before they happened."[322] (See Table 19-1.)

Table 19-1. Changes in Electricity Prices. (1998-2002) (%)

	Entire Period 1998-10/02	NETA in Operation 3/01-10/02
Change in		
Wholesale price	-40	-20
Commercial and industrial price	-25	-18
Domestic price		
Switched customers	-17	$\left.\begin{array}{c} \\ \end{array}\right\}$ -1 to -3
Unswitched customers	- 8	

Source:
Comptroller and Auditor General, *The New Electricity Trading Arrangement*, 2003.

The House of Commons committee reading the report also asked why price to suppliers fell more than the price they charged. All anyone seemed to know was that the suppliers had 1.5% margins when regulated (but some made bigger profits) and two suppliers had 4.2- 4.3% margins after deregulation, which caused the committee to comment that "Ofgem has very limited information on suppliers' profit margins..."[323] It also noted that:

> In Ofgem's view NETA has ... delivered benefits in excess of £1.5 billion ... based on the assumption that NETA causes prices in the wholesale market to fall to the price that a new entrant would charge in generating electricity to recover its costs, rather than earning a set price above this level as under the old arrangements.[324]

Not a bad return, if credible, for a market with £7.5 billion turnover. Implementing NETA cost the market entities £39 million and everyone else in the industry £580 million, so, by those OFGEM calculations, NETA paid for itself in five months.

The electricity intelligentsia pounced on the claims and asserted that:

• The generating oligopolists deliberately propped up pre-NETA prices to sell their power stations at peak values. Wholesale prices fell when dominant generators lost

322. *Ibid.*
323. Committee of Public Accounts, *The New Electricity Trading Arrangements in England and Wales*, Second Report of Session 2003-2004, House of Commons, 1 December 2003, p. 4.
324. _____, *op. cit.*, p. 7.

market power as they divested capacity. New capacity put pressure on prices, too. It had nothing to do with NETA.

- To avoid balancing charges, generators kept plants operating unnecessarily and suppliers ordered more electricity than needed. The system ran "long," each plant ran less efficiently, and the industry produced more CO_2 than it would have, otherwise.

- Decision making remained inflexible. OFGEM continued to control it indirectly by entrusting the balancing mechanism to ELEXON, a so-called subsidiary of National Grid (which had a minimal control over it). The Electricity Act of 1989 made no provision for a balancing mechanism, but gave OFGEM supervision of National Grid. Therefore, OFGEM, overseeing ELEXON via National Grid, had final say over modifications of NETA and dominated the discussions.

- NETA disadvantaged small generators that could no longer sell into a Pool but had to find their own customers. Combined heat and power (cogeneration) plants, the most efficient producers, suffered because they had to align electricity production to the needs of the industrial firms that took their heat, rather than the schedule of the electricity market. They risked imbalance charges if their host's needs got in the way of their sales obligations. Windmills of course, would have trouble scheduling production.

- Wholesale prices no longer provided adequate signals to maintain required peak capacity.

The complex NETA arrangements cost more to run than the Pool. On top of all that, wholesale prices rose sharply from spring to winter of 2003. OFGEM did not boast about that.

From 1999 through early 2001, National Power and PowerGen divested 12,500 MW (16% of industry capacity), more than required to gain approval to purchase distribution and supply operations. Did they want to reduce exposure to generation? Did they know something? At the same time, all generators put 3,650 MW of new plant into service and took 2,500 MW of old plant out of mothballs. Capacity increased but so did demand. Those changes, combined with the fluctuation in fuel costs, cannot explain all pre- and post-NETA price fluctuations. Something else must have happened. (See Table 19-2.)

Table 19-2. Before and After NETA Implementation (1998-2002)

	1998	2002	% change
	[1]	[2]	[3]
Operations			
Maximum load at peak (GW)	56.3	61.7	9.6
Generating capacity at peak (GW)	73.3	76.7	4.6
Electricity sales (TWh)	303.5	319.8	5.4
Big Two generating capacity (GW)	25.0	10.0	-60.0
Prices			
Coal fuel (£/tonne)	30.17	29.66	-1.7
Natural gas fuel (p/therm)	20.05	18.53	-7.6
Electricity to ultimate customers (p/kWh)	5.32	4.49	-15.6
Wholesale electricity (p/kWh)	2.58	1.63	-37.0

Sources:
Digest of United Kingdom Energy Statistics, various issues.
Ofgem and Offer, various publications.

The Big Two transferred plant to Eastern in 1996. Yet, according to economist Andrew Sweeting, "generators exercised more market power, as measured by the Lerner index, in the late-1990s than in 1995 or 1996."[325] Their profit margins rose.

Oligopolists try to keep prices low enough to discourage competitors from entering the market. But what if the Big Two kept prices high to attract buyers for their generating plants, that is, to sell to the suckers and get out at the top of the market? Maybe Labor was right, in that those high prices also attracted unnecessary gas-fired units to the market and thereby ruined the prospects for the coal industry.

NETA raised risks for generators without retail customers. But the government would not permit the Big Two to acquire customers unless they divested generation. Both needed money to expand internationally and build a UK customer base. National Power, in 1997, borrowed in order to pay out an enormous dividend. Moody's downgraded the company's bonds "in the expectation that business risk will increase and cash flow quality deteriorate..."[326] Debt as a percentage of total capital jumped from 24% in 1996 to 52% in 1997, according to Moody's. PowerGen piled up debt to buy East Midlands and the biggest utility in Kentucky, LG&E Energy. PowerGen's debt ratio rose from 28% in 1998 to

---

325. Andrew Sweeting, "Market Power in the England and Wales Wholesale Electricity Markets 1995-2000." M.I.T. Center for Energy and Environmental Policy Research, 04-013 WP, August 2004, p. 7.
326. Moody's Investor Services, *Moody's Electric Utility Sourcebook*, October 1997, p. 267.

65% in 1999, Moody's calculated. Selling UK power plants would bring in badly needed cash for the indebted generators.

There was no shortage of buyers either trying to expand an existing operation or to get in on a worldwide energy boom before the opportunities disappeared. Half the generation capacity sold by PowerGen and National Power went to the boomers in 1999.

Edison Mission Energy, a heavily leveraged subsidiary of California-based Edison International, bought 4,000 MW of coal-fired plants, as part of an international expansion that tripled assets in one year. The UK plants were to provide, according to Moody's, 15% of "future cash flows," although by October 2000, the rating agency added that "Uncertainty surrounding the changes in regulatory guidelines have negatively impacted ... expected cash flows..." [327]

AES, an American firm with decentralized management that listed "having fun" as a priority and a heavy debt load (70% of capitalization), bought the enormous Drax station (4,000 MW). AES put up 10% and borrowed the rest. Moody's said that "overbuilding" of competitive gas generators posed the greatest threat to Drax, but AES Drax (as rebranded) "has a hedging contract that protects a significant portion of its revenues..." [328] Unfortunately, the contract was with TXU Europe, which, subsequently bet the wrong way on prices, failed, and left AES Drax to face lower power prices without any protection. AES then walked away from Drax.

To complement its nuclear fleet, British Energy picked up 2,000 MW of coal-fired plants from National Power and sold its recently purchased supply business. (In 2002, the government had to rescue the company from near bankruptcy brought on, in part, by those transactions.) Scottish Power and London Electricity, with customers but limited generation in England and Wales, bought the remaining plants.

Frenzy pervaded the power market. Power producer stocks in the United States rose 100% in 1999 and another 40% in 2000 in a flat market. The world needed electricity. Deregulation meant big profits. Enron led the way, like a Pied Piper. Bankers told insular, inexperienced utility executives to expand or end up as "road kill." British power executives may have been smart enough to reduce exposure to the UK generation market, but not that smart because they put the money from the sales into other markets that tanked later on.

Once the dominant generators sold plants, either they shifted from supporting the price to garnering customers through competitive pricing, or they just plain lost control of price due to sale of those facilities. Maybe something other than the new market, then, accounted for the drop in wholesale prices that took place concurrently with the opening of NETA?

---

327. _____, *Moody's Power Company Sourcebook*, October 2000, p. 323.
328. _____, *op. cit.*, p. 231.

Joanne Evans and Richard Green asked: did wholesale prices fall due to NETA, increased competition or because generating supply rose more than demand? From a "static" study of April 1996 - March 2002 data, they concluded that more competition and too much supply – not NETA – depressed prices. Then they noticed a drop in profit margins six months before NETA's introduction unexplained by either competition or supply. Perhaps "it was only the imminent prospect of new trading rules that allowed prices to fall." [329] Before NETA, generators eschewed price cutting because competitors would notice (everyone could see Pool prices) and retaliate. With NETA in operation, generators could make private deals. And, within a few months of NETA's start "there was no longer a sufficient period" to impose "punishment strategies" on price cutting competitors. [330] Perhaps generators believed NETA would keep them from maintaining "high prices under the new trading rules" and they needed to "switch from a high-price to a low-price equilibrium, even while still trading in the Pool." [331] Wholesale prices fell, then, because NETA precluded easy collusion between generators?

Retail prices dropped less than wholesale. Monica Guiletti, Luigi Grossi and Michael Waterson analyzed generator and supplier operating margins, the "spark spread" (difference between the generator's fuel cost per kWh and the price per kWh at which it sold electricity) and the "supplier margin" (retail price less wholesale price less distribution costs). Spark spread as a percentage of price fell sharply after NETA's introduction, but supplier margin as a percentage of price rose. Adding the two margins, total margin as a percentage of price did not change. What generators lost suppliers picked up. And the same firms owned both businesses. [332] As the three researchers put it, "Despite its stated intention of reducing prices ... the net effect of NETA ... instead merely rearranged where money was made in the system." [333]

Why, around NETA's inception, did generators purchase suppliers at a "going rate" of roughly £200 per customer? [334] If they borrowed the £200 at 6%, they had to earn £12 pre-

---

329. Joanne Evans and Richard Green, "Why did British electricity prices fall after 1998?," Research Memorandum 35-2003, Business School, University of Hull, 2003, p. 12."

330. _____, op. cit., p. 14

331. _____, op. cit., p. 15.

332. Here is how to keep the total margin constant. Customer paid £250 for electricity, before NETA. Of that amount, Generator made a £25 profit on electricity sold to Supplier, and Supplier made a £10 operating profit on sale to Customer. After NETA went into operation, Generator made only a £15 profit because NETA had reduced Generator's opportunity to overcharge, but Supplier (now owned by Generator) earned a £20 profit:

	Before NETA	After NETA
Generator profit	£25	£15
Supplier profit	10	20
All other costs	215	215
Customer bill	250	250

333. Monica Giuletti, Luigi Grossi and Michael Waterson, "Price transmission in the UK electricity market: was NETA beneficial?," draft December 2008, University of Warwick, p. 1.

334. Ian Graham and Simon Flowers, "PowerGen," Merrill Lynch, 9 February 2001, p. 11.

tax just to pay the interest. Consumers switch suppliers, move or die, so investment in the consumer has a limited life. Gas supply customers stuck around for less than 10 years and electricity customer turnover was running around eight years. The supplier writing off investment in the customer over 10 years, would have to earn another £20 per year. Adding it up, to break even the supplier would have to clear £32 per year in operating profit (before interest, amortization of investment and income taxes). The average household spent £250-300 per year on electricity. According to filings with OFGEM, the average supplier in 2002 earned an operating profit of less than 5% (financial reports for earlier years showed similarly low numbers), or about £14 per customer. Why would generators make what looked like a losing investment? Possibly because:

- They planned to market an array of products to customers. Selling gas, for instance, might add £10 per year to operating profit, bringing margin close to break even.

- Expecting to lose control of generating pricing, they wanted to reassert price control via supply. With a firm grip on a customer base and a smaller generating fleet tailored to that base, they could maintain profitability. Integration provided at least some protection when generating prices fell after NETA. Generators without supply affiliates collapsed.

- They wanted to hedge against decline in generation prices. With retail prices steady, retail profits rise when generator prices fall, and vice versa. Combination, then, reduces but does not risk eliminate risk.[335] Investment in supply looked more like an insurance policy than a business opportunity.

- Combinations yield economies. Mergers reduce costs per customer. A larger, more diverse customer mix makes for more predictable operation, a benefit considering NETA's penalties for missing targets. Owning generating assets may provide a better hedge against risks than financial instruments. Perhaps there were economic reasons for vertical integration in the industry, after all.

By late 2002, the Big Six controlled three quarters of domestic supply. Four of them owned just enough capacity to meet the needs of their domestic and small commercial customers and they bought power in the market to supply large customers. The fifth had more capacity than needed for its small customers. The sixth and largest supplier, Centrica, trading as British Gas, had inadequate generating capacity, forcing it to purchase the bulk of its needs on the wholesale market. Four large generators had no affiliated supplier company with small customers. With no clout in the NETA market, they folded, sold out at a low price or required government bailout.

---

335. In 2002, Moody's cited a power company "largely protected from falling power prices due to a volume hedge between the company's generation portfolio and its ... customers." A year later, the bottom fell out of the market and the hedge failed to provide the expected protection. Moody's Investor Services, "Innogy plc," *Moody's Power and Energy Company Sourcebook*, October 2002, p. 193.

Unison, a trade union, and the National Right to Fuel Campaign, hired Cornwall Energy, a consulting firm, to investigate prices in the NETA/BETTA market. At publication, in January 2008, the joint sponsors of the study declared with refreshing forthrightness:

> The average family is being ripped off... companies are just passing increases in wholesale costs straight through to consumers. But many of these companies are setting the wholesale price themselves... Nearly 30% of the extra expenditures has gone in increased profits.[336]

They deconstructed household bills. From 2003 to 2006 prices rose more than costs. Generating margin per kWh shot up from 0.49p out of a total price of 6.86p to 2.29p out of a price of 10.46p. (Industrial prices rose from 3.0p to 6.7p per kWh.) Higher fuel costs accounted for about 1.0p and increased generating margin 1.8p of the 3.6p rise in price per kWh. By 2008, price had risen another 2.0p for domestic and 1.8p for industrial customers, with higher fuel costs accounting for roughly 1.0p of the increase. Admittedly the margin in 2003 was low (far below the 0.9p margin of 1998), but margin after 2003 certainly made up for the previous shortfall. Margins recovered from the post NETA trough to a level higher than before. Generators increased their share of the electric bill (See Table 19-3.)

Ending the duopoly and implementing NETA and BETTA weakened generator control over wholesale prices. But merger of suppliers and generators into a handful of firms replaced a duopoly controlling part of the market with an oligopoly in control of most of the market. NETA and BETTA produced short-lived benefits for consumers.

---

336. Unison, "Report Exposes Golden Hole in Energy Firms' Accounts – Inquiry Demanded into Missing Billions," press release, Jan. 22, 2008, no pagination.

Table 19-3. Household Electricity Costs (2003 - 2006)

	2003 £millions	2003 p/kWh	2006 £ millions	2006 p/kWh
	[1]	[2]	[3]	[4]
Generating fuel costs	1,815	1.61	3,020	2.63
Generating operating margin	557	0.49	2,635	2.29
Total generating charges	2,372	2.10	5,655	4.92
Transmission and distribution	1,800	1.60	2,032	1.77
Other supply, VAT and losses	2,541	2.25	3,337	2.91
Other expense and supplier margin	1,020	0.91	983	0.86
Total other expenses and margins	5,361	4.76	6,352	5.54
Total expenditure on electricity	7,733	6.86	12,007	10.46

Source:
Cornwall Energy, "Gas and Electricity Costs to Consumers," Paper for the National Right to Fuel
Campaign, January 2008, pp. 26-33.

Shuffling Cards

After Stephen Littlechild retired, the mergers and acquisition business burgeoned:

- Most American players exited. Continental Europeans filled the vacuum, taking over major generators, distributors and suppliers.

- Fifteen electric distribution utilities fell under control of eight management groups, only two of which belonged to a domestically owned and publicly traded UK company.

- Six firms came to dominate the supply and generation markets.

The buyers paid more than regulated asset value, undoubtedly because the companies earned more than cost of capital and buyers may have viewed regulatory cost of capital as a low and possibly fictitious bar that they could easily hurdle, but also because they simply viewed the utilities as cash cows to leverage up in order to produce incremental earnings to impress gullible shareholders. The transactions should have concentrated regulatory minds on this question: what constituted a return adequate to attract capital for investment needed to serve the public? The obvious answer was: less than what the companies earned.

From NETA's inception through 2011, 31 major transactions occurred. In 2001 and 2002, the Germans moved into the generating sector. (See Table 19-4A.) In 2003-2006, firms with no electricity experience entered (Table 19-4B). In 2007-2011, French, Spanish, Irish and Hong Kong companies mopped up much of what was left (Table 19-4C).

By 2010, a fully formed, largely foreign-owned Big Six controlled the market: British Gas (Centrica, UK), EDF Supply (EDF, France), E.ON UK (E.ON, Germany), npower (RWE, Germany), Scottish Power (Iberdrola, Spain) and SSE (SSE, UK).

Table 19-4A. Major Non-Government Utility Transactions (2001-2002)

Date	Buyer	Asset Purchased or Merged
5/01	British Energy and Cameco, Canadian uranium miner	Operating lease on Bruce nuclear station, from Ontario Power Generation, a Canadian utility
12/01	Centrica, UK gas retailer	Enron Supply, US supply arm of bankrupt Enron, US gas and power firm.
12/01	American Electric Power	Fiddlers Ferry and Ferrybridge UK power stations, from Edison Mission.
1/02	National Grid	Niagara Mohawk Power, a US utility
5/02	RWE, German utility	UK REC Northern's supply business, from US utility MidAmerican, a successor to Cal Energy and owner of Northern.
5/02	Vivendi, French water utility	UK water utility Southern Water from ScottishPower.
5/02	RWE	Innogy, UK generator, formerly National Power
7/02	E.ON, German utility	PowerGen, UK generator
7/02	EDF	SEEBOARD, the UK REC, from American Electric Power
8/02	Centrica	Direct Energy, North American retail supplier from Enbridge, a Canadian pipeline.
10/02	National Grid	UK gas transmission utility Lattice
10/02	E.ON	TXU Europe, bankrupt UK REC and supplier, from TXU.

Note:
Date of transaction closing.

Sources:
Electricity Association, Policy Research, 21st October 2002, "Who owns whom in the UK electricity industry" Digest of U.K. Energy Statistics, Energy Networks Association, "Gas fact sheet 01," *Energy Quote,* UK Business Park, "UK Activity Report," Company and press report

Table 19-4B. Major Non-Government Utility Transactions (2003-2006)

Date	Buyer	Asset Purchased or Merged
2/03	Cameco	British Energy's share of Bruce nuclear operating lease
3/03	Cargill and Goldman Sachs, American trading firms	Teeside, bankrupt UK power station owned by Enron
4/03	Moyle Holdings, subsidiary of non-profit, Northern Ireland-based Mutual Energy	Moyle Interconnector, transmission line from Northern Ireland to Great Britain, from Viridian (formerly Northern Ireland Electricity)
10/03	E.ON	UK REC Midlands Electricity, from US utility First Energy (successor to GPU)
12/03	FPL, US utility	AmerGen, US operations of British Energy
7/04	Scottish and Southern Energy	UK power stations Fiddlers Ferry and Ferrybridge from American Electric Power
10/04	Various investment firms	National Grid sells some regional gas networks acquired in Lattice merger
12/05	Public markets	Creditors of bankrupt Drax power station sell it to investors
12/06	Arcapita, Bahrein investing group	Viridian, holding company for Northern Ireland utility

Notes and Sources:
See Table 19-4A.

Table 19-4C. Major Non-Government Utility Transactions (2007-2011)

Date	Buyer	Asset Purchased or Merged
1/07	MidAmerican, US utility	Pacificorp, US utility purchased from ScottishPower
4/07	Iberdrola, Spanish utility	ScottishPower
8/07	National Grid	Keyspan, a US gas utility
11/07	North West Electricity Networks, Australia /US investment consortium	UK REC Norweb from United Utilities
1/09	EDF	British Energy
11/09	Centrica	20% of British Energy from EDF
10/10	Cheung Kong Infrastructure, Hong Kong conglomerate	UK Power Networks REC (formerly London, Eastern and SEEBOARD RECs) from EDF
12/10	Electricity Supply Board, Irish utility	Northern Ireland Electricity from Arcapita
2/11	GDF Suez, French power generator	International Power, UK generator spun off from National Power
5/11	PPL	UK REC Central Networks formerly Midlands and East Midlands) from E.ON

Notes and Sources:
See Table 19-4A.

Seven groups (six foreign owned) now owned the UK's electricity distributors (Figure 19-1).

Figure 19-1. Ownership of Distribution Networks (2012).

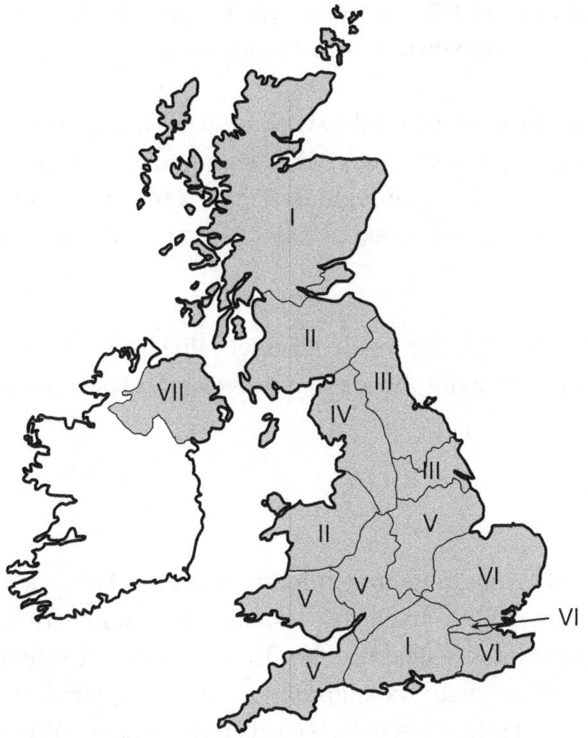

Note:
I. Scottish and Southern Energy – SSE, UK
II. SP Energy Networks – Iberdrola, Spain
III. Northern Powergrid – Colonial FirstState, Australia and US
IV. Electricity North West – Berkshire Hathaway, US
V. Western Power Distribution – PPL, US
VI. UK Power Networks – Cheung Kong, Hong Kong
VII. Northern Ireland Electricity – ESB, Ireland

Three 2010-2011 deals highlighted the gap between private market and regulatory asset value.

In July 2010, Bahrein-based Arcapita announced the sale of Northern Ireland Electricity's network and maintenance and construction operations to ESB (Electricity Supply Board of Ireland) for £1.2 billion (120% of regulatory asset value). Arcapita had purchased NIE's parent, Viridian, in 2006, and tried to dispose of it in 2008.

That same month, EDF sold EDF Energy Networks to Cheung Kong Infrastructure, a unit of a Hong Kong conglomerate. The £5.8 billion price included payment for EDF's contracts to service Heathrow and Gatwick airports, the London Underground and the Channel Tunnel. Credit Suisse estimated that Cheung Kong paid 118% of regulated asset value and noted that "… the private market continues to see significant value in UK infrastructure assets given the visible and stable cash flows they provide, as well as stable regulatory environment and strong inflation link."[337] Profit on the sale offset EDF's losses in the United States and EDF needed the cash to pay down debt and fund nuclear construction. EDF preferred to own customers and power plants, not distribution lines.

In May 2011, PPL, an American utility with network assets in England, purchased Central Networks from E.ON, a German utility trying to rationalize far flung holdings. PPL paid £4.0 billion, about 142% of regulated asset value according to a broker's calculation. PPL, seeking regulated assets to balance its unregulated operations, had to beat out other bidders.

Regulated utilities sell far above regulated asset value because they earn more than cost of capital. Buyers who pay up for those properties clearly expect that situation to continue.

National Grid

National Grid tried to export its transmission model to America. It purchased New England Electric System (NEES), an electric transmission and distribution utility, as a beachhead on America's shores, a deal which raised revenues by one half. America's utilities, however, would not part with transmission assets and quarreled over management and ownership of regional grids so National Grid could not assemble an American transmission company.

National Grid financed the NEES deal with funds stashed away from Energis, formed in 1992 to string a fiber optic communications network along transmission right of way. NG took Energis public when telecommunications stocks were hot. Energis' market value reached £10 billion in 2000, making it one of the biggest London-listed companies before it collapsed in 2001 and filed for bankruptcy in 2002. National Grid, before the end, though, had extracted enough cash to jump start its American expansion.

Management negotiated long term deals with NEES' regulators – 20 years in Massachusetts. National Grid would lower prices 2%, freeze them for five years, then raise them in the following five years in line with prices charged by neighboring companies plus a

---

337. Credit Suisse, "European Power Breakfast, UK Regulated Utilities: CKI announces bid for EDF's UK Networks," 30 July 2010, p. 4.

charge to recoup the original 2% reduction. In the final ten years, the utility would share cost savings with customers. Rhode Island regulators signed a similar arrangement.

Merrill Lynch analyst Ian Graham wrote, "The deal can be extremely lucrative, but only if NEES aggressively cut costs …"[338] Making the deal more attractive, the company could take income tax deductions on interest paid on the deal's debt in both the US and the UK.

In September 2000, NG moved on Niagara Mohawk, an upstate New York electric and gas distribution and transmission utility recovering from a near-death experience caused by expensive power contracts that followed a disastrous nuclear saga. To get approval, NG agreed to cut electric prices 8%, freeze them for 10 years, and share with customers re-turn on equity over 11.75%. NG would keep gas rates flat until December 2004 but could pass on commodity costs to consumers.

The negotiated rate plans started with returns on utility equity of 11% in Massachusetts, 10.5% in Rhode Island and 10.6% in New York but with sufficient cost reductions, NG could earn 15-16%. (US regulatory decisions in 2000-2002 averaged 10.2% return on eq-uity.) National Grid, however, having paid more than the utilities' book equity, could not earn 15-16% on its investment. In fiscal years 2005 and 2006, National Grid USA, the par-ent of the American utilities, earned 6.8% on its equity investment. The utilities, them-selves, earned an average 11.2% return on the much smaller regulated equity.

Return on utility book equity fell to 9% in 2007 and 7% in 2008. National Grid worked outside the conventional American regulatory framework in the misguided hope that regulators would embrace incentive regulation and utilities would entrust their transmission assets to it. Management miscalculated, and, in retrospect, overpaid for its American beachhead.

Borrowing to finance big acquisitions required a larger balance sheet and more cash flow to support loans. A big acquisition in the UK would solve the problem. The UK regulator kept demanding more cost reductions, too. So it made sense to do a merger that would both enlarge the balance sheet and cuts costs by taking out duplicate expenses. But what utility was big enough to make a difference to National Grid? Lattice was.

In 1997, British Gas plc split into Centrica and BG plc. In October 2000, BG spun off pipe-lines and distribution into Lattice plc. The market disliked Lattice. Its stock traded below regulated asset value. In April 2002, NG and Lattice announced a merger that valued Lattice shares at only 4% above market and 14% below regulatory asset value. One analyst said, "… I was a little bit surprised at Lattice selling itself so cheaply. It looks like a cosy deal."[339]

---

338. Ian Graham, "National Grid," Merrill Lynch, 28 February 2000, p. 12.
339. Sophie Barker, "Grid and Lattice form utility supergroup," Telegraph.co.uk, 23 April 2002.

The merger more than doubled National Grid's size, cost-cutting produced savings and there was an added bonus. The Securities Exchange Commission (SEC) of the United States, that desultory regulator of utility holding companies, required its charges to maintain common equity at no less than 30% of capital. National Grid barely met that standard. Merrill Lynch analysts wrote, "The merger restores the SEC ratios to around 40% ... The merger creates debt capacity of ... up to $12 billion ..."[340] National Grid could buy another American company.

OFGEM thought the merger would not reduce competition because "the networks deliver substantially different products."[341] It would increase "system security and performance" and reduce operating costs, and OFGEM would deliver those benefits to customers via "the price control process."[342] On the other hand, the merger would pave the way for more acquisitions in the United States, which could divert management attention from the British networks. OFGEM liked to compare utilities for benchmarking purposes and even considered whether "to compensate customers for the detriment associated with loss of a comparator caused by ... merger..."[343] but decided that it did not compare NG and Lattice enough to warrant concern.

British Gas agreed that the merger would produce operating synergies and benefits for consumers but in the system operation and balancing markets it would "open up opportunities for arbitrage arising from the links between the two markets through gas fired electricity generation..."[344] The new combination could manage networks and maintenance to produce greater revenue and share information that would give it a trading advantage. Putting both networks under one parent would increase the risk to security of supply, too, if the parent failed.

But no matter. OFGEM could handle any problems. OFGEM, the Office of Fair Trading and the Secretary of State approved the creation of National Grid Transco. The deal moved ahead despite a row about the alleged extra-marital affairs of National Grid's chief executive, which the British press, characteristically, blew up into "speculation the scandal could capsize a £13 bn merger with Lattice."[345] The board backed the chief executive, no other firm put in a competing bid, and the deal closed in October 2002.

---

340. Philip Green and Simon Flowers, "National Grid Transco," Merrill Lynch, 4 November 2002, p. 14.
341. Office of Gas and Electricity Markets, *Proposed merger of National Grid Group plc and Lattice Group plc to create National Grid Transco plc, A consultation paper*, May 2002, p. 9.
342. *Ibid.*
343. Office of Gas and Electricity Markets, *Proposed merger of National Grid Group plc and Lattice Group plc to create National Grid Transco plc*, A consultation paper, May 2002, p. 18.
344. British Gas, "Proposed Merger of National Grid Group plc and Lattice Group plc to create National Grid Transco plc, Response from British Gas," May 2002.
345. Terry Macalister, "National Grid board backs Urwin," Guardian.co.uk, 29 July 2002.

Then, in 2004, National Grid Transco raised more cash while reducing its exposure to British regulation by selling four local gas distribution networks. That provided funds to reduce debt and pay a large dividend to shareholders.

National Grid, in February 2006, proposed to purchase Keyspan, a natural gas and electric utility serving New York and New England. National Grid paid 160% of book value for the stock (260% after deducting intangibles). The transaction increased the company's investment in the United States by more than one half.

This time, National Grid escaped an onerous regulatory settlement by claiming that the deal would spare Keyspan's customers from a 10% rate hike and assuring that prices through 2017 would rise at roughly 1.0-1.5% per year, half the expected rate of inflation. National Grid closed the transaction in August 2007. Keyspan's assets, without multi-year deals to drag them down, earned close to allowed returns on regulated equity. But an 11% return on book equity is only 7% on a purchase price equal to 160% of book equity. By the beginning of 2010, ten of National Grid's 12 American subsidiaries had filed for or planned to file for rate relief to revise rate arrangements made or in force at time of merger with National Grid.[346] For the fiscal years 2012-2016, National Grid's average earned regulated return on equity was only 8.9%.

National Grid became a transatlantic utility. No amount of operating expertise and clever multi-year price planning, though, could extract it from the grip of the American regulatory process. In the United Kingdom, National Grid turned a monopoly over electric transmission into a monopoly over electric and gas transmission, creating the UK's largest publicly traded, domestically owned utility, and the only one with international reach.

---

346. National Grid used a time-honored means to boost reported returns on American investments. It used leverage. Here is how it works. Regulators allow Utility A to earn a 10% return on its stockholders' equity of $100, meaning a profit of $10 after paying income taxes. Company B decides to take over A, and pays $130 to the shareholders in A, in order to induce them to sell. Since B paid $130 to buy a company that earns $10, it earns only 7.7% on its investment ($10/$130). The regulator does not raise the allowed profit just because B paid more than the book value on which the regulator set the return. That 7.7% return would not please B's shareholders, so B's management decides to borrow $100 of the $130 required to purchase A. It will pay 6% interest on the borrowing ($6). B, furthermore, can deduct that interest expense, when calculating income tax back home. Assuming a 33% tax rate, it will save $2 in tax payments. Thus, B will report $10 of additional after tax income from A, less $6 of interest costs, plus $2 income tax saved on other income back home. That is, it can claim that the acquisition increased after tax income by $10-$6+$2 = $6. Since B invested only $30 of stockholder money (the rest being borrowed), it will report that the investment in A produces a 20% return on equity ($6/$30).

The seeming alchemical process can have negative effects. If B keeps piling up debt, investors and rating agencies will view that debt as riskier, which will force B to pay higher interest rates for all its debt. The stock of B will look riskier, too, because of the large debt obligation that has to be paid before stockholders get their returns and that will reduce the stock price. Finally, some American regulators take note of the borrowing that financed the equity investment and reduce the overall return to reflect the lower cost of capital that they believe the borrowing produces.

Nuclear Fumbles

When John Major's government floated British Energy, the company paid dividends that exceeded earnings. Its income statement made provisions for back-end nuclear cycle costs. Whether the company could fund those provisions, though, depended on how well it invested any funds that it set aside assuming that it had the resources to do so. Its long term contracts to sell 25% of output to Scottish utilities expired in 2005. Pool price swings could affect profits but City analysts still predicted rising dividends and earnings through 2005, even under "price war conditions." The National Audit Office asserted in 1998 that the government "created British Energy as a robust company, obtaining a high degree of assurance that the company would have the capacity to meet its liabilities."[347]

British Energy expanded abroad despite problems at home. In 1997 it partnered with Philadelphia-based PECO Energy, setting up AmerGen Energy to buy and run nuclear power plants divested by American utilities. PECO's chief executive, Corbin McNeill, was one of those nuclear navy types the electric industry turned to when in trouble. He had a track record for nuclear rehabilitation and a team that had demonstrated its skills. British Energy had little experience operating US-type reactors and no relevant regulatory expertise.

Next year, British Energy tried to buy London Electricity to obtain a customer base, but Électricité de France beat it to the draw. So, instead, in 1999, British Energy bought the supply arm of South Wales Electricity (SWALEC) to add customers ahead of NETA's opening, and the Eggborough power station to gain the flexibility required to avoid NETA's balancing charges.

In mid -2001, the company reversed course, selling SWALEC and mortgaging Eggborough to fund an 82% interest in Bruce Power Limited Partnership, established to secure an operating lease on Ontario Power Generation's enormous and troubled Bruce nuclear power station. The partnership planned to improve Bruce's unhappy operating performance (although British Energy lacked experience with Canadian CANDU reactors) and to prosper by selling Bruce's power into the huge northeastern electricity market. Within a few years, analysts predicted, Bruce would account for one fifth of British Energy's earnings.

British Energy's home market tanked before foreign markets took off. Shortly after the Eggborough purchase, wholesale power prices collapsed in the run up to NETA. In March 2001, British Energy's chairman warned the government that the company needed help. In July the company halved its dividend. British Energy had another headache. When it sold SWALEC in August, it could not unload that firm's principal liability, a contract to buy power from Enron's Teeside power station at a fixed price that exceeded the going market

---

347. National Audit Office, Department of Trade and Industry, Report by the Comptroller and Auditor General, *The Sale of British Energy*, HC 694 Session 1997-98, 8 May 1998, Contents inside cover.

rate. British Energy not only did not have customers to whom it could sell electricity but it also had to pay higher-than- market prices for power it did not need. Electricity prices declined into the winter of 2001-2002, to below British Energy's break even point. The company continued to pay a dividend, though. In August 2002, the second Torness unit went out of service (the first unit's outage began in May). The stock fell 30%. The company said 2002 output would fall 5% below the predicted level. Then management decided to refuel the Dungeness B station ahead of schedule and that caused investors to wonder whether there were problems at Dungeness as well.

Bad news piled on. The company lost £337 million pretax in the six months ended September 30, 2002. The Teeside contract liability ballooned to £349 million. The decommissioning fund fell in value. The company had to repay £110 million of debt in March 2003. As of late August, it had only £78 million cash. It had a £610 million bank line but directors would not trigger the loan because they could not see how the company could repay it – a triumph of honesty over commercial necessity. The directors decided that British Energy could not meet its obligations in September 2002. The government came to the rescue with a £410 million credit line that expired in three weeks. British Energy stock fell to 13% of the 1996 offering price. Three weeks later the government increased the credit line to £650 million and in November, with a restructuring plan in the works, extended the loan to March 2003.

British Energy sold its North American investments. After reorganization, British Energy's creditors ended up with new debt and 97.5% of the shares and existing shareholders 2.5%. (They had lost 97.5% of their original investment by January 2005, when the new shares began to trade.) British Energy wrote down its generating stations by £3.738 million, to one quarter of former worth, in order to reflect values in the current market. The government helped by setting up a Nuclear Securities Fund responsible for decommissioning and liabilities for uncontracted spent fuel and operating waste. In return, British Energy deposited a bond in the fund, agreed to make fixed payments into it, and to pay the fund 65% of operating cash flow (the "cash sweep"). The fund could, at any time, convert its right to the cash sweep into 65% of the company's shares. In June 2007, the government converted part of the sweep into shares and then sold the 29% of the shares it received, and the balance when Électricité de France bought British Energy in early 2009. The government calculated that the proceeds of the sale exceeded the money required for fund purposes by £3.6 billion. The rescue paid off.

How did the government let British Energy get into such a mess? The company produced one-fifth of the nation's electricity, most of it carbon free. Officials, however, went on endlessly about British Energy being an ordinary commercial entity which had to stand on its own, but the government was bound by international law to take responsibility for nuclear waste. The failure of British Energy would put the government directly back into the nuclear power business.

The National Audit Office (NAO) investigated. Its report, issued in February 2004, was introduced by a self-exculpatory press release which began with the declaration that Sir John Bourn, head of the NAO "reported ... that contingent risks he and the Public Accounts Committee had identified following British Energy's privatisation had materialized. The Department for Trade and Industry had clearly recognized the risks ..."[348] That was true, in the sense that NAO's 1998 report contained warnings about risks (as did the prospectus) and admonished the government to keep the risk factors under review, especially those involving the Nuclear Decommissioning Fund and its investment policies. For instance:

> ... despite the transfer of nuclear liabilities ... future Governments remain exposed to any residual uncertainty about the capacity of British Energy to meet its nuclear liabilities.

> Many of British Energy's nuclear liabilities are large ... It is ... not possible to be certain about the capacity of British Energy to meet all of these costs...[349]

> Liabilities ... could fall to the taxpayer if British Energy does not have sufficient revenues ... to meet these liabilities...[350]

But, on the whole, NAO had categorized British Energy as a "robust" company. NAO's 1998 report did not dwell on the possibility of the event that caused British Energy's undoing, the collapse of electricity prices, which other government policies helped to precipitate.

Parliament's Committee of Public Accounts, in 1999 advised the government to "monitor carefully the company's ongoing ability to meet ... liabilities without recourse to taxpayers"[351] and to maintain ownership of its Special Shares (golden shares) in British Energy beyond their 2006 expiration date because "it is still likely that many uncertainties and risks about the ability of the company to finance ... long-term liabilities will remain."[352] The Department of Trade and Industry duly promised to monitor the company and to review the special shares' status in 2006.

The NAO's 2004 report and British Energy's own financial reports laid bare governmental dealings with the company. In the spring of 2000, after electricity prices fell, company officials warned the Department of Trade and Industry that, if prices fell another

---

348. National Audit Office, "Press Release – Risk Management: The Nuclear Liabilities of British Energy plc," 8 February 2004, p. 1.
349. National Audit Office, *The Sale of British Energy*, p. 25.
350. _____, op. cit., p. 26.
351. National Audit Office, Report of the Controller and Auditor General, *Risk Management: The Nuclear Liabilities of British Energy plc*, HC 264 Session 2003-2004, 6 February 2004, p. 56.
352. _____, *op. cit.*, p. 58.

20%, "it would threaten the viability of British Energy and raise the issue of its ability to meet liabilities."[353]

The company requested changes in the reprocessing contract with government-owned British National Fuels (BNFL) that accounted for over 30% of expenses. BNFL said no, unless British Energy restructured all its obligations (that is, went bankrupt). The company tried to get its customers excused from paying the recently imposed Climate Change Levy on the grounds that nuclear power did not contribute to global warming, but the Treasury would have none of that, fearful that it would have to make concessions to other energy producers whose product did not contribute to climate change. British Energy asked for lower property taxes, arguing that the newly imposed rates (April 2000) taxed its units more than those of other electricity producers, which the NAO confirmed. The Department of the Environment, Transport and Regions "believed that British Energy had been treated fairly."[354]

Those three items, together, might have covered the year's expected cash deficit. The NAO commented that "the Department in discussions with other departments whose policies affected British Energy did not specifically draw their attention to the risks posed by British Energy's liabilities." [355] Then again, the Department also stood aside from negotiations between BNFL and British Energy because it did not want to interfere in purely commercial negotiations, and it seemingly managed the NETA process without understanding what a price change beyond its calculations might do to British Energy's ability to meet its nuclear obligations.

Perhaps the government was slow to see the problems because the company continued to pay a dividend, an indication to it of solvency. And the government would not want to interfere in the market or favor one company over another, either. City financial analysts seemed less gloomy than management, too. Maybe, electricity prices would rebound to solve the problem. The government hired a banker for advice in January 2002 and learned, eventually, that if British Energy failed, the Nuclear Decommissioning Fund would have to get in line with other creditors and BNFL could lose its biggest customer. Finally, in September 2002, the rescue began.

On February 11, 2004, Sir Robin Young, KCB, Permanent Secretary, Department of Trade and Industry, and Mr. Adrian Montague, CBE, Chairman of British Energy (and one of the UK's most prominent business leaders and at the time also Deputy Chairman of Network Rail, Chairman Designate of Infrastructure Investors and Chairman of Michael Page International) arrived at the House of Commons to testify about British Energy's rescue. They fell into the unfriendly hands of Gerry Steinberg, Member for the City of Durham,

---

353. _____, *op. cit.*, p. 27.
354. NAO, *op. cit.*, p. 25.
355. *Ibid.*

who declared the privatization "a complete disaster."[356] Sir Robin noted that the NAO described British Energy as "robust" at privatization. MP Steinberg then asked Chairman Montague "How many days a week do you put in British Energy?... Do you think you should be full time?"[357]

Steinberg then questioned the government's lack of effective action after the spring of 2000 warnings, saying "Anybody ... with any common sense would have ... thought, bloody hell, something is wrong here..." to which Sir Robin answered, "As I said ... we had hugely upped our monitoring ...This was not an emergency... You will remember also that we cannot, as a department, intervene on behalf of a single private company, it is not our policy and that would be unfair –," at which point MP Steinberg interrupted, "This is not a normal private company, is it?... but you treated it as though it was Marks and Spencer or an ordinary private company, that is the whole point."[358] Sir Robin replied that the government could not aid a private company" without getting permission from the European Commission, and if we do help in any way, the first people to complain would be its competitors."[359]

Steinberg also objected to paying a carbon tax on nuclear power: "It is like saying my mother when she was alive who could not drive should pay a car tax," to which Chairman Montague replied, "I do not understand your analogy."[360]

In short, British Energy could not insulate itself from the market, did not build a customer base and became entangled in contracts (Teeside and BNFL) that fixed costs at high levels. Paying dividends after the problems emerged may have sent a perverse message to the government: no emergency, because the company can still afford to pay them. Branching abroad did not help at home. The government, in turn, bought into the fiction that British Energy was a normal company that could fend for itself, with the right to do whatever it pleased, because all shareholders and creditors were consenting adults.

If British Energy foundered, the government would have ended up with the nuclear waste, certainly a reason to prevent collapse rather than pick up the pieces afterwards. Although the government made a case that international rules and domestic competition issues prevented it from acting, it could not make a case that it tried hard to modify or get around the obstacles.

356. "Risk management: the nuclear liabilities of British Energy plc (HC 354-i)." Public Accounts Committee, 11 February 2004, uncorrected transcript of evidence. Gerry Steinberg web page.
357. *Ibid.*
358. *Ibid.*
359. *Ibid.*
360. *Ibid.*

After the rescue, British Energy sold North American operations. In April 2005, it hired Bill Coley – veteran executive of Duke Power from the era when that North Carolina company had a reputation as the best nuclear operator in the country – as chief executive officer. In the spring of 2006, prices rose, earnings rebounded and British Energy paid its first dividend since the rescue. The government leaped into action. The Nuclear Liabilities Fund converted sweep payments into company shares and sold them with the proceeds set aside to decommission existing nuclear stations. This move reduced the company's need to reserve funds for the cash sweep and improved its balance sheet by raising the equity component of capitalization.

Nuclear power returned to the spotlight in the fall of 2005. Tony Blair's government seemed to favor more nuclear power, although energy minister Malcolm Wicks warned: "Don't think this government's in the business of saying we're going to become heavy subsidisers of nuclear energy because we aren't." [361] The government wanted to reduce carbon emissions, had no permanent nuclear waste storage policy, the operating licenses for all but one existing nuclear expired by 2023, oil and natural gas prices had risen and the country increasingly depended on fossil fuels from unstable regions. Unless it planned to rely on windmills and solar power, entirely, the United Kingdom would have to do something.

Back in 2003, Blair described new nuclear plants as an 'unrealistic option.' [362] Environmentalists warned against a nuclear subsidy. Insiders, though, insisted that the nuclear option did not require legislation. British Energy had valuable assets needed for a nuclear program: power plant sites, experienced employees and connections to the transmission network. The company needed partners to develop new generating stations on those sites.

British Energy, in 2007, had roughly 11,000 MW of capacity valued on the books at £2 billion. Free cash flow (after dividends and capital expenditures) ran about £100 million per year. Building a 1,000 MW nuclear station might cost £2 billion spread over 5-10 years. British Energy directors would have to bet the company to do one project. Delays or cost overruns could ruin the company, something that Bill Coley must have understood from his days at Duke Power. British Energy did not have the scale to prudently undertake one facility, no less the four that some in government had in mind. The company needed partners or a deep-pocketed owner.

In January 2008, Gordon Brown's government proposed a nuclear power building program, in order to reduce carbon dioxide emissions and improve energy supply security. But John Hutton, business secretary, assured the public that "We are not giving planning

---

361. Jean Eaglesham, "Britain to decide on new nuclear power stations next year," *Financial Times*, September 29, 2005, p. 3.
362. Christopher Adams and Jean Eaglesham, "Blair review to look at private investment in nuclear plants," *Financial Times*, November 30, 2005, p. 4.

permission to new power stations … and we are not going to subsidise them. "[363] The Tory spokesman approved, as long as the government provided no subsidy.

The government's white paper, *Meeting the Nuclear Challenge*, lauded nuclear power because it produced no greenhouse gas emissions, was affordable compared to other carbon-free alternatives, dependable, safe and improved the nation's energy diversity and "it would be in the public interest to allow energy companies the option of investing in new nuclear stations…"[364] and the government should encourage their construction.

The UK would have to build 30-35 GW of new capacity over two decades while 22 GW of old power stations (including 10 GW nuclear) would close "based on published lifetimes."[365] The report optimistically estimated nuclear construction costs of £1,250 per kW for a 1,600 MW station, or £2.8 billion completed including cost of capital during construction.

The government would develop criteria for nuclear power sites, pursue generic designs for plants, assure waste management plans that protect taxpayers and establish a strong Nuclear Installations Inspectorate. For nuclear projects to go forward, the market would require "a clear signal… This White Paper gives that signal."[366]

Policy affirmations do not launch construction projects. Anyway, the Scottish Executive said that it was unlikely to approve a nuclear project and the Welsh Assembly declared a station in that region "unnecessary."[367] The government in London could prevent construction, but as far as taking affirmative action, it was pushing with a string.

The white paper declared that new nuclear power stations would bring fuel security and environmental benefits to the nation, but:

> It will be for energy companies to fund, develop and build new nuclear power stations in the UK…[368]

> Their decisions will be affected by their view on the underlying costs of new investments, their expectation of future electricity, fuel and carbon prices, expected closures of existing power stations and the development time for new power stations.[369]

---

363. Jim Pickard and Ed Crooks, "UK gives new plants green light," *Financial Times*, Jan. 11, 2008, p. 5.
364. Department for Business Enterprise & Regulatory Reform, *Meeting the Energy Challenge: A White Paper on Nuclear Power*, January 2008, p. 10.
365. _____, *op. cit.*, p. 13,
366. _____, *op. cit.*, p. 35.
367. _____, *op. cit.*, p. 122.
368. _____, *op. cit.*, p. 37.
369. _____, *op. cit.*, p. 119.

And just in case the potential builders missed the point:

It is not intended that incentives will be provided through the fiscal regime to invest in nuclear power generation in preference to other types of electricity generation.[370]

The market, then, had to support nuclear investment or there would be none. But the market had never supported nuclear power. The government always propped it up.

The press said that British Energy was talking to potential partners. Maybe Centrica? It had 15 million customers but neither the experience nor scale to tackle British Energy, a company its own size. Only a determined leviathan could swallow British Energy.

In September 2008, in the midst of a global financial crisis, Électricité de France (EDF) offered to buy all British Energy shares for £12.5 billion. EDF operated 58 nuclear reactors. The French government, owner of 85% of EDF and 79% of nuclear engineer Areva, wanted to make France a leading commercial nuclear power, with those two national champions spearheading the effort. A few days before, EDF lost a bid to buy Constellation Energy, a nuclear-oriented US utility that suffered a disastrous trading reversal and sought succor from Warren Buffett. EDF ended up with a consolation prize, 50% interest in Constellation's nuclear fleet. Between the two deals, EDF secured ownership in 12 more nuclear stations.

The Times of London wrote that EDF planned to construct four reactors at British Energy sites using French technology, that the acquisition would "ensure a steady flow of exports for France's nuclear and engineering industries," while a Labor parliamentarian warned that the combined group could force up prices, and the story ended with a lament:

Britain led the world in civil atomic power ... but a decision ... to pursue ... the Advanced Gas Cooled Reactor proved a costly mistake when the rest of the world adopted a different technology ...[371]

In November 2009, Centrica bought 20% of British Energy from EDF for £2.3 billion, paid by selling its Belgian retail business to EDF for £1.2 billion and the balance in cash. The deal valued British Energy at £11.5 billion, less than EDF paid for it. Centrica agreed to take 20% of British Energy's uncontracted power generation as well as to buy 18 TWh at market prices during a five-year period beginning in 2011 (about 8% of Centrica's annual needs). Then EDF put its electricity networks (RECs) up for sale, which would help pay

---

370. _____, *op. cit.*, p. 154.
371. Robin Pagnamenta, Lewis Smith and Adam Sage, "Anger as France becomes a nuclear power – in Britain," TIMESONLINE, September 25, 2008.

for the nuclear push. EDF's business plan centered on customers and generators, not regulated delivery.

Centrica, EDF and British Energy together in 2009 produced 31% of the UK's power with 26% of generating capacity, and served 44% of the retail customers. The deal concentrated the business into fewer hands and sold still another major firm to a foreign owner.

Regulation

The handover of OFGEM from Callum McCarthy to Alistair Buchanan in 2003 did not, on the surface, change much on the regulatory front and the generation/supply oligopoly continued to do its thing, undeterred by occasional harsh words.

Pricing orders ratcheted down the risk premium and benchmarked expenses and capital spending to pressure utilities to lower costs. The inferior utility would, henceforth, earn lower returns. The new regime, however, frankly discussed how much the utility could earn if it beat the standards, and showed a concern, if not an obsession, with attracting investment to modernize, meet future demand and reduce carbon dioxide emissions. The fact that companies rarely appealed after declaring how tough the findings were, though, indicated that OFGEM's bark was still far worse than its bite. (See Table 19-5.)

Table 19-5. Electric Regulatory Decisions (2004-2009) (%)

Year	Sector or Company	Po	X	Pre-tax Real Cost of Capital	Treasury Indexed Bond Yield	Risk Premium
	[1]	[2]	[3]	[4]	[5]	[6]
2004	Distribution	1.3	0.0	6.90	1.7	5.2
2005	National Grid	6.0	0.0	6.25	1.7	4.5
2006	Transmission	5.0	-2.0	5.05	1.5	3.6
2009	Distribution	-8.3	5.6	4.70	1.1	3.6

Notes by Column:
2. Initial year price increase or decrease (-) from previous price level before application of RPI-X formula.
3. X factor price change above or below rate of inflation.
5. Average of annual yields of 2.5% Index Linked Treasury Stock 2016 and 2% Index Linked Treasury Stock 2006 for 2004 and 2005, and of the 2.5% issue alone for 2006 and 2009.
6. Pretax real cost of capital minus Treasury indexed bond yield.
2 and 3. 2009 estimated by author from decision text and tables.
1 to 6. Author's estimate of average decisions for multiple utilities.

Sources:
Ofgem for orders and Bank of England for interest rates.

Squeezing out big savings became more difficult. In the transmission sector, the regulator and National Grid negotiated annually on a reward/penalty for beating or missing a target for "incentivized balancing costs." In the early years, the arrangement coined money for National Grid. After a few years, National Grid had a harder time beating the targets and with the introduction of BETA and intake of more wind power, the numbers became too unpredictable for the scheme to work, and it was modified in 2011. (See Table 19-6.)

Table 19-6. Transmission Incentives (2002-2011) (£ million)

Fiscal Year Ended March 31	Expense Target	Actual Expense	Maximum Incentive	Maximum Penalty	Incentive/ Penalty Paid to National Grid
	[1]	[2]	[3]	[4]	[5]
2002	382	263	46	-15	46
2003	367	286	60	-45	49
2004	340	281	40	-40	32
2005	320	289	40	-40	12
2006	378	427	40	-20	-4
2007	No target	495	–	–	–
2008	430–445	451	10	-10	-1
2009	530-545	827	15	-15	-15
2010	586-616	417	15	-15	15
2011	511-566	282	15	-15	15

Notes:
All figures in current £.
National Grid and Ofgem could not agree on an incentive plan for fiscal 2007. Ofgem proposed £390-410 as target. National Grid had forecast £451 expense.

Source:
Ofgem, *National Grid Electricity Transmission Operator Incentives from 1 April 2011*, 10 June 2011, p. 4.

By 2010 OFGEM shifted from paring costs to encouraging capital spending on the network in order to boost security and reduce carbon emissions. OFGEM launched the RPI-X@20 review of policies, and sought new arrangements to bring more generating capacity on line, capacity of the right sort. National Grid's May 2009 seven-year forecast showed no shortage of capacity, true, but European rules to reduce carbon emissions could force closure of power stations.

Prices mattered, too. In 2008, a House of Commons committee issued a report on "fuel poverty" which fretted about rising energy prices, but:

> ... our concern is to maintain a public policy environment in which UK energy prices ... are as low as possible, but also one in which the other crucial objective of energy policy – environmental sustainability and security of supply – are delivered. We are aware of the urgent need to bring forward investment ...[372]

Electric companies needed prices high enough to attract capital. But did consumers pay too much because only six firms controlled 55% of generation output and almost three quarters of price setting plant. Suppliers claimed that they lost money so do not blame them for high prices. They just bought from generators that set prices. Of course, they owned the generators.

The committee asked why only a small percentage of those who switched to obtain lower prices actually pick lower cost suppliers, while a larger percentage picked higher cost suppliers? Why did competition from outside suppliers not induce incumbents to lower their prices to the competitors' level? Why did suppliers charge those with pre-paid and credit meters more than the costs required to serve those generally low income customers?

Researchers Chris M. Wilson and Catherine Waddams Price measured switching's benefit in terms of "consumer surplus."[373] Switchers achieved only 26-39% of possible gains, and 27-38% of them "actually *reduced* their surplus"[374] due to "high search costs"[375] (time and effort spent finding the right product) and "difficulties involved in comparing highly complex non-linear tariffs"[376] (consumers could not figure out the pricing schedule) rather than high pressure sales tactics or consumer ignorance about their own consumption patterns.

The House committee concluded that it had not heard evidence suggesting active collusion, but:

> It is clear ... that in a... market dominated by six big players, it is easy for those players to make informed judgements about the behaviour of their competitors. This can distort competition, without any active collusion occurring. The regulator therefore needs to remain very watchful.[377]

---

372. House of Commons, Business and Enterprise Committee, *Energy prices, fuel poverty and Ofgem*, Eleventh Report of Session 2007-2008, Vol. 1, HC 293-1, 28 July 2008, p. 3.
373. Chris M. Wilson and Catherine Waddams Price, "Do Consumers Switch to the Best Suppliers?," University of East Anglia, May 2006, p. 2. They defined consumer surplus as "the net utility derived from the firm's chosen price-product combination as valued by any individual consumer."
374. _____, *op. cit.*, p. 1.
375. _____, *op. cit.*, p. 2.
376. _____, *op. cit.*, p. 3.
377. House of Commons., Business and Enterprise Committee, *op. cit.*, p. 41.

OFGEM investigated, and in 2010, "… noted that the Big 6 suppliers pricing patterns tracked one another closely" [378] and "Ofgem remains concerned about whether the energy supply market is working in the best interests of consumers."[379]

Measuring electricity profit margins, though, was tricky and complicated by need to allocate profits between offerings in gas and electric packages. The government expressed puzzlement but not enough to require a uniform system of accounts other than an unaudited Consolidated Segmental Statement beginning in 2009 that did not provide the level of detail that American regulators required. The combined margin for gas and electric supply and the margin on generation sold to domestic customers tended to move in opposite directions, more evidence that owning a supply business provided a hedge for the generator. Even with the partial ability to offset bad profits in one business with good profits in the other, the generator/supplier combinations showed earnings records that would discourage new investments in either competitive supply or generation other than by the hopelessly optimistic. (See Table 19-7.)

Table 19-7. Net Margin on Sales to Domestic Electricity Customers (2000-2012) (£billions and %)

Year	Electric Domestic Revenue (£)	Gas Domestic Revenue (£)	Total Domestic Revenue (£)	Generation Margin (£)	Supply Margin (£)	Total Margin (£)	Total Margin as % Total Domestic Revenue
	[1]	[2]	[3]	[4]	[5]	[6]	[7]
2000	7.1	5.2	12.3	0.5	2.2	2.7	22
2001	7.2	5.5	12.7	0.3	1.8	2.1	17
2002	7.2	5.8	13.0	0.2	2.1	2.3	18
2003	7.3	6.0	13.3	0.3	2.0	2.3	17
2004	8.7	7.9	16.6	0.6	1.8	2.4	14
2005	9.2	7.8	17.0	1.3	0.2	1.5	9
2006	11.0	9.6	20.6	2.4	0.6	3.0	15
2007	11.9	9.5	21.4	2.3	0.8	3.1	14
2008	13.6	11.5	25.1	2.5	0.4	2.9	12
2009	13.8	12.4	26.2	2.3	0.8	3.1	12
2010	13.5	12.6	26.1	1.4	1.6	3.0	11
2011	13.9	11.7	25.6	2.3	1.3	3.6	14
2012	14.9	15.0	29.9	2.1	1.6	3.7	12

Notes:
Columns 1-3 – Revenues for entire domestic (residential) electricity and gas market of UK.

378. Ofgem, *Electricity and Gas Supply Market Report*, 22 February 2010, p. 4.
379. _____, *op. cit.*, p. 7.

Column 4 – Estimated domestic market share of earnings before interest and taxes (EBIT) of electric generation operations of Big Six.

Column 5 – Combined electric and gas supply EBIT for Big Six.

Column 6 – Total of electric and gas generation and supply margins.

Column 7 – Total Big Six EBIT margin as percentage of entire UK domestic revenue.

Sources:

Columns 1-3 – DECC, *Digest of UK Energy Statistics*, Table 1.7, "Domestic Sector," various years.

Columns 4-6 – Ofgem, "Energy Supply Probe," 6th October 2008, Figure 8.1, p. 100 (for 2000-2007).

_____, "State of the Market Assessment," 27th March 2014, p. 6 (for 2009-2012).

_____, "Electricity and Gas Supplier Market Report," 22 Feb. 2010, pp. 24,25.

_____, "Making the profits of the six largest energy suppliers clear," 25 November 2013.

_____, "Electricity and Gas Supply Market Report," 19 December 2011, p.11.

_____, "Financial Information Reporting: 2010 Results," p. 8.

NERA, "Energy Supply Margins Update December 2010."

Then OFGEM made headlines. Financed by charges on regulated companies, it now looked like any City business: board chairman, chief executive officer and staffers titled "partner." In 2009, on the heels of a parliamentary expense account scandal, OFGEM's expense reports "were slipped out without notice on the watchdog's website,"[380] as a news story put it. Top executives racked up charges for overnight stays in London, a newspaper subscription to the *Financial Times*, drinks for the poverty conference team, and £5 for a meal at OFGEM's canteen, despite salaries of £145,000 for the part time chairman and £265,000 for the chief executive. (American regulators would die to collect those salaries.) Newspaper columnist David Hughes wrote "Ofgem's greedy bosses make MPs look models of self-restraint ... Just when you thought you were outrage-proof at the venality of our public servants, out of the woodwork pops Alistair Buchanan, the notably feeble energy 'watchdog.'"[381]

The expense account peccadillos mattered less than whether OFGEM met the objective that decorated its press releases, in the blue band surrounding the word "Ofgem" in the orange lozenge: "Promoting choice and value for all gas and electricity customers." The Utilities Act of 2000 diluted OFGEM's role, giving the Competition Commission power to veto license modifications. The Enterprise Act of 2002 gave the Office of Fair Trading jurisdiction over anti-competitive behavior and, in the words of Oxford economist John Bower, "removed the primary role that Ofgem ... played since 1990, namely, promoting competition."[382] Why blame OFGEM?

The industry's oligopolistic structure rendered "choice" into "not much of a choice," but was OFGEM the responsible party? As for rising prices, perhaps the industry had sim-

---

380. Jon Swaine, "Ofgem chief Alistair Buchanan claimed £50 for 'fuel poverty' drinks," Telegraph.co.uk, 25 August 2009.

381. David Hughes, *Daily Telegraph News Blog*, August 25, 2009.

382. John Bower, "Why Ofgem?," Oxford Institute of Energy Studies, August 2003, p. 4.

ply run out of prudent ways to save money and the new regulator realized that game was over, that the implementation of intelligent networks and carbon dioxide reductions would require investment, and prices had to rise in order to pay for them. Shooting the messenger has a long tradition.

Environment

Margaret Thatcher, a chemist by training, understood the role of greenhouse gas emissions in global warming. Her government set targets for their reduction. Privatization produced a net emission reduction of 54 million metric tons (coal-fired emissions declined 77 million metric tons, offset in part by a 23 million metric ton increase from replacement fuels). More reductions took place later after the Great Recession depressed the economy's output.[383] (See Table 19-8.)

Table 19-8. Annual Emissions of Carbon Dioxide in the UK (1990-2011) (million metric tons)

	1990	1995	1997	2000	2005	2010	2011
	[1]	[2]	[3]	[4]	[5]	[6]	[7]
Electricity generation							
Coal	170	116	93	90	103	80	82
Other fuels	34	47	57	69	70	77	62
All electricity	204	163	150	159	173	157	144
Other sectors	388	391	402	394	381	341	315
Total economy	592	554	552	553	554	498	459

Note:
Carbon emissions from burning coal as a generation fuel estimated by author.
Tonne = metric ton.

Sources:
1. Department of Energy & Climate Change, Statistical Release, "UK Climate Change Sustainable Development Indicator: 2010 Greenhouse Gas Emissions, Provisional Figures and 2009 Greenhouse Gas Emissions, Final Figures by Fuel Type and End- User," 25th March 2011.
2. _____, Statistical release, "UK Climate Change Sustainable Development Indicator: 2012 UK Greenhouse Gas Emissions, Provisional Figures and 2011 Greenhouse Gas Emissions Final Figures by Fuel Type and End Uses," 28th March 2013.
3. _____, "Historical Coal Data: Availability and Consumption 1853 to 2012
4._____, _Digest of UK Energy Statistics_, various issues

---

383. Government data may not take into account emissions that occur during the production and transportation of the fuel itself, so they may understate emissions and distort comparisons of carbon emissions by fuel type. Burning natural gas produces lower emissions than burning coal, but how much less is another matter.

The Electricity Act of 1989's Non-Fossil Fuel Obligation (NFFO) encouraged nuclear and renewable energy. The NFFO expired in 2002, having accomplished little. The renewable component of electric generation rose from 1.8% in 1990 to 2.1% in 1997 and 2.9% in 2002 (but reached 6.6% in 2010.) The RECs also had to offer energy saving advice to customers and the Energy Saving Trust spent over £100 million per year to help consumers save energy.

The Climate Change Levy that almost sank British Energy taxed output from conventional resources (including large hydro and nuclear) but not from coal bed methane, combined heat and power (cogeneration), or other renewables. The tax also funded energy efficiency projects.

The Renewable Obligation introduced in April 2002 actively supported renewables as opposed to taxing fossil fuels. The government issued Renewables Obligation Certificates to producers of electricity generated by one of a specified list of resources, one certificate per kWh produced (although offshore wind earned two certificates and sewage gas only one half a certificate per kWh). Licensed electricity suppliers had to buy those certificates in increasing amounts, rising from 3% of kWh sales in fiscal 2002/2003 to 15.4% of sales in fiscal 2015/2016 (lower requirements in Northern Ireland). Suppliers that did not buy renewable electricity had to fulfill requirements by purchasing the required number of certificates at a government-set price.

Renewable producers met targets. Their output added about 5% to the electric bill. The Energy Efficiency Commitment (2002) required suppliers to help electric and gas customers reduce energy usage, with the goal of reducing greenhouse gas output by the equivalent of 0.4 million metric tons of carbon dioxide by 2005.

The *Energy White Paper* of 2003 foresaw the low carbon future required to combat climate change, a decline of domestic energy supplies and a need to update energy infrastructure. *The Stern Review: The Economics of Climate Change*, issued by the Chancellor of the Exchequer in 2006, analyzed the impact of climate change on the economy and proposed action to mitigate it. The government committed to a 15-18% reduction (from 1990 levels) of carbon dioxide emissions by 2010, a modest goal considering the reductions that had already taken place.

Legislation followed. The Climate Change Bill, published in 2007, bound the UK to reduce carbon dioxide emissions 26-32% below 1990 levels by 2020 and 60% by 2050. The Energy Bill of 2008 supported carbon capture and sequestration and nuclear decommissioning, and strengthened the Renewables Obligation. In 2009, the Chancellor proposed a carbon budget that, using 1990 as a base year, would reduce greenhouse gas emissions 32% by 2018-2022, and 80% by 2050. The former goal even looked plausible. Greenhouse gas emissions had already fallen 26% from 1990 to 2009.

Right before the electorate ousted Labor, OFGEM issued *Project Discovery*, which set a new direction for electricity policy. The February 2010 document emphasized the need for "secure and sustainable energy."[384] It made five principal points:

- The electricity industry has to replace old facilities with plants that emit little or no greenhouse gases, a project requiring "unprecedented levels of investment" at a time of "increased risk and uncertainty."[385]

- Unless investors see certainty in the pricing of carbon emissions, they will delay making decisions to build low- or no-carbon generation.

- Wholesale market price signals "do not fully reflect the value that customers place on … security" so generators lack the "incentives to make additional peak energy supplies available and to invest in peaking capacity."[386] (Economists called this the "missing money" problem.)

- International interdependence creates risk, not only of losing a transmission link, but also greater dependence on foreign gas, as the UK's own North Sea production declined.

- Electric and gas prices will rise, possibly becoming unaffordable to some customers, and will reduce the competitiveness of British industry.

The Big Six, OFGEM thought, would build new generation, possibly jointly, which boded ill for competition. The Big Six competed like investment managers who keep portfolios as similar as possible, so clients cannot fire them for underperforming. OFGEM said they "sought to benchmark … against each other in order to minimise the risks of their energy costs deviating materially from the average."[387] They would not raise their costs to meet government objectives unless all the others did. The government had to twist arms or make an offer too good to refuse.

OFGEM put in a plug for demand side response. Convince customers to reduce demand at peak times, as an alternative to serving them with expensive power. Privatization had sidelined those efforts. Competitive suppliers, after all, made for money selling more – not less – electricity.

Project Discovery even strayed from Thatcherite orthodoxy into the moral arena familiar to readers of Adam Smith's *The Theory of Moral Sentiments*. In the old days, electric firms strove to meet standards of public service – something inherent in the job. In the new era, they aimed to maximize their own profits, even at the expense of network users as a

---

384. OFGEM, *Project Discovery: Options for delivering secure and sustainable energy supplies*, Ref, 16/10, 3 Feb. 2010.
385. _____, *op. cit.*, p. 1.
386. _____, *op. cit.*, p. 2.
387. _____, *op. cit.*, p. 17.

whole. As OFGEM put it, "the reputational risk driver ... is diluted." [388] With moral compass replaced by market mechanism, the regulator must provide an enforceable directive or an economic incentive to assure that good things happen.

OFGEM cut through dogma, looked ahead and argued that the existing market would not deliver what the country required in a timely fashion. OFGEM had changed direction.

Conclusion

After the elimination of the Pool, prices fell. Then the Big Six oligopoly replaced the Big Two and prices rose. Utilities started to run out of the impressive operating efficiencies that made price cap regulation so attractive. Customers could choose their suppliers and enjoy all the protection and benefits that competition and light-handed regulation provided. Then the government focused on climate change and deviated from hands-off, market-based principles espoused from time of privatization. The regulator took note of a big defect in short term market pricing: it did not attract long term investment. On May 7, 2010, voters ejected Labor and brought back the Tories, the original advocates of market-based solutions, and made way for still another reorganization of the electricity industry.

---

388. _____, *op. cit.*, p. 21.

Chapter 20
The Coalition Changes the Course (2010-2015)

The electricity market must be designed for the desired generation mix of the future. [389]

– Vincent de Rivaz and Sam Laidlaw

David Cameron's Tories, in a Coalition with the Liberal-Democrats (Lib-Dems), promoted lower carbon emissions, reliable electricity supply, renewables and nuclear power and turned the electricity market upside down.

DECC Sets Policy

Green-leaning Chris Huhne (Lib-Dem) became Secretary of State for Energy and Climate Change. The Lib-Dem platform called for carbon-free electricity by 2050 and no new nuclear plants. The Tories opposed nuclear subsidies and were pale green at best.

The Coalition pledged to reduce "greenhouse gas emissions by at least 80% by 2050. " [390] The private sector, it argued, needed help to raise the funds needed to meet the goals.

With 16 of 17 operating nuclear plants scheduled to close by 2025, the Coalition would "aim to remove unnecessary barriers to new nuclear … without providing public subsidy." [391] Builders, a delusional DECC declared, could begin construction in 2013-2014 and have a plant running by 2018. [392] Three consortia announced plans for 15,600 MW of new capacity at five sites. Restarting the nuclear program would require "government support … public acceptability … regulatory certainty … market certainty." [393] But no government money. [394]

---

389. Vincent de Rivaz and Sam Laidlaw, "Far from being a drain on the public purse, new power plants will be a major boost to UK economy," *The Telegraph, http:// www.telegraph.co.uk/finance*, p. 2, 14 December 2010.

390. Department of Energy & Climate Change, *2050 Pathways Analysis*, July 2010, p. 1.

391. _____, op. cit., p. 167.

392. As an alternative to building new nuclear stations, the government could, where feasible, extend the operating lives of existing stations. In 2012, UK the government extended the operating lives of two plants. In the US, the Nuclear Regulatory Commission has extended the operating licenses of numerous reactors from expiration after 40 years of operation to 60 years, and some utilities have begun to discuss extensions to 80 years.

393. Department of Energy & Climate Change, *Pathways Analysis*, July 2010, p. 168.

394. By International Financial Standards, the UK government already had a big nuclear liability and "the epic task of nuclear decommissioning is … likely to cost the UK £61 billion…" (Adam Jones, "Tank drivers may think twice at UK plc's accounting revolution," *Financial Times*, June 28, 2011, p. 16.)

OFGEM Launches RIIO

In October, OFGEM unveiled RIIO ("Revenue = Incentives + Innovation + Output"). "If Britain's energy ... companies are to deliver the networks needed for a sustainable energy sector, the way we regulate them needs to change," and utilities must deliver "sustainable energy and long term value..." [395] OFGEM would set deliverables, lengthen the price control term to eight years (with mid period reviews) to foster long term thinking, allow third parties to own and operate large projects and create new incentives for efficiency and innovation. RIIO was:

> ... a price control framework that encourages network companies to deliver in response to commercial incentives with the potential to earn higher returns and face less intrusive regulatory scrutiny if they innovate and outperform.. [396]

> Utilities that did not deliver "... will see real ... downside, including below average returns ... and potential licence revocation." [397]

In December 2012, RIIO-T1 set National Grid's prices for 2013-2021. [398] "Under RIIO ... Companies are required to develop and submit well-justified business plans ..," [399] and "The objective of RIIO is to encourage network companies to play a full role in the delivery of a sustainable energy sector, and to do so in a way that delivers value for money..." [400]

The first case, RIIO-T1, for National Grid, set numerous objectives each with its own payoff. [401] Customers would share some savings "from the year they are made." [402] OFGEM would review deliverables annually and could reopen decisions. It calculated a cost of capital that incorporated a fixed return on equity and an interest rate tied to a lagged

---

395. OFGEM, "RIIO: A new way to regulate energy networks," Financial decision, October 2010, introductory section, no pagination.

396. _____, *op. cit.*, p. 4.

397. _____, *op. cit.*, p. 11.

398. OFGEM, *RIIO-T1: Final Proposals for National Grid Electricity Transmission and National Grid Gas, Final Decision – Overview document, 17 December 2012.*

399. _____, *op. cit.*, "Overview," first page (no numbering).

400. _____, *op. cit.*, p. 1.

401. Setting numerous specific objectives moves regulation farther from the light-handed concept that the regulator should set general goals and management should figure out what to do to reach the goals. As Lou Gerstner, the man who saved IBM famously wrote, ""People do what you inspect, not what you expect." (*Who Said Elephants Can't Dance*, 2002). Utility managers will seek to meet all goals as ordered by RIIO. Doing so becomes the new *modus operandi* for the rational manager. Whether doing so provides better or more efficient service to the consumer depends on how well the regulator has set the requirements. If customers want something different or managers can come up with better ideas matters less than doing what the inspector wants them to do.

402. _____, *op. cit.*, p. 14.

bond index. [403] The order projected that NG would maintain an investment grade on its bonds and earn an after tax real return on equity of 7% (about 10% on nominal basis, in line with US utility returns). But a broker wrote that NG could "achieve c 13-14% post-tax nominal equity returns in the UK." [404] Prices would rise but OFGEM's overview did not spell out how much (8% per year looked likely from the numbers).

RIIO effectively buried RPI-X. Yet, if it attracted capital and forced utilities to promptly share savings with customers, would anyone mourn the decease of RPI-X?

Generating Market Reform

Late in 2010, chief executives of E.ON UK, EDF Energy, Centrica and Scottish and Southern told the *Sunday Telegraph* that they wanted the government to "finally force up the price of electricity enough to make low-carbon nuclear power achievable." [405] The "industry needs to find ways of attracting more capital." [406]

Chris Huhne opposed a subsidy but £40 billion worth of new nuclear stations would not materialize without one. Citigroup's Peter Atherton presciently commented: "The Government's definition of a subsidy is literally a bag of cash delivered ... to each nuclear power plant ... What's going to happen will be an economic transfer of risk from company to consumer. Of course it's a subsidy." [407] Wholesale power sold for 5p/kWh, break even for new nuclear plants was 6p/kWh but they needed 8 p/kWh "to be attractive for companies to make the start..." [408] However, as one green energy investor conceded, nuclear costs "... still compare very favorably with renewables." [409]

---

403. Previously, regulated utilities had devised strategies to appropriate the benefits of interest rate changes for their shareholders. The RIIO-T1 order forced them to pass on benefits, with a delay, to consumers. Considering that the order came out at a time of extraordinarily low interest costs and that most observers expected interest rates to rise, this innovation looked as if might benefit the company more than the consumer, assuming that OFGEM did not find offsets elsewhere in the RIIO framework to prevent that from happening. There was no economic logic, though, to varying the interest component of cost of capital but not the equity component, because the cost of equity would vary with interest rates as well. The formulation seemed a compromise designed to assure that OFGEM was half right no matter the direction of interest rates and cost of capital.
404. Mark Freshney, Guy MacKenzie and Vincent Gilles, "National Grid," Credit Suisse, 20 March 2013, p. 4.
405. Rowena Mason, "Britain's power chiefs reveal nuclear blueprint," 13 November 2010, http:// www.telegraph.uk/ finance, p.1.
406. _____, *op. cit.*, p. 3.
407. _____, *op. cit.*, p. 2.
408. *Ibid.*
409. _____, *op. cit.*, p. 4.

Nuclear builders required "the certainty of fixed waste fees."[410] A few days after the *Sunday Telegraph* interview, the government capped fees per plant at £1 billion, supposedly three times the likely costs (this after the government contributed £4 billion to previously unfunded nuclear waste liabilities and excused generators from contributing to the waste disposal facility.) Chris Huhne proclaimed "The biggest energy shakeup in 25 years: We need to unlock private investment on an unprecedented scale and ensure the low-carbon revolution at the lowest cost to consumers."[411] Where the market failed DECC would step in.

The *Electricity Market Reform Consultation* of December 2010 predicted that electricity sales would double by 2050, one quarter of generating plants would close by 2030 and renewable energy would furnish 30% of power by 2020. Electric companies would have to spend £120 billion through 2020, twice as much as in the previous decade. "Without reform, the existing market will not deliver the … long-term investment, at the price we need."[412] The report did not specify cost but a 5-10% per year addition to the electric bill looked like a good estimate.

The consultation paper proposed:

- Taxing carbon emissions to reduce the profitability of burning high carbon fuels.
- Contracts for differences to provide stable prices for low carbon generators.
- Capacity payments to induce generators to install and operate plant required at peak periods.
- Emissions performance standards for new power stations.

Those proposals would head off "looming" difficulties. Without action, capacity margin would fall from 20% to 10%. The UK, depending more on intermittent renewables that require backup for when they go off line, needed higher – not lower – capacity margins. Meeting emission goals required a "largely decarbonised" generating sector by the 2030s.[413]

No-carbon generators, with their high capital (expensive to build) and low operating (no fuel) costs, are "not well suited to the UK market where gas is the marginal plant."[414] Gas plants, with low capital and high operating costs, set market prices, which vary with the price of gas. Fluctuating market prices, creating unpredictable and possibly mini-

---

410. Rowena Mason, "UK taxpayers face unlimited nuclear waste bills if costs spiral," *http://telegraph.co.uk/finance*, 8 Dec. 2010, p. 2.

411. Chris Huhne, "The biggest energy market shake-up in 25 years,"16 Dec. 2010, *http: // www.telegraph.co.uk/ finance*, p. 1.

412. Department of Energy and Climate Change, *Electricity Market Reform Consultation Document*, The Stationery Office, December 2010, p. 5.

413. _____, *op. cit.*, p. 16.

414. _____, *op. cit.*, p. 24.

mal profit margins at non-gas plants would scare off investors. [415] The government would have to fix that.

A stock broker described the policy as "positive for UK renewables and clean power generators and bad for dirtier operators..."[416] The government planned a price floor for carbon emissions that would increase over time and a UK Green Investment Bank.

The Carbon Price Floor, effective in April 2013, taxed carbon emissions at the sum of the EU trading price for carbon permits plus a variable tax to equal £16/metric ton initially, £30 in 2020 and £70 in 2030. Industrial customers objected. Environmentalists said the floor would raise prices without reducing emissions because big carbon emitters would move abroad and emit the same amount as before. Coal prospects looked bleak, the biggest coal-fired plant switched from coal to wood pellets and in July the largest coal miner went bankrupt.

In June 2011, DECC issued a six part national policy statement. The country had to meet emission targets. Market prices did not attract necessary investment. Out of 85 GW of existing capacity, 22 GW would close by 2020 (12 GW due to environmental regulations). The UK had to "ensure that developers deliver the required levels of investment..."[417] The government would consider "further interventions in energy markets ... to ensure that developers come forward..."[418] and help consumers cut usage.

Decarbonization, DECC claimed, would increase electricity demand, not a problem because the UK had vast renewable resources (40% of Europe's wind), but all those windmills would change network flows and require more transmission lines. Renewables would account for 33 GW of the 59 GW of new capacity needed through 2025, but they "are not capable on their own of meeting our future needs ..." in part "because of their inherent intermittency."[419]

---

415. How does gas set price? As an example, gas generation fuel costs 8p per kWh generated. The gas generator bids to sell at 10p per kWh. If gas price falls to 2p, the generator bids to sell at 4p per kWh. Either way, it makes an operating margin of 2p per kWh to cover fixed costs and profit. (Admittedly, in a perfectly competitive market, generators could not operate that way, but when a half dozen of them know what the others are doing, they can set prices in a non-competitive manner.) The windmill has no fuel costs but requires 8p per kWh to cover fixed costs and earn a profit. With gas price at 8p, the gas generators set price at 10p, the wind generator sell its output at 10p, and collects a 10p margin for fixed costs and profit. With gas price at 2p, gas generators set price at 4p, and the windmill's operating margin falls to 4p.— less than its fixed costs and required profit. (The same analysis applies to a nuclear plant with its low fuel cost. It has to earn a high margin to cover high fixed costs and provide a profit.) The windmill's unpredictable margins raises its cost of capital. The market price may not reach a level high enough to provide that return. Investors may require a non -market assurance of profitability.
416. Alan D. Wells, CFA, *et, al.*, "European Utilities: UK Renewable Energy," Morgan Stanley, April 4, 2011, p. 1.
417. Department of Energy and Climate Change, *Overarching National Policy Statement for Energy*, Version for Approval (EN-1), June 2011, p. 12.
418. _____, *op. cit.*, p. 13.
419. Department of Energy and Climate Change, *National Policy Statement for Nuclear Power Generation* (EN-6), Volume II, June 2011, p. 4.

"The Government believes that energy companies should have the option of investing in new nuclear power stations "[420] although it "will need to be satisfied that effective arrangements exist or will exist to manage and dispose of the waste…, "[421] an odd comment because "… the Government recognises that it has a responsibility to deal with long-term higher activity waste management…"[422] No breaks for nuclear, let the market decide.

The policy statement, predictably, came up with the same four solutions. Shortly thereafter, in July, DECC issued a three report roadmap for the electricity market.

New Role for OFGEM

The *Ofgem Review* featured the usual bromides. By "promoting competition" in generation and retail supply and "replicating competition" in the network, "economic regulation" delivered greater efficiencies, lower consumer prices, and investment in energy infrastructure and services. OFGEM, however, admitted that existing regulation might not be "capable of meeting the challenges of the future."[423]

Government sets goals, "policy outcomes" that OFGEM must deliver.[424] Suppliers must meet those "secure., sustainable and affordable" goals.[425] And "… even where competition is the best approach, it may produce imperfect outcomes … Nor will markets necessarily deliver in line with wider public interest objectives… and regulatory intervention … can correct these failures."[426]

OFGEM's had its marching orders: reinforce the grid, create a competitive transmission network for offshore wind power and improve competition in wholesale markets, monitor consumer markets and encourage installation of smart meters to help customers curb usage at peak periods. Smart meters in all homes would collect timely price information and give customers greater control of usage so "It will be paramount to keep consumers engaged in the market."[427] That is, make customers pay attention, this time around, or the meters will do them little good. Finally, to emphasize who runs the show, the DECC expressed "confidence that the regulator's decisions would be aligned with the Government's strategic policy framework."[428]

---

420. _____, *op. cit.*, Volume I,p. 1.
421. _____, *op. cit.*, Volume II, p. 13.
422. _____, *op. cit.,*, Volume. II, p. 16.
423. Department of Energy and Climate Change, *Ofgem Review Final Report*, July 2011, p. 5.
424. _____, *op. cit.*, p. 6.
425. _____, *op. cit.*, p. 8.
426. _____, *op. cit.*, p. 9.
427. _____, *op. cit.*, p. 14.
428. _____, *op. cit.*, p. 17.

A Roadmap and More

DECC's *Renewable Energy Roadmap* boasted of "… the best wind, wave and tidal resources in Europe … Our challenge is to bring costs down and deployment up."[429] The Green Investment Bank would support renewables and the Green Deal reduce energy consumption. Renewable generation would quadruple between 2010 and 2020. Renewables, though, cost more than the best conventional alternative, the combined cycle gas turbine (CCGT). (See Table 20-1.)

Table 20-1. Levelized Cost per kWh (2010 p/kWh)

	Power Plant Going into Service in	
	2010	2020
Offshore wind	14.9-19.1	10.2-17.6
Onshore wind	7.5-12.7	7.1-12.2
Biomass	9.4-16.5	9.3-15.6
Combined cycle gas turbine	7.6-7.9	8.7-9.1

Note:
Levelized cost is average cost over the lifetime of the power plant.

Source:
DECC, *UK Renewable Energy Roadmap*, July 2011, p. 18.

Planning our electric future: a White Paper for secure, affordable and low-carbon electricity predicted that the reforms would reduce electricity costs in 2010-2030 by £9 billion and the average annual residential electric bill would rise from £485 to £642 with the reforms in place and to £682 without them (all in 2009 prices).

The Energy Bill

Chris Huhne, faced with a charge of perverting the course of justice over a traffic offense, resigned from DECC in February 2012. Ed Davey, Lib-Dem Under-Secretary of State for Employment Relations, Consumer and Postal Affairs at the Department for Business, Innovation and Skills, took over. He steered the Draft Energy Bill into Parliament in May 2012 with aspirational language ("to deliver secure energy … and drive ambitious goals"), threat ("blackouts could become a feature of daily life") and doubt that the market could "incentivise this investment efficiently."[430] The Energy Act 2013 became law in December 2013.

---

429. _____, *UK Renewable Energy Roadmap*, July 2011, p. 3.
430. H.M. Government, *Draft Energy Bill*, May 2012 (London: UK Stationery Office, 2012), p. 5.

Economic liberals derided the bill. *The Economist*, denounced "fundamentally flawed" policy and argued that "renewable energy is a means to many worthwhile ends ... But too often it becomes an end in itself ... as it has in Britain."[431] Despite claims that the reform would achieve its goals "while minimising costs to the consumer,"[432] the bill looked more like a boost for nuclear and renewable energy than a plan to economically reduce carbon dioxide emissions. The *Financial Times* saw "a Gosplan-like reabsorption of the market into the belly of Whitehall" as opposed to the "liberalized model that has assured ... low prices ... for 20 years."[433]

Dieter Helm described the bill as "opening the flood gates to lobbyists in search of subsidies" and a "complex morass."[434]

The chemical industry attacked the bill. It would make UK industry less competitive. Spokesman Alan Eastwood declared "the whole approach smacks of the medieval religious conviction that self-flagellation is the route to salvation."[435] The government caved, omitted a carbon tax from the bill and excused industry from decarbonization costs. Complaining helps.

Alex Henney decried the technical ineptitude of civil servants: "Their shortcomings are exacerbated by the frequency with which they (and ministers) change jobs; their naive marketism; and the interest some have in greenness at the expense of reality and cost."[436] As an example, he estimated the capital cost of an unregulated UK nuclear project at 11.2% vs 7.8% for a regulated US project. That difference would raise the price of power from 5.6p to 8.0p. British consumers, then, will pay more because "The DECC has seemingly rejected" the regulated approach to develop nuclear power "because it clings to naive delusions that this development is in unexplained ways market related."[437]

Chiara De Bo and Massimo Florio, Italian economists, flagged the bill as a complete turnaround in thinking, the realization that "existing arrangements do not provide the right price signals " to attract investment for new generating capacity, so the government will have to play an active role in attracting capital, "a role that deeply alters the previous

431. "Poles Apart, *The Economist*, July 14, 2011, p.14.
432. Department of Energy and Climate Change, *Planning our electric future: a White Paper for secure, affordable and low-carbon electricity*, July 2011, p. 15.
433. "UK Energy Reform," *Financial Times*, May 26/27, 2012, p. 8.
434. Dieter Helm, "Energy proposals losing their spark," *Financial Times*, June 1, 2012, p. 8 Those were prescient words. Ed Davy, after leaving office, commented that "The amount of lobbying that goes on is tremendous, from every single part of the energy industry." Not exactly a surprise. A small energy firm argued that "success in gaining a CFD has been based on a company's ability to lobby, not an ability to compete in the market. This favours larger generators." That comment sounds more like the truth than sour grapes. (Quoted in Ashley Thomas, "Utilities," Societe Generale, 27 May 2015, p. 6.
435. James Shotter, "Proposed carbon floor sparks chemical outcry," *Financial Times*, June 25, 2012, p. 16.
436. Alex Henney, "The Collapse of the Government's Electric Generation Policies," 11 May 2012, p. 3.
437. _____, *op. cit.*, p. 14.

market-based approach."[438] Perhaps, they said, the EU's energy policy (based on the old UK model) was also "inadequate to solve long term investment challenges."[439] The Energy Act, then, heralded a change in belief about what markets could accomplish, a theological conversion of sorts.

DECC scaled hyperbolic heights to extol the bill as "Essential ... to power low-carbon economic growth, to protect consumers, and to keep the lights on," noting that expected electric industry spending "dwarfs the investment needed for the Olympics," and "will stimulate supply chains and support jobs in every part of the country, capitalising on our engineering prowess and our natural resources, cementing the UK's place at the forefront of clean energy development."[440]

The law was "An Act to make provision for the setting of a decarbonisation range... for ... reforming the electricity market for the purposes of encouraging low carbon electricity generation or ensuring security of supply..."[441]

The Secretary would set a target for electric generation's carbon intensity from 2030 onwards, taking into account "scientific knowledge... technology ... economic circumstances ... impact on ... competitiveness ... fiscal circumstances... social circumstances... the structure of the energy market... difference in circumstances between England, Wales and Scotland... ," international circumstances and "duties ... under ... the Climate Change Act 2008..."[442]

Electricity Market Reform (EMR) would deliver the low carbon, secure electric future.

Contracts for Differences (CFDs), the key part of EMR, would encourage "electricity generation which in the opinion of the Secretary of State will contribute to a reduction in emission of greenhouse gases."[443] The Secretary would issue CFDs with set strike prices to selected low carbon generators. All suppliers had to buy those CFDs. An intermediary, the counterparty, would assure compliance with terms. As with other CFDs, the buyer pays the low carbon generator the difference when market price falls below strike price, the low carbon generator pays the buyer the difference when market exceeds strike price.

---

438. Chiara Del Bo and Massimo Florio, "Electricity Investment: An Evaluation of the New British Energy Policy and Its Implication for the European Union," March 20, 2012, p. 2.

439. *Ibid.*

440. Department of Energy & Climate Change, "An Energy Bill to Power Low-Carbon Economic Growth, Protect Consumers and Keep the Lights On," Press Notice 2012/151, 29 November 2012, p. 1.

441. *Energy Act 2013*, 18th December 2013, p. 1.

442. *Energy Bill* (HC 135), pp. 2-3.

443. *Energy Bill* (HC Bill 100), p. 2.

Initially, the Secretary could allocate CFDs and bilaterally negotiate with generators planning "bespoke" projects for which "generic terms are unsuitable." [444] He could issue 15 year contracts and set different strike prices for different resources, keeping in mind the need to "maintain a level playing field." [445] The Secretary could turn away applications once the cost of the CFD portfolio became too high. [446] Whether consumers got "the best deals" [447] depended on the Secretary's judgment.

DECC predicted that suppliers would pass on CFD costs to the consumer because the consumer is "the ultimate beneficiary of the transition to a low carbon electricity system ..." [448]

Until the CFD program began, the Secretary could sign a contract ("Investment Instrument") to advance construction of a low carbon generator in case "there is a significant risk that the low carbon energy generation ... will not occur or will be ... delayed..." [449]

To choose and pay for capacity needed to assure reliability, the law created the Capacity Market. DECC would forecast capacity needs. The System Operator (National Grid) would hold periodic auctions for the purpose of "providing electricity or reducing demand for electricity." [450] National Grid would select the best proposals from generators or demand side managers to "deliver energy in periods of system stress." [451] Existing plants could secure one-year contracts, refurbished ones up to three years and new plants longer. All suppliers would pay for the contracts and could pass on costs to customers.

The Emission Performance Standard (EPS) prohibited new power stations operating under base load conditions from emitting more than 450 g of carbon dioxide per kWh (40% lower than existing coal stations), a standard that no new conventional coal station could meet. DECC could suspend the standard when "there is an electricity shortfall, or a significant risk of an electricity shortfall." [452]

Because the UK lacks a liquid wholesale market in which participants can "quickly and easily buy or sell power at a price that reflects supply and demand fundamentals," [453] the

---

444. Department of Energy & Climate Change, *Supplementary Memorandum to Delegated Powers and Regulatory Reform Committee on Part 2 (electricity market reform) of the Energy Bill*, 17 June 2013, p. 3.
445. H. M. Government, *Draft Energy Bill*, p. 28.
446. Given that renewable costs were declining rapidly, this policy seemed designed to fill the quota with high cost renewables and blocking off markets for later, lower cost renewables.
447. _____, *op. cit.*, p. 12.
448. Department of Energy & Climate Change, *op. cit.*, p. 6.
449. H..M. Government, *op. cit.*, p. 70.
450. *Energy Bill* (HC 135), p. 13.
451. Department of Energy & Climate Change, *op. cit.*, p. 51.
452. *Energy Bill* (HC 100), p. 40.
453. H.M. Government, *Draft Energy Bill,*, p. 34.

Secretary, can modify licenses to facilitate "participation in the wholesale electricity market..." and to promote "liquidity in the market."[454]

To promote "competition in domestic supply," the bill requires" licence holders to change the domestic tariff or other terms ... so as to reduce the costs to ... customers..."[455] and it encourages the Secretary to promote "clear and easy to understand" tariffs.[456]

The law established the Office of Nuclear Regulation (ONR) to oversee nuclear health and safety, defined as "protecting persons against risks of harm from ionising radiations from ... nuclear sites"[457] all under "appropriate Secretary of State controls ..."[458] Unlike its US counterpart, the ONR reports to a politician rather than holding status as an independent agency and does not oversee financial qualifications of nuclear owners, thereby perpetuating the notion that nuclear power plants are ordinary commercial ventures.

"The Secretary of State will set the outcomes... Ofgem ... will ... act in the manner best calculated to further the delivery of the policy outcomes..."[459] DECC will revise its strategy periodically to bring "greater clarity and coherence to the roles of both Government and Regulator"[460] and issue a Strategy and Policy Statement to set "strategic priorities"[461] no less than once every five years.

Just before introducing the Energy Bill, Ed Davie presented a tariff simplification plan. Suppliers would provide four schedules (two standardized and two based on special criteria) and indicate the cheapest rate. The proposal did not guarantee lower prices, just a less confusing price list. Suppliers could always reduce price discrepancies by jacking up the cheapest rates.

Hard Times and Bad Timing

Coalition energy policy raised electric bills in the midst of the Great Recession. OFGEM warned that prices would rise with or without the new policy. It tried, in the fall of 2011, to cap prices. It threatened to investigate the Big Six, simplify tariffs and require generators to auction part of their output to new competitors. Cabinet members had mixed feelings. Energy Secretary Chris Huhne wanted action, but Chancellor George Osborne charac-

---

454. *Energy Bill* (HC 100), p. 20.
455. *Energy Bill* (HC 135), p. 92.
456. *Energy Act 2013*, p. 109.
457. *Energy Bill* (HC 100), p. 48.
458. *Draft Energy Bill*, p. 47.
459. _____, *op. cit.*, p. 41.
460. _____, *op. cit.*, pp. 12-13.
461. _____, *op. cit.*, p. 93.

terized any plan to lower UK carbon emissions before European competitors acted as economically suicidal.

The December 2011 Carbon Plan limply began: "Even in these tough times, moving to a low-carbon economy is the right thing to do."[462] The financial community said "new nuclear is simply too expensive."[463] The Tories still favored nuclear power. But not natural gas. Burning it produced carbon dioxide, drilling for it caused earthquakes, gas prices were unstable and its suppliers unreliable. They had a shale gas revolution in the United States. They would not have one in the United Kingdom.

The Carbon Plan's best case scenario to 2050 projected a drastic shift away from fossil fuels (Table 20-2). Higher cost plants would replace existing ones. MARKAL, the plan's number crunching model, used point estimates and assumed perfect foresight, forward looking and rational consumers and perfectly competitive markets – not exactly real world conditions. The Coalition's 40 year plan looked like an aspirational framework.

Table 20-2. Carbon Plan Best Case

	2010 [1]	2050 [2]
Generating capacity (GW)		
Fossil	70.8	28
Nuclear	10.9	33
Renewable and other	8.5	45
Total	90.2	106
Generation (GWh)		
Fossil	283	60
Nuclear	62	210
Renewable and other	36	220
Total	381	490
Net electricity supplied (GWh)	361	470

Notes
Author's estimates for generation in 2050.
All 2050 fossil fueled units use carbon capture and sequestration processes.

---

462. Department of Environment and Climate Change, *The Carbon Plan: Delivering our low carbon future, Presented to Parliament pursuant to Sections 12 and 14 of the Climate Change Act of 2008*, December 2011, p. 1.
463. Catherine Airlie and Matthew Carr, "U.K.'s Cheapest Way to 2050 Carbon Goal Would Triple Nuclear," *Bloomberg News*, Dec. 1, 2011, 12:07 PM EST.

Sources

2010 data from DECC, *Digest of UK Energy Statistics 2011*, pp. 138, 141, 143, 144.

2050 data from DECC, *The Carbon Plan*, 2011, pp. 19, 122.

Nasty Politics and Consumer Backlash

Late in 2013, energy prices rose. Labor leader Ed Milliband promised in September that if he won the May 2015 election he would freeze domestic gas and electric prices for 20 months. John Major, former Tory prime minister, called for a tax to capture windfall profits that energy companies would earn in the coming winter and use the money to aid the poor. Industrial customers demanded an end to the carbon floor. Then, according to the press, in November prime minister David Cameron pledged to roll back green taxes and told his minions to "get rid of the green crap" that drove up energy bills. [464] Story denied, but somebody got a message out.

Alistair Buchanan departed OFGEM in June 2013. Andrew Wright, a senior staffer took over on an interim basis. News reports said that "the regulator's new leadership is determined to get tougher with the industry in the face of soaring fuel bills and mounting public anger." [465] Wright took himself out of the running for the top job a few days after a parliamentary committee berated OFGEM as a "toothless tiger" and he admitted, "Consumers have a perception that the market is not working well and that's something we agree with." [466] OFGEM selected Dermot Nolan, a Yale educated economist and Irish energy regulator as its new chief executive. He took over in March 2014, a month when OFGEM's latest limp motto, "Making a positive difference **for energy consumers**," looked as if it needed a revision.

In March 2014, OFGEM, the Competition and Markets Authority and the Office of Fair Trading issued a "State of the Market Assessment" that found "low levels of customer satisfaction," [467] "**weak competition**" and "**possible tacit coordination**." [468] The average customer, it said, could save £100 a year by switching to the best tariff of a competitor, without explaining how the confused consumer could penetrate the thicket of fixed vs variable rates, tariff expiration dates, locational issues and exit fees to achieve the savings. (See Table 20-3.)

---

464. Rowena Mason, "David Cameron at centre of 'get rid of all the green crap' storm," 21 November, 2013, *theguardian.com*.

465. Terry Macalister, "Electricity networks told to reduce costs by regulator', 22 November 2013, *theguardian.com*.

466. Jennifer Rankin, "Ofgem's acting chief executive pulls out of race for top job," 29 November 2013, *theguardian.com*.

467. OFGEM, "State of Market Assessment," 27 March 2014, p. 10.

468. _____, *op. cit.*, p. 11.

Table 20-3. Comparable Household Bills (2014)(£)

	Average	High	Low
Ten best fixed rate deals	1,017	1,045	991
Ten best variable price deals	1,137	1,176	1,058
Big Six representative bill	1,412	1,491	1,384

Notes:

Standard consumption package.

Source:

Kate Palmer, "Latest predictions: how to beat energy bill rises," Telegraph.co.uk, 20 August 2014.

Despite admitting that the Big Six earned "slightly below the estimated cost of capital,"[469] the study bafflingly could not "conclude whether these profits are excessive.[470] The report belatedly also noticed that "for a given change in wholesale costs, retail prices respond quicker when costs increase than when they fall. This could provide an indication that the market is not strongly competitive."[471]

In June, OFGEM referred the energy market to the Competition and Markets Authority for investigation, with results expected conveniently after the 2015 election. OFGEM's new chief executive said that the upcoming study would "clear the air… rebuild consumer trust … and certainty investors have called for."[472]

OFGEM kept up the pressure. The draft of its eight-year price plan for the distribution networks required first-year price reductions of roughly 15%, followed by small annual price changes. Although OFGEM used a 6.0% real after tax return on equity to set prices, its "plausible ranges"[473] showed the companies earning 11% on average if they met operating targets. (On the company's books the 6% translates into a nominal return of over 10% and the 11% to over 20%.) The draft also called for an 8% cut in capital and operating expenditure budgets which the companies could overcome, OFGEM said, by employing smart grid technology. OFGEM proudly declared that it had knocked £12 off the average annual household bill.

Late in the summer of 2014, the Big Six agreed to return £153 million of unclaimed deposits to customers who had switched suppliers. The heat was on. Show respect for customers!

---

469. _____, *op. cit.*, p. 6.

470. _____, *op. cit.*, p. 7.

471. _____, *op. cit.*, "Supplementary Appendices," p. 7.

472. OFGEM, press release, "Ofgem refers the energy market for a competition investigation," June 26, 2014.

473. OFGEM, "RIIO-ED1 Draft determinations for the slow-track electricity distribution companies," 30 July 2014, p. 44.

By December Britons read warnings that the lights might go out, reserve margins fall to 2% by late 2015, coal fired stations close due to European environmental rules, and National Grid, the system operator, would pay owners of mothballed gas fired stations to put the units back in service and industrial users to get off the grid during peak periods.[474] At year end, NG held the first auction to sign up capacity.[475] The price came in at half the expected level. Experts worried that generator owners would close down plants and worsen the capacity picture.

Fuel prices collapsed in early 2015. Politicians demanded that electricity prices reflect those lower costs. In February, DECC launched "Power to Switch," telling consumers to reach for £2.7 billion in potential savings by ditching their current suppliers in favor of lower priced competitors.[476] (Low-priced independent suppliers had about 10% of the market.) The Big Six could have lowered retail prices to reflect lower costs, but a key part of their business model was to make good margins in supply when they made poor margins in generation. If they only could make poor margins, why would they or anybody else build the power plants the government claimed were needed? With a CC investigation underway, a critical report issued by Parliament and both major parties bashing energy firms almost as vigorously as bankers, it looked as if the economically liberal framework for the energy business had broken down beyond repair.[477] Power plant builders in the future would require guarantees, CFDs or whatever the government came up with to go forward with projects. The brouhaha over margins demonstrated the imprudence of any other strategy.

This cast of uncertain politicians and changing regulators sent mixed signals. Previously they aimed to decarbonize, increase security and attract capital. They shelved those con-

---

474. Mark Johnson, "Keeping the lights on," *The Economist*, The World in 2015, p. 108.
475. The two capacity auctions held through September 2016 priced capacity below expectations. Neither auction led to a commitment to build new capacity. Bidders simply declared existing units available for capacity service. The auctions each added about £1 billion (2%) to the average electric bill. With reserve margins predicted to be well under 10% through 2020, the industry needed new generating facilities, not reclassification of existing plant to make an extra profit. The auction looked more like a bailout of under utilized plant than an inducement to add capacity.
476. Kevin Burke, "British Households could save £200 a year by switching energy supplier," energybiz, March 15, 2015, energybiz.com.
477. The Energy and Climate Committee's March 2015 report was a model of polite criticism. It argued that the government had put too much emphasis on adding generating capacity rather than demand side response and that policy raised customer costs. The committee raised the issue of conflict of interest in National Grid's roll as advisor on the need for capacity, operator of transmission and owner of lines that would carry the generation from the new capacity. Most of the capacity contracts, it noted, went to existing (generally fossil-fueled) stations, so the capacity market, by promoting fossil-fuel generation, negates the benefits of the CFDs that aim to promote renewables. By agreeing to so many renewable contracts early on, the government may have locked out better deals for consumers in the future. (See fn 477.) Finally, small market players have difficulty operating in the new market. But, the committee concluded, the government rolled out the overall program smoothly, but needed to work on the parts. House of Commons, Energy and Climate Change Committee, *Implementation of Electricity Market Reform*, eighth report of Section 2014-25, HC 684, 4 March 2015.

victions when electric bills rose. They really signaled that no matter a government's leanings, it keeps hands off energy markets only when prices decline.

Nuclear Morass

The nuclear saga continued:

Three consortia planned nuclear stations at existing generating sites:

- EDF/ Centrica – The two owners of British Energy planned 6.4 GW of new capacity at Hinkley Point (units C and D) and Sizewell, using French state-owned Areva's technology.
- NuGen – French generator GDF Suez (renamed Engie in 2015), Spanish utility Iberdrola (owner Scottish Power) and SSE planned 3.6 GW at Sellafield.
- Horizon – German utilities E.ON and RWE planned 6 GW at Wylfa and Oldbury.

GDF Suez was the world's largest generator and EDF the world's biggest nuclear operator. A good start. Then in 2011, SSE exited NuGen, deciding to invest in renewables instead. In March 2012, E.ON and RWE dropped out of Horizon because Germany's decision to prematurely close nuclear stations in the wake of Japan's Fukushima disaster would slash the cash flows they required to undertake the project. In December, news leaked out that Centrica would not exercise options for 20% of Hinkley Point and Sizewell.

Horizon attracted Russian, French, Chinese and Japanese attention. Finally, in November, 2012, Hitachi, a Japanese firm, bought it. In early 2014, another Japanese giant, Toshiba, took over NuGen after the government announced its subsidy deal with EDF.

EDF had cash demands back home. It needed partners for Hinkley Point and chose a Chinese state company. "Energy security" takes on a new meaning when foreign state-owned firms control the nation's entire nuclear sector.

Hinkley Point's price tag rose from £4.5 billion to £7 billion per unit in May 2012. A City analyst said the plant required 16.5 p per kWh sold to make a profit while EDF said no more than 14.0p. Wholesale power at the time sold for 5.0 p.

Southern Company's Vogtle nuclear construction project, provides an instructive comparison. At the time, it looked as if Vogtle at completion could cover all costs plus profit by charging 7.8P/kWh while EDF required 14.0p. to do the same. If both projects cost roughly the same and operated similarly, only cost of capital could account for the difference. Did EDF require a return on investment (pretax operating income) two-and-a-half times higher than Southern? Did operating in Britain rather than Georgia (USA) demand such a high risk premium? Or were all nuclear cost estimates sheer fantasy? (See Table 20-4.)

Table 20-4. Nuclear Power Cost Estimates (p/kWh)

	Southern	EDF
	[1]	[2]
Fuel	0.6	0.6
Operations and maintenance	0.8	0.8
Depreciation	1.4	1.4
Miscellaneous taxation	0.6	0.6
Total costs of production	3.4	3.4
Pretax operating income	4.4	10.6
Total price	7.8	14.0

Notes

1. Conversion rate of £1.00 = US $1.60.
2. Southern refers to its Plant Vogtle 3&4 projects under construction, with total construction cost raised by 10% to EDF estimate for Hinkley Point project, on assumption that Southern estimate is too low.
3. EDF refers to Hinkley Point project awaiting approval. Assumes same costs of production as Vogtle.
4. Forty-year depreciation.

Sources

1. Southern Co. financial reports.
2. Seth Borin, Todd Levin and Valerie M. Thomas, "Estimates of the Cost of New Electricity Generation in the South," Working paper # 54, Georgia Tech School of Public Policy, March 26, 2010.
3. Max Chang, David White, Ezra Hausman, Nicole Hughers and Bruce Biewald, "Big Risks, Better Alternatives," Synapse Energy and Economics, October 6, 2011.

To do a politically acceptable deal EDF would have to stand down from 14.0p and the government give something in return, preferably something without an easily calculable cost. In October 2013, EDF upped the price tag for the two units to £16 billion (in 2012 prices) and delayed the operating date to 2023. It agreed to sell power at 9.25p per kWh, indexed for inflation, with adjustments allowed for a windfall profit tax, changes in law and operating costs for 35 years. (Wholesale prices at the time were roughly 6p per kWh.) EDF would take construction cost risk. Other terms were not made public.[478] The Treasury would guarantee (for a fee) £10 billion of project loans. Nuclear builders waiting in the wings liked the deal.

---

478. It was subsequently revealed that the government had given EDF guarantees covering nuclear waste disposal and insurance, allowed a delay in opening to 2033 and assured EDF of a payment of £22 billion if the UK shut the plant down before 2060. (Damian Carrington, "Hinkley Point C nuclear deal contains £22 bn 'poison pill' for taxpayer," *theguardian*, 18 March 2016.)

The government would sign if the "deal is fair, affordable and value for money.. and in line with the ... policy of no public subsidy for new nuclear where similar support is not available to other forms of low carbon generation."[479] The deal would produce jobs, too, the press releases said, and the nuclear program would reduce electric bills by 2030.

Critics objected to price indexation for a facility whose costs were largely fixed. Environmentalists argued that more nuclear power in the UK would depress coal prices and thereby encourage more coal burning elsewhere, so net carbon emissions would not decline.

The European Union modified the terms in October 2014, raised the Treasury's loan guarantee fee and required the project to share benefits with consumers if over the 60 year operating life return exceeded or cost fell below expectations. The price tag rose to £24.5 billion including inflation and interest costs (£34 billion including all capital costs). The government claimed that nuclear would cost less than other decarbonization measures. Highly touted giant projects always seem to cost less than alternatives until the construction begins.

Subsequently, EDF agreed to take over Areva's troubled nuclear engineering division and searched for more partners. The UK's Chancellor of the Exchequer went to China to encourage more Chinese investment. EDF's CFO resigned, reportedly due to an internal dispute over whether EDF could afford the project. EDF's board voted to go ahead in mid 2016 after a contentious, split vote. Then the new prime minister, Theresa May, put off a decision, to the great consternation of the Chinese government, which had plans to build reactors throughout Britain, after which she gave the project a go-ahead.

The EDF design had not been put into service anywhere. Two other projects of this design were over budget and behind schedule. Hinkley Point looked like a high stakes pursuit from which political sponsors could not find a graceful exit, a nuclear Moby Dick, perhaps.

Originally, EDF claimed the project would earn a 10% on capital. If EDF could borrow 65% of funds at 3.5%, shareholders could earn at least 22%.[480] Cost overruns, though, could erode or eliminate profitability and cost overruns were endemic to nuclear construction. Only State controlled entities that put non-financial objectives ahead of risk adjusted return on investment can take those kinds of chances.

Whether consumers get a good deal depends on whether better alternatives were crowded out of the market by the nuclear plant and on how well the government bargained with EDF, and the initial indications are that bargaining was not the government's strong point.

---

479. Department of Environment & Climate Change, *op. cit.*, p. 53.
480. Both UK government (the guarantor of much of the project's debt) and EDF debt yielded less than 3% in early 2015. Interest rates could rise during the duration of the construction.

Policies and Consequences

Tories restructured the UK's electricity system in part because they believed that the market reflects the judgments of millions of participants with intimate knowledge of events around them, so it makes better decisions than bureaucrats operating within the sheltered walls of government. The market has the information, not the bureaucrats. The market as constituted, however, did not make decisions that attracted capital to generation investments so the government decided to do so itself and send the bill to consumers. The market had been deregulated in part to put investment risks on investors, not consumers. The new policy put the risk directly back on consumers. The Tories returned all significant decisions that affected the long term to the government or the regulator, leaving consumers with the almost meaningless choice of which supplier to use.

The government selected projects, locked in long-term sales for those projects and thereby crowded out alternatives. The three nuclear sites, alone, could generate one-fifth of the UK's power. Anyone with a better idea had to compete for a shrinking slice of a static market.

Policy makers predicted higher prices despite the arrival of new technology that would improve network efficiency and give customers better control of usage. Breaking the Big Six's grip might lower prices but also their incentive to invest. Tariff simplification might yield better consumer decisions but also cause suppliers to raise prices in order to replace profits previously made by confusing customers. The UK could, also, openly re-regulate, simplify life for all and reduce electric company cost of capital but the odds of that happening are lower than a Papal nullification of priestly celibacy. It might solve a problem but does not merit consideration.

The Coalition extended regulatory reach. The minister decides fuel mix, price and capacity. Paying for those decisions will constitute an increasing part of the electric bill. Firms will move facilities out of the competitive market into safer ones, especially when the government offers unduly high returns and financing.[481] Felling trees in Louisiana and shipping wood pellets manufactured from them to the UK for burning may or may not reduces carbon emissions in the most economical manner but the government likes the idea. Switching an old power station from selling electricity to providing capacity does not create a new power station but it does create a more reliable stream of revenue for its owner.

---

481. As an example, Drax Group moved quickly to take advantage of the new regime. In December 2012, the owner of the UK's biggest coal-fired power station, announced that it had "secured financing to support the Group's transformation into a predominantly biomass-fueled electricity generator." Drax, Press Release, "Biomass Financing Secured," Dec. 20, 2012, p. 1. UK Green Investment Bank, a government agency, provided the funds.
    In December 2014, E.ON followed suit, announcing that it would get rid of all fossil and nuclear investments and stick to distribution and renewables. Early in 2015, the EU questioned the size of the Drax subsidy. Trying to substitute for the market is not an easy business.

The executives who make the decisions do so based on whether the government makes that decision more profitable on a risk adjusted basis than the alternatives, not on its cost effectiveness in reducing emissions. Regulated projects (by contract or old fashioned regulation) looked more attractive than competitive ones. Managers quickly take note and act accordingly. As Deep Throat advised, "Follow the money."

The Coalition first relegated natural gas to the back burner. UK production had peaked and Russian supplies were undependable. Yet gas-fired electric generation produces half the greenhouse gases per kWh as coal and intermittent renewables require back-up from quick-starting gas generators. Environmentalists, however, want to eliminate rather than reduce emissions. The Coalition revised its views late in the game, but opposition to drilling remains high and British politicians may never permit a shale gas revolution even if Britain has the gas.

Conclusion

The pro-market Coalition opted to decarbonize electricity production and to attract long term investment by introducing layers of mandates and regulation. Generators reacted by choosing assured rather than competitive prices and maneuvering existing plant into favored categories in order to obtain the best price. The government manifested more certainty about electricity's future than warranted, fretted about running short of supply even though electricity sales dropped 13% from 2010 to 2014 (to the lowest level in 17 years) despite a 7% growth of the economy. Demand in the future might not reach anticipated levels, either, thanks to more efficient customer usage, electricity storage or on-site generators. New technologies, whose costs have been dropping rapidly, might enable some customers to cut the ties to the grid, entirely. Lower demand would reduce the need for grid-based generating capacity that the government wants in service to assure reliability. Grid-based generating facilities and transmission lines needed so badly, according to the plans, could turn into tomorrow's stranded assets. Should anybody bet the ranch on a forty year projection?

The first regulator, an economist, tried to channel the industry via a theoretical regulatory framework. The second, an economist with financial experience, tightened the screws. The third regulator, an accountant, bent the regulatory framework to reflect reality, and in doing so, buried much of the work of his predecessors. The fourth regulator, just on the job, launched still another investigation of the market.

The Coalition shook up the industry, the third shake up (fourth if NETA and BETTA count separately) in 20 years although consumers might not have noticed except for the change in prices. Tories who celebrated F.A. Hayek's teachings reintroduced barely disguised central planning. After Tories decisively won the May 2015 election, and kicked the Lib-

Dems out of government, they set to work dismantling support for expensive, subsidized, domestic renewable projects while wooing a Chinese state enterprise to pour vast sums into the UK's expensive, subsidized nuclear effort.

Then the Tories ejected David Cameron after the UK voted to exit the European Union, and his successor, Theresa May, quickly declared, "It's just not right that two thirds of energy customers are stuck in the most expensive tariffs."[482] She reaffirmed her commitment to Hinkley Point, which will raise everyone's electric bill, folded DECC into a new government department and set 2025 as the phase out date for coal-fired generation. All of which demonstrates that UK governments can write whatever energy policy they want, with the only consistent aspect being that customers will foot the bills.

---

482. Theresa May, "Speech to Conservative Policy conference," Birmingham, UK, 5 Oct. 2016.

Chapter 21

Twenty Four Years Privatized: UK vs USA (1990-2015)

The liberal argument is in favor of making the best possible use of the forces of competition as a means of coordinating human efforts... [483]

– F. A. Hayek

Change the guiding principles from government ownership and central planning to competition and light regulation. Then, said the true believers, competition will force down costs and the competitors will pass on the savings to consumers. They were partly right.

Selling Stock Ownership

The government wanted money. It collected billions from the share sales. It wanted to enlarge the shareholder population, too ("shareholder democracy"). Individuals owned 28.2% of UK shares in 1981. [484] From 1982 through 1989, the government sold shares in British Petroleum, British Telecom, British Gas, British Airways, British Steel and the water industry of England and Wales. Individual ownership of shares, however, fell to 20.6% in 1989, 16.5% in 1997, the year after the privatizing British Energy, and 11.5% in 2010. As the UK Shareholders' Association conceded, small shareholders "clearly subsequently disposed of their holdings rather than become long term stock market investors as was once hoped." [485]

Of the original 17 electricity firms floated, by 2010 only SSE still traded on the stock exchange. [486] Small investors cashed out with everyone else. In 2007, ten million of them still owned UK shares, with an average holding of £24,000. In 2010, around eight million of them owned £26,000 of shares, on average. Privatization did not give birth to peoples' capitalism.

---

483. F. A. Hayek, *The Road to Serfdom* (Chicago: University of Chicago Press, 2007), p. 85.
484. Office for National Statistics, "Percent of total equity: Individuals Not seasonally adjusted Updated on 9/12/2009," Series DEYI: SRS.
485. UK Shareholders' Association, "UK Stock Market – Background Information and Statistics, UK Stock Market Statistics, Last revised July 2007," no pagination.
486. National Grid was sold later and it has remained a publicly traded company.

Where Have All the Savings Gone?

After privatization, costs fell thanks to staff reductions and termination of the coal contracts, with some of the savings reaching consumers. (Costs declined after nationalization, too, and customers collected the benefits, but that is a different story.)

Break the electric bill into three components per kWh sold:

- Fuel – Generators can control fuel mix but not fuel price. They can determine how to consume fuels more efficiently.

- Regulated profits – Regulators set prices. Good managers can beat regulatory targets.

- Other expenses and unregulated profits – Functioning competitive markets should drive down those expenses and profits.

All three components of the bill fell during nationalization (1950 - 1990 data) and even faster in the first decade of privatization. After that, the "other expenses and unregulated profit" category rose. (Renewable expenses including taxes accounted for one quarter of the increase in "other" from 2000 to 2010 and perhaps two-fifths of the increase from 2010 to 2015.) The sharp drop in fuel costs after 2008 simply gave everyone an opportunity to take a piece of the savings before consumers could get to them. Twenty five years after privatization, customers paid more, in real terms, than before. (See Table 21-1.)

Table 21-1. Components of the Real Price of Electricity in the UK (1950-2015) (2003 p/ kWh)

	1950	1970	1990	2000	2001	2008	2010	2015
	[1]	[2]	[3]	[4]	[5]	[6]	[7]	[8]
Fuel	4.64	2.64	2.36	1.31	1.44	2.71	2.34	1.31
Regulated profit	1.38	1.12	0.86	0.78	0.81	0.69	0.68	1.10
Other expenses and unregulated profit	5.83	4.56	4.60	3.13	2.68	4.76	4.79	6.56
Total	11.85	8.32	7.82	5.22	4.93	8.16	7.81	8.96

Notes:

1. 1990 (last nationalized year), 2001 (NETA begins), 2008 (highest gas prices), 2010 (end of Labor government) 2015 (terminal year).
2. GDP deflator used.
3. All data based on financial accounts of nationalized industry before privatization, and prospectuses and company accounts after privatization, all adjusted to calendar year basis.
4. Fuel costs before privatization from the accounts of the nationalized industry. After privatization, natural gas expenses from government statistics, nuclear fuel expenses from company accounts, or

based on fuel costs of similar companies. Other fuel expenses based on reported volumes of use and price indices in government reports.

5. Profit defined as after tax income available to compensate capital, that is, interest expense plus net income. All data from accounts of the nationalized industry and from the financial accounts of the electricity companies, after privatization. Price defined as total value of electricity sold to ultimate customer, as reported by DECC divided by kWh sold.

Sources:

1. Electricity Council, *Handbook of Electricity Supply Statistics 1987* (London: Electricity Council, 1987).
2. Alex Henney, *A Study of the Privatisation of the Electricity Supply Industry in England & Wales* (London: EEE Limited, 1994).
3. UK Department of Energy and Climate Change, *Digest of United Kingdom Energy Statistics, Table 1.7* (various issues) and QEP table 3.2.1.
4. _____, *Energy Trends* (various issues). Table 5.1.

In post-war USA, expenses and prices fell rapidly until unanticipated nuclear and legally mandated purchased power costs hit. The drops then resumed from the early 1990s through the crash of Enron in 2001, after which they and turned up. (See Table 21-2.)

Table 21-2. Components of the Real Price of Electricity Sold in the USA [1,2] (1950-2015) (2005 ¢/kWh)

	1950	1970	1990	1992	1996	2000	2001	2008	2010	2015
	[1]	[2]	[3]	[4]	[5]	[6]	[7]	[8]	[9]	[10]
Fuel	1.91	1.23	2.11	1.89	1.53	1.66	1.66	3.05	2.22	1.64
Regulated profit	3.01	1.73	1.98	1.83	1.65	1.05	1.24	0.94	0.89	0.94
Other expenses and unregulated profit	7.45	3.58	5.01	5.14	5.08	4.97	5.14	5.10	5.86	6.19
Total	12.37	6.54	9.01	8.91	8.26	7.68	8.04	9.09	8.97	8.77

Notes:

1. 1992 (passage of Energy Policy Act which opened way for wholesale generation competition), 1996 (states begin to restructure local markets). 2001 (wholesale generation market collapses and Enron goes bankrupt), 2008 (natural gas prices peak), 2010 (for comparison with UK), 2015 (terminal year).
2. GDP deflator.
3. For utility and no-utility generation, based on filings made to Federal Energy Regulatory Commission (FERC) and Energy Information Administration (EIA).
4. Electric utility operating income (after tax income available from operations to pay interest charges plus net profit). Based on filings made to FERC, adjusted upward for estimated operating income of public power agencies (Federal, state, cooperative and municipal utilities).
5. Total revenue from sale of electricity to ultimate customers divided by kWh sales.

Sources:

1. Edison Electric Institute, *Statistical Yearbook of the Electric Utility Industry* (various issues).
2. U.S. Energy Information Administration, *Electric Power Monthly* (various issues).
3. _____, *Electric Power Annual* (various issues).

From 1950 to 1970, costs and prices fell in both countries. From 1970 to 1990, expensive nuclear programs in both countries and independent power contracts in the US raised costs. Deregulation and privatization (1990-2010) took place under ideal conditions. With most nuclear projects completed, and enough capacity to meet needs, the industries could reduce work force, close down old coal plants and end expensive independent power contracts. Highly efficient, low-cost gas generators went into service. All stars were aligned for consumers to reap the benefits. But they did not. Operating savings were offset by other expenses or disappeared into someone else's pockets. (See Table 21-3.)

Table 21-3. Comparison of Annual Rates of Change of the Components of the Real Price of Electricity (1950-2014) (%)

	1950-1970	1970-1990	1990-2010	2010-2015
	[1]	[2]	[3]	[4]
UK				
Fuel	-2.7	-0.5	0.0	-10.1
Regulated profit	-1.0	-1.4	-1.1	11.1
Other expenses and unregulated profit	-1.2	0.1	0.2	7.4
Total price	-1.5	-0.3	0.0	3.7
USA				
Fuel	-2.1	2.7	0.2	6.0
Regulated profit	-2.6	0.7	-3.7	1.1
Other expenses and unregulated profit	-3.6	1.7	0.7	1.1
Total price	-3.1	1.7	-0.1	-0.5

Sources and Notes:
See Tables 21-1 and 21-2.

Comparison Shopping

Imagine performing an experiment. Take identical countries, restructure electricity markets in some and leave others as a control group. Then compare. Policymakers, though, did not experiment. They restructured and hoped for the best. The UK restructured ahead of other European countries and more thoroughly than the USA, where only 14 states and the District of Columbia remain restructured after seven states reverted to regulation.

Competition should force sellers to first reduce costs and then prices, in order to capture or retain customers. Assume that regulated utilities lack incentive to reduce costs or prices, except when prompted by regulators who rarely have the information, competence or staff needed to keep a sharp eye on utility costs and pricing. Therefore, prices in competitive markets should decline over time relative to prices in regulated markets, all other things being equal.

All other things are not equal, but the UK, USA and Continental countries have similar economies, climates, fuel mixes and programs to promote renewable resources. They all started from the same place, a regulated electricity industry in 1990. The price impact of deregulation, if any, should show, to some degree, in inter-country comparisons.

In 1990, UK utilities charged residential consumers slightly less and industrial customers slightly more than consumers in the original European Union 15 (EU 15). Since then, UK prices declined modestly relative to those in the EU 15, not so much because the government restructured the industry but because it slapped lower taxes on electric bills. Compared to the USA, though, prices in the UK and EU 15 remained embarrassingly high. (See Tables 21-4 and 21-5.) Privatization never changed that part of the picture.

Table 21-4. UK Electricity Prices as % of Prices in EU 15 and USA (1980-2015)

	1980	1985	1990	1995	2000	2005	2010	2015
	[1]	[2]	[3]	[4]	[5]	[6]	[7]	[8]
UK domestic prices with taxes included as % of:								
EU 15 mean price	103	90	99	80	94	83	79	104
EU 15 median price	109	83	97	76	91	83	83	107
USA average price	133	124	141	147	127	159	158	187
UK industrial prices with taxes included as % of:								
EU 15 mean price	120	102	104	91	102	91	105	126
EU median price	126	105	114	92	100	88	107	138
USA average price	108	88	152	145	136	149	157	211
UK domestic prices with taxes excluded as % of:								
EU 15 mean price	–	–	103	94	107	93	99	146
EU median price	–	–	99	92	100	100	103	156
USA average price	–	–	144	140	124	150	158	189
UK industrial prices with taxes excluded as % of:								
EU 15 mean price	–	–	–	94	107	93	113	150
EU median price	–	–	–	92	100	100	113	165
USA average price	–	–	–	140	139	145	162	184

Notes:
1. EU 15: Austria, Belgium, Denmark, Finland, France, Germany, Greece, Ireland, Italy, Luxembourg, Netherlands, Portugal, Spain, Sweden and the United Kingdom.
2. Purchasing power parity exchange rates for household customers.
3. UK domestic excluding taxes 1990 estimated by author.

Sources:

International Energy Agency, *Electricity Information 2001 with 2000 data*, Tables 32 and 36.

U.K. Department of Energy & Climate Change, "Quarterly Energy Prices," March 2011, Tables 5.3.1 and 5.5.1.

U.K. Department of Trade and Industry, "Quarterly Energy Prices," June 2007, Tables 5.3.1 and 5.5.1; March 2016, Tables 5.3.1,5.4.2, 5.5.1, and 5.6.2.

See Table 21-2 for U.S. sources.

Table 21-5. UK Electricity Prices Ranked in EU 15
(1= lowest price, 15 = highest price) (1980-2015)

	1980	1985	1990	1995	2000	2005	2010	2015
	[1]	[2]	[3]	[4]	[5]	[6]	[7]	[8]
Domestic price								
Tax included	9	5	5	4	6	5	4	10
Tax excluded	–	–	–	7	11	9	7	15
Industrial price								
Tax included	14	9	11	6	8	7	9	14
Tax excluded	–	–	–	8	8	7	11	14

Notes and Sources:

See Table 21-4.

In America, some states deregulated, some that deregulated re-regulated, and some remained regulated throughout. States deregulated to reduce high prices but prices did not decline relative to those in re-regulated and regulated states. Some blamed an over-reliance on natural gas as a fuel (because gas price rose sharply) in deregulated states but re-regulated states burned gas in the roughly same proportion without the same price inflation. The increasing cost of renewable energy contracts might explain the puzzling performance if deregulated states took relatively more renewables, but they did not. Whatever the virtues of restructuring, it did not produce significant monetary benefits to consumers.[487] (See Figure 21-6.) Nor did it improve the industry's ability to raise low-cost capital because the industry had no difficulty raising capital previously, other than during the bleakest period of nuclear construction. Did deregulation improve service quality and product offerings? Probably not.

---

487. Two University of California economists came to a similar conclusion and declared that "the greatest motivation for restructuring was rent shifting not efficiency improvements," which seems a polite way of saying that it was all was about how to redistribute profits, not how to give consumers a better deal. (Severin Borenstein and James Bushnell, "The U.S. Electricity Industry after 20 Years of Restructuring," Energy Institute at Haas, September 2014, p. 1.)

Table 21-6. Average US Electricity Price to Ultimate Customers by State as Percentage of US National Average (1992-2015) (%)

	1992	1996	2001	2008	2010	2015
States that deregulated	117	123	111	130	126	121
States that re-regulated	100	97	96	95	95	98
States that remained regulated	87	89	88	90	93	97

Notes:
See Table 21-2 for explanation of years chosen.
Sources:
1. "Average Retail Prices of Electricity to Ultimate Customers by State," Energy Information Administration, *Electric Power Monthly* (various issues).
2. Leonard S. Hyman and Richard E. Schuler, "Electricity Restructuring, Consumer Prices and the Cost of Capital," HICSS-47 Conference, Jan. 2014, and subsequent calculations by author.

Privatization and restructuring barely affected UK consumers, other than those who made a bundle on the stocks, not a stellar result, but better than that suffered by some Americans.

Global Warming and Coal

In 1990, the UK and USA both relied heavily on coal and nuclear power and emitted similar levels of carbon dioxide per kWh generated. UK generators reduced greenhouse gas emissions by shifting from coal to gas, and by 2000 UK emissions per kWh had fallen by one third while US emissions per kWh barely budged. After 2000, UK generators had less scope to reduce fossil fuel emissions and began to rely more on renewables. The Americans got to work when the shale gas revolution allowed them to replace coal with cheap gas. Then they pushed more renewable generation into the grid. In both countries, dropping coal from the generating mix made the biggest impact on carbon emissions. (See Tables 21-7 and 21-8 .)

Table 21-7. Reduction in Annual Carbon Dioxide Emissions in UK and USA (1990-2015) (% and million metric tons)

	UK 1990-2000	UK 2000-2010	UK 2010-2015	USA 1990-2000	USA 2000-2010	USA 2010-2015
	[1]	[2]	[3]	[4]	[5]	[6]
% change						
Real GDP	28.4	14.9	10.9	39.7	18.0	10.4
Electric generation	18.0	1.1	-11.0	26.1	8.4	-1.2
Change in annual carbon dioxide emissions – million metric tons						
Total carbon dioxide emissions	-41.6	-57.7	-110.2	876.4	-246.3	-339.0
Electric generation emissions	-45.8	-2.5	-54.5	476.1	-32.4	-400.0
Breakdown of change in carbon dioxide emissions from electric generation – millions of metric tons – from replacement of:						
Coal by gas generation	-42.1	-3.7	—	—	-59.0	-167.0
Oil by gas generation	-4.1	—	—	—	-16.7	—
Coal or oil by nuclear or renewable generation	-6.5	—	-24.0	—	—	-151.0
Gas by nuclear or renewable	—	—	-28.0	—	—	—
Change in volume and other	6.9	1.2	-2.5	476.1	43.3	-82.0

Sources:

1. Measuring Worth, "What Was the U.K. GDP Then," various dates.
2. _____. "What Was the U.S. GDP Then," various dates.
3. Joseph V. Spadaro, Lucille Langlois and Bruce Hamilton, "Greenhouse gas Emissions of Electricity Generation Chains: Assessing the Difference," *IAEA Bulletin*, 42/2/2000, p. 19.
4. U.K. Department for Business, Energy & Industrial Strategy, "Final UK greenhouse gas emissions national statistics 1990-2015," 7 February 2017.
5. U.K. Department of Energy and Climate Change, *Energy Trends* (March 2011), Table 5.1, "Fuel used in electric generation and electricity supplied," and subsequent dates of issuance.
6. _____, Statistical Release, "UK Climate Change Sustainable Development Indicator: 2010 Greenhouse Gas Emissions, Provisional Figures and 2009 Greenhouse Gas Emissions, Final Figures by Fuel Type and End-User, 31st March 2011," Tables 7 and 8, and subsequent dates of issuance.
7. U.K. Office for National Statistics, Real time GDP database.
8. U.S. Energy Information Administration, *Electric Power Monthly*, June 2009, Tables 3, 4, 58 and 59.
9. _____, *Electric Power Monthly*, May 2011, Table 1.1, 2.6,7.1-7.6.
10. _____, *Emissions of Greenhouse Gases in the United States 2009*, March 2011, Tables 6, 7, 12.
11. U.S. Environmental Protection Agency, *Inventory of U.S. Greenhouse Gas Emissions and Sinks: 1990-2008*, Washington, DC, April 15, 2010 (EPA 430-R-10-006), Table ES-2.
12. _____, "DRAFT Inventory of U.S. Greenhouse Gas Emissions and Sinks: 1990-2015", Feb. 14, 2017.
13. U.S. Energy Information Administration, *Monthly Energy Review*, Tables 7.1, 7.2a, 12.1 and 12.6, various dates.

Table 21-8. Electric Industry Emissions of Carbon Dioxide in UK and USA (1990-2015) (Metric tons per GWh)

	1990	2000	2010	2015
	[1]	[2]	[3]	[4]
UK emissions (metric tons per GWh)				
Total electricity output	641	421	410	332
Fossil fueled electricity output	824	562	546	618
USA emissions (metric tons per GWh)				
Total electricity output	602	603	552	473
Fossil fueled electricity output	869	852	782	754

Notes and Sources:
See Table 22-7.

Many American politicians viewed climate change as a left-wing plot connected to a war against coal. Without political pressure to act, American utilities embraced carbon dioxide reduction slowly. Americans stuck with coal longer.

International Standing

British, Americans and Continental Europeans took different approaches to electricity reform. The British regularly changed ownership, regulation, incentives and structure. Americans deregulated in some states and not in others, leaving much of the industry as before. On the Continent, large utilities (many state-controlled) dominated, regulation changed more slowly and competition and vertical dis-integration came later than in the UK.

Industry restructuring can produce dramatic results. Natural gas deregulation led to increased supply and lower prices. Financial deregulation produced a plethora of new products, some toxic. After airline deregulation, millions of new fliers flocked to airports, attracted by low prices and the blandishments of new carriers. Restructuring electricity, in contrast, had little discernible impact on consumers.

The UK generation market grew relatively slowly in 1990-2010, at 0.9% per year, compared to 1.5% in the USA and 1.1% in Germany-France-Italy combined. After 2010, UK and US generation actually declined while it barely grew on the Continent. Without any growth, the British had no need to install modern equipment to satisfy rising demand. Building a big project without a government assurance for its output became a risky business because it might be left without a market. On the other hand, low growth reduced the risk of damage from poor regulation because bad regulation discourages investment capital, and without a growing market the industry does not require new investment. No growth has advantages and disadvantages. (See Figure 21-1.)

Figure 21-1. Electricity Production (TWh) (1960-2015)

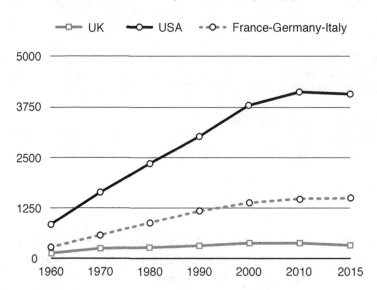

Sources and Notes:
All data from Appendix, Table P.

Electricity usage per capita or per unit of GDP first fell in the UK at the time of the 1970s energy crisis, ahead of the USA and the Continent, but not because British industry faded faster between 1960 and 1990, because it did not (Figures 21-2 and 21-3). After the Great Recession (Figure 21-4), despite industrial recovery in the USA, electricity demand there tapered off, anyway. Energy conservation took hold everywhere.

The sales slowdown, compounded by the Great Recession, created an uncertainty foreign to electricity planners who always believed in eternal growth. Yet this is an old industry, and it may have reached the top of the S-curve, with little growth and even decline ahead.[488] If flattened demand represents a temporary dislocation caused by the Great Recession, then the Coalition government correctly anticipated the need for massive investment in the future. If it is the next phase of a secular trend, the government will have saddled consumers with the carrying costs on unnecessary plant and investors with the risk that future politicians will not require consumers to pay for that plant when the time comes.

---

488. Leonard S. Hyman, "All You Need to Know," *Public Utilities Fortnightly*, November 2013, p. 14.

Figure 21-2. Electricity Production per Capita (1960-2015) (kWh)

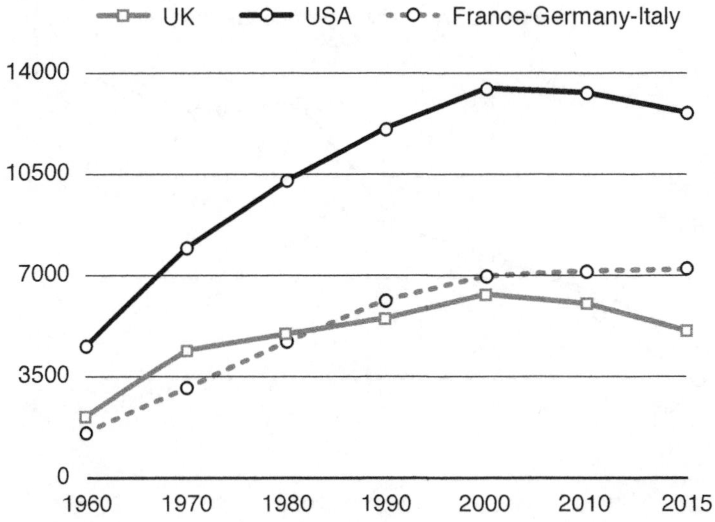

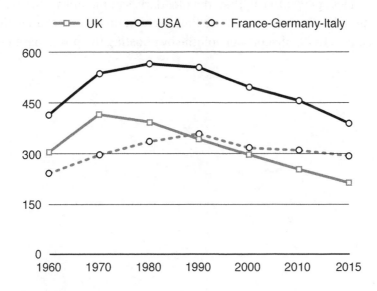

Sources and Notes:
All data from Appendix, Table P.

Figure 21-3. Electricity Production per $ Real GDP (1960-2015)
(million kWh per billion 1990 $ of GDP)

Sources and Notes:
All data from Appendix, Table P.

Figure 21-4. Industrial Production Index (1960-2015) (1990 = 100)

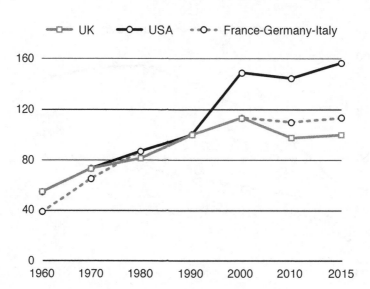

Sources and Notes:

All data from Appendix, Table P.

High electricity prices may have contributed to the relatively fast shrinkage of the UK's industrial base after 2000 (Figure 21-5). Still, the country's modern, services-oriented economy required less power to run than the old, dark, satanic mills. The newly minted electric companies, despite incentives to sell more could do little to change the picture. They could, of course, try to offset lower volume by boosting the profit margin.

Figure 21-5. Average Price of Electricity (1960-2015) (¢ per kWh)

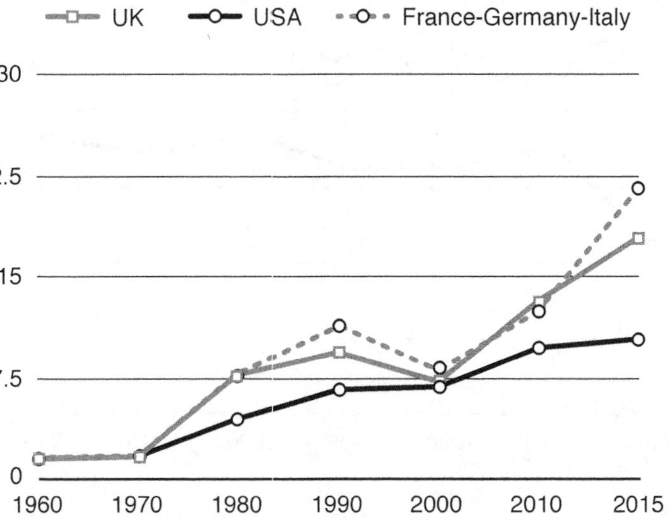

Notes:

1960-2015 data for UK and USA, 1980-2015 for France-Germany-Italy.

Composite price for France-Germany-Italy weighted by total generation.

Source:

All data from Appendix, Table P.

Gas Fades

The UK's natural gas industry had only one growth market left: sale to electric genera-tors. America's gas industry revived when it exploited vast shale gas reserves which drove down gas prices. The UK may have significant shale gas reserves, but nobody will know without drilling, an activity that provoked intense opposition from those who prefer their moors pristine. Ironically, even with more gas supplies, sales to electric generators, could falter, as the electricity industry makes greater use of renewables and attempts to restart nuclear power.

The growth in the natural gas market (excluding sales to electric generators) peaked in the USA in the 1970s and in the 1980s in the UK (Figure 21-6). Appliances became more efficient. Consumers learned to use less gas. It might require a substantial price reduc-tion to halt that trend. (See Figures 21-7 and 21-8.)

Figure 21-6. Gas Consumption (Excluding Sales to Electric Generators) (1960-2015) (billion therms)

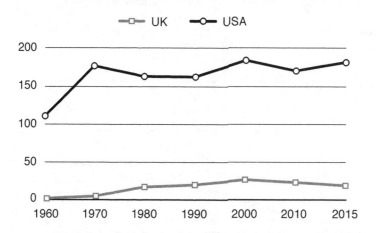

Sources and Notes:

All data from Appendix, Table I.

Figure 21-7. Gas Consumption (Excluding Sales to Electric Generators) per $ of Real GDP (1960-2015) (therms per 1990 $1,000 of real GDP)

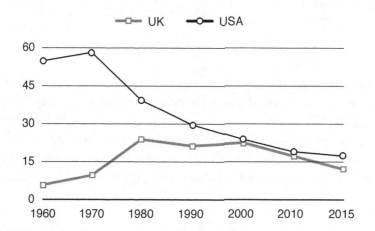

Sources and Notes:

All data from Appendix, Table J.

Figure 21-8. Gas Consumption (Excluding Sales to Electric Generators) per Capita
(1960-2015) (therms)

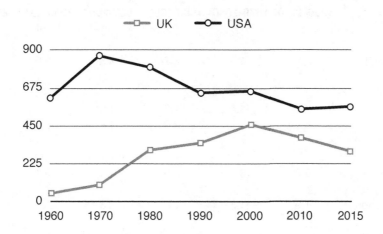

Sources and Notes:

All data from Appendix, Table K.

UK Growth Rates

Energy usage grew twice as fast as the economy before the energy crisis of the 1970s and half as fast afterwards. In the early 2000s, electricity sales fell despite growth in GDP, presenting a dilemma for investor-owned, profit-seeking electric companies whose owners wanted growing earnings and dividends. Once they ran out of growth and cost reductions, raising prices was the only way left to give shareholders what they wanted. (See Table 21-9.)

Table 21-9. Annual Average Rates of Growth in UK (1960-2015) (%)

| | 1960-1970 | 1970-1980 | 1980-1990 | 1990-2000 | 2000-2010 | 2010-2015 |
	[1]	[2]	[3]	[4]	[5]	[6]
Real GDP ($1990)	2.9	1.9	2.7	2.5	2.0	1.3
Electricity (kWh)						
Generation	6.2	1.4	1.2	1.6	-0.2	-2.4
Sales	6.6	1.5	2.0	1.8	-0.9	-1.6
Gas (therms)						
Total sales	8.0	9.4	1.6	6.3	0.1	-6.0
Sales excluding to						
electric generation	7.9	11.5	1.6	2.8	-1.7	-3.9
Real prices ($1990)						
Electricity (all sales)	-1.4	1.5	-1.3	-3.0	3.1	3.5
Electricity (residential)	-1.5	-0.3	1.9	-3.5	3.4	3.9
Gas (all sales)	-4.0	-4.5	0.0	-8.0	7.2	10.0

Notes and Sources:
All data from Appendix, Tables A, B, D, H, I and L.

Operating More Efficiently

In the first privatized decade, UK electric companies reduced high reserve margins, and raised load factor They closed old, coal-fired clunkers, replaced them with modern gas-fired generators, slashed staff, reduced heat rate and fixed the mess at the nuclear plants. They outdid the Americans on most counts. After a decade of squeezing, though, they had more difficulty continuing the pace, especially after consumers began to buy less electricity. Could they squeeze out more? As companies run out of big savings, price cap regulation starts to resemble rate of return regulation in disguise. (See Table 21-10.)

Table 21-10. System Operations (1990-2015)

	1990	2000	2010	2015
	[1]	[2]	[3]	[4]
Annual generation per kW of capacity (kWh)				
UK	4,366	4,752	4,257	4,162
USA	3,898	4,401	3,954	3,560
Reserve margin (%)				
UK	36.0	23.3	36.3	36.2
USA	25.6	18.7	26.4	NA
Nuclear % of capacity				
UK	17.0	15.9	12.7	11.7
USA	13.9	12.2	9.7	9.0
Nuclear % of generation				
UK	19.3	22.9	16.2	20.9
USA	19.0	19.8	19.6	19.4
Heat rate (BTU/kWh)				
UK	10,017	8,686	8,714	8,445
USA	10,366	10,231	9,756	9,189
Load factor (%)				
UK	62.2	67.4	64.6	68.2
USA	60.3	61.2	59.0	NA

Notes:
1. US load factor for 2010 estimated by author.
2. UK government does not calculate heat rate. Author's estimates based on fuels used.
3. NA — not available.

Sources (not numbered in table):
1. EEI, *Statistical Yearbook of the Electric Power Industry*, Table 33.2, "Generation Per Kilowatt of Installed Nameplate Capacity," Table 4.6, "Installed Generating Capacity in the United States," Table 2.1, "Capability, Peak Load and Kilowatt Hour Requirements," Table 4.8, "Consumption of Fossil Fuels for Electric Generation."
2. UK DECC, *Digest of United Kingdom Energy Statistics*, Table 5.10, "Plant loads, demand and efficiency," Table 5.6, "Electricity fuel used, generation and supply," Table 5.7, "Plant capacity – United Kingdom," Table 5.4, "Fuel used in generation."
   US EIA, *Electric Power Annual*, Table 4.2, "Net Internal Demand, Capacity Resources, and Capacity Margins by North American Electric Reliability Corporation Regions, Summer," various issues.
   _____, *Electric Power Monthly*, Table C1, "Average Heat Content of Fossil-Fuel receipts," Table 1.1, "Net Generation by Energy Source: Total (All Sectors)." Various issues.

Conclusion

When privatizing, the government said, "Decisions about the supply of electricity should be driven by the needs of customers."[489] The industry, however, probably collected more benefits than consumers.

Thanks to privatization, the British did get cleaner air. If the electricity industry had just stopped burning high priced British coal, and cut staff to reasonable levels, consumers would have come out ahead. The government, though, had more in mind than consumer benefits. It had other goals: to rid itself of one of the last vestiges of the socialist experiment, and to raise money without raising taxes, and in those goals it succeeded.

---

489. Secretary of State for Energy, *Privatising Electricity*, Presented to Parliament by the Secretary of State for Energy by Command of Her Majesty (London: HMSO, February 1988.

Part Five

Political Power and Power Politics:
Real World Lessons
and the
Triumph of Technology

Chapter 22
Power, Politics, Consumers and Technology

What is studied is a system which lives in the minds of economists but not on earth.[490]

– Ronald Coase

In the two centuries stretching from Humphry Davy to Ed Davey the British operated and regulated energy utilities, possibly the longest run of industry supervision by one government in modern history. Is that historical record relevant? After all, in those oft-repeated, and immortal words, "This time is different."

Is It Different?

True, consumers own more appliances, and companies operate more efficiently, but the basic model remains: central station delivers power over transmission lines to load centers where distribution lines take it to customers. In the future, on-site power sources, electricity storage, electric cars functioning as portable power and storage devices, or even wireless transmission of electricity may change the game but until then the business remains the same old network with the same old product.

The government retained its dominant role for 130 years, except for a brief break during which Thatcherites pretended to defer to the primacy of the market, after which the government reverted to its role of industry decision maker in camouflaged form. That role grew as sales growth petered out. Until new uses for or means of producing electricity materialize, prudent electric company managers will not invest based on wishful thinking. They will, instead, set strategy to take advantage of government environmental, regulatory and security directives, not to steal a march on competitors, or make a grab for non-existent growth. The success or failure of previous government policies should tell a lot about how similar policies will work.

If human nature remains unchanged, past reactions to incentives, prohibitions, sales pitches and regulations should indicate how people will react to similar stimuli today. Otherwise, social science research would amount to useless storytelling of a one-off nature.

---

490. Ronald Coase, "1991 Nobel Lecture: The Institutional Structure of Production," in Oliver G. Williams and Sidney G. Winter, eds., *The Nature of the Firm: Origins, Evaluation and Development* (NY: Oxford U. Press, 1993), p. 229.

Samuel Insull or Charles Merz, coming back to life, could easily resume work. Resurrected customers, too, would recognize the set up, if not all the appliances. Revived politicians would grasp the issues and interfere with the market with the same alacrity as before. Like a performance of *Julius Caesar* in modern dress, costumes change but not words. True, Mrs. Malaprop warned, "We must not anticipate the past,"[491] but how many ways can managers and policymakers slice and dice a kiloWatt-hour before repeating themselves?

Business schools teach case studies for a reason.

Twenty Lessons

One hundred thirty years of innovation, politicking, lawmaking, marketing, bureaucratic maneuvering and meddling offer practical lessons for those who plan to regulate, deregulate, invest in, buy from or sell to the electricity industry, or to protect or fleece its customers.

Lesson 1: The customer gets the leftovers

Legendary management consultant Peter Drucker declared, "The customer is the business."[492] Adam Smith put it less succinctly:

> Consumption is the sole end purpose of all production; and the interest of the producer ought to be attended to so far as it may be necessary for providing that of the consumer. The maxim is so perfectly self-evident, that it would be absurd to attempt to prove it.[493]

Not self-evident to British policymakers. Parliament elevated the parochial interests of municipalities over those of consumers. Laborites nationalized the industry because they believed in scientific socialism and public ownership and Tories privatized it because they believed in the virtues of private ownership. Neither had evidence that customers would benefit. They just knew. Both used electric bills to subsidize special interest, too. Someone gained, maybe the miner or the generator owner, but not the consumer. Lord Weir spearheaded the one reform specifically designed to help consumers, and even then special interests appropriated benefits before they reached the public.

---

491. Richard Brinsley Sheridan, *The Rivals*, Act IV, Scene II.
492. Peter F. Drucker, *Managing for Results* (NY: Harper & Row, 1964), p. 91.
493. Adam Smith, *An Inquiry into the Nature and Causes of the Wealth of Nations* (NY: The Modern Library, 1937), p. 625.

During deregulation in America, bankers, lobbyists and lawyers, salivating at the prospects, negotiated with outgunned public representatives to transfer regulated assets into unregulated hands and to foist charges on customers to pay for assets supposedly devalued ("stranded" in regulatory parlance) by competition. Deregulation may have transferred more value between industry players than it transferred from producers to consumers.

When governments reorganize industries, they allocate benefits to the best-represented special interest groups rather than ordinary consumers.

Lesson 2: Regulation never goes away

Electricity buyers and sellers function within a network whose rules are set by a regulator.

Reformers separated generation and supply from the regulated networks. They did not eliminate the interaction between regulated and unregulated activities. The carrying capacity of regulated transmission affects the value of unregulated power plants because they require transmission to reach markets. Demand for unregulated power determines need for regulated transmission. Regulators supervise pricing and set market shares for renewables. They determine need for capacity to assure reliability. Those actions affect the market for unregulated power. Market players adjust their activities to take advantage of regulation within a semi-deregulated market, to position their assets for the highest risk adjusted returns.

In deciding whether to deregulate, Alfred E. Kahn, the economist of deregulation, argued that:

> The final consideration is the dynamic character of the industry's technology and market. The more fluid an industry is the more the technology and future course of development are subject to rapid, unpredictable change, the more unregulated competition recommends itself as the proper institution of social control. [494]

Three factors changed the electric industry at time of privatization and deregulation: opening up abundant natural gas supplies, development of computation and communications systems that could quickly handle numerous transactions and introduction of large scale, efficient, gas-burning generators. Those developments opened the market to competitive generators but those entrants did nothing that an incumbent could not have done.

---

494. Alfred E. Kahn, "The Passing of the Public Utility Concept: A Reprise," National Economic Research Associates, May 1983, p. 5.

Kahn also warned that:

> ... the partial introduction of competition into a thoroughly-regulated industry creates distortions ... which can ... be resolved ... only by extending the freedoms ... to all competitors ... And that ... is the historical process in which we find ourselves inevitably engaged ... unless ... we encounter another unforeseeable ... conjuncture ... [495]

The industry may have reached that conjuncture. When the market cannot attract needed investment the government has to do so by regulation or by mandate. Instead of inevitably expanding, the competitive sector may end up shrinking in the face of a determined expansion of the regulated sector.

Not only will regulation not go away but it could drive out competition without the introduction of new sources of electricity that can operate independently of the grid.

Lesson 3: No regulatory formula works forever

Regulation should evolve to fit circumstances. Regulators, though, are bad at spotting inflection points. The British hung on to the sliding scale long after it had ceased to work. Americans stuck with a 6% rate of return for decades. The British have to go to Parliament to change rules. Americans, at least, can make changes state by state.

Rules suitable for periods of stable prices may work less well during inflation or deflation. American rate of return regulation had a successful six decade run because costs kept falling so regulators could simultaneously reduce prices and allow utilities to earn returns high enough to attract capital. When costs turned up, the process turned contentious. The British restructured electricity at a time when the industry suffered from high operating costs and excess capacity. UK regulation worked perfectly for an industry that could slash costs and needed little new capacity. That framework required an overhaul when conditions changed.

During the first eighty years, electricity costs fell steadily. Only fools would invest in new electric plant knowing that future investors operating with lower costs could undercut them and that consumers would gladly desert the old supplier in favor of the new, lower cost one. The regulator, then, had to protect investors from competitors and consumers in order to assure that any investment took place. When the cost curve turned up and new plant started to cost more than old, the regulator no longer had to protect previous investors in order to get money into the industry. At that point, consumers needed

---

495. _____, "Competition in the Electric Industry is Inevitable and Desirable," in N.Y. State Energy Research and Development Authority, *The Electric Industry in Transition* (Arlington, VA: Public Utilities Reports, 1994), pp. 26, 27.

protection, but regulators already had moved on to other tasks. If renewable energy and fuel costs continue to decline, the regulator may have to protect legacy conventional and even existing renewable electricity producers from the impact of the new technologies and from irate consumers who will want cost declines reflected immediately in prices.

What works until it stops working. After that, look for a new formula.

Lesson 4: Unnaturally contrived markets favor quick witted market gamers

Normal markets evolve to meet customers' needs. Operators and regulators learn from experience. They revise rules to improve products or services. If the market performs poorly traders can take their business elsewhere.

The electricity industry, whether regulated or not, required a mechanism to buy, sell or trade power, so in the 1920s it developed power pools whose purpose was to help the co-operating electric companies achieve a lower price for their customers. Governments in the past two decades imposed designs on electricity markets, set rules relatively quickly and made everyone trade in the markets. In those new markets, competing generators (each trying to maximize its own profit) would somehow achieve a lower price for consumers than cooperating regulated utilities. The change in structure changed the incentives.[496] As was egregiously demonstrated in California, profit maximizing competitors act differently than cooperative utilities in a market and sharp-elbowed, morally-challenged traders will threaten customer reliability to make a buck and can run rings around system operators. English market players were just not as blatant. The UK government got rid of the Pool because it offered too much opportunity for price collusion.

When a contest ensues between market operators promulgating and enforcing rules and manipulators looking for a new opening, the enterprising market manipulator can outwit the system operator and regulator for long enough to make a killing and then move on to the next tactic, and the public will surely pay the bill.

Keeping morally challenged manipulators out of the market is as difficult as keeping squirrels away from bird feeders.

---

496. Poorly designed financial incentives often achieve perverse results because they cancel out rather than supplement moral imperatives that already produce good results without financial incentives. In the old days, the pool participants and the pool operator worked together to maintain reliable and reasonably priced service. That was their job and they took pride in doing it. With the introduction of the competitive market, the pool participant got paid to maximize profit and let the pool operator worry about how to maintain system reliability. And if the participant could increase profit by gaming the reliability rules, it would do so. For an explanation of the principles behind the argument, see Samuel Bowles, *The Moral Economy* (New Haven: Yale University Press, 2016).

Lesson 5: The regulatory dilemma – to protect or not protect the incumbent from competition – will not go away until regulation does

The regulatory bargain requires regulators to protect utility investors from forces that threaten their profitability in return for a limitation on that profitability. The bargain reduces the regulated firm's risk and cost of capital, which translates into lower electricity prices. (Roughly one-fifth of the total electric bill goes to pay for cost of capital and associated taxes.) A regulator who breaks that bargain raises regulatory risk and cost of capital.

Regulators could make alternative bargains. They could make promises good until modified for pressing reasons. Consumers might then benefit from unexpected technological events or from correction of previous regulatory errors. Making the deal temporary adds to risk incurred by the regulated firm, so regulators would have raise the return and permit faster asset depreciation. Utilities could recover their money before the bargain changes, after which they could take their chances on obsolescence the way other companies do.

Then again regulators could emulate the American solution once competition appeared on the horizon. They could foist charges on consumers to pay utilities for the expected loss of asset value ("stranded assets") caused by the introduction of competition or technological change. Imagine the bogus claims. Regulators, too, could delay the onset of competition until consumers compensate the utility for the expected loss of value. But regulators cannot save the obsolete utility. Trolley car lines failed. No regulator could extract payments from former riders who chose to take a bus or their own car to work.

Regulation lowers the utility's cost of capital by placing the risk directly on the consumer. Competition raises the electric company's cost of capital by returning the risk to the company. In the long run, though, the customer may pay indirectly by compensating all risk takers with a higher return on capital, but in the short run, competitive markets remove from particular customers the risk that they get stuck paying for an egregiously bad local project.

The added efficiencies of the competitive market must exceed the added cost of capital or consumers could end up paying more for the same service. Otherwise, why bother?

Lesson 6: Politicians like to choose technologies and make customers pay for them

Politicians confuse means with ends. They focus on renewable energy rather than on the reduction of greenhouse gas accumulations that warm the climate. The former is a means, the latter is the end. Although promoting renewables should reduce greenhouse emissions, doing so may prevent more efficient reduction or removal methods from getting to market because the government has already made its selection. Taxing carbon emissions might force consumers and producers to seek the most economical way to dodge the tax and perhaps find a better solution than the government's. Politicians, however, dislike taxes and like visible projects that imply action, windmills, for instance rather than energy saving equipment buried in a building.

After World War II, politicians made British coal the fuel of choice and supported it through cozy contracts between nationalized coal and electric industries. Customers paid for the contracts. Then politicians promoted nuclear power and customers footed the bill for that expensive effort. Several times, in fact. Then they promoted windmills and wood pellets.

Forcing the electric company or consumers to buy a possibly uneconomical product misleads the citizenry into thinking the product has commercial merit. More likely, the policy props up a failing industry, hides the subsidy in the electric bill and certainly distorts the market.

Too many politicians, deep down, have little faith that the market will find the right answer and colossal faith in their own abilities to find it, and the wisdom of imposing their solutions on captive customers. Mrs. Thatcher and her close associates declared their faith in competitive markets, but not when it came to nuclear power, which may just prove the point.

Most politicians are not rocket scientists, and it shows.

Lesson 7: Nuclear power is not and never was an ordinary commercial venture

Governments insure nuclear ventures. They have to pick up the pieces when safety or waste disposal arrangements fail. Conventional nuclear units are so massive and the cost of construction or operating accident so great that only the largest firms can prudently build and operate them, often in joint ventures (to spread the risk) that turn projects into diplomatic as well as engineering feats. Twenty-first century nuclear builders, to limit the risks to them from cost escalation and changing market conditions during that long construction period, now demand government backing, including debt guarantees. Small, modular, factory-manufactured units designed for assembly in the field could change the picture, reducing financial and market risks, but no such units are ready for the market.

Financial analyst and fund manager William I. Tilles pointed out another peculiar risk.[497] A nuclear mishap anywhere can affect public perception anywhere, as when the Japanese accident at Fukushima caused the Germany to close down its nuclear plants. The actions of the sloppiest nuclear operator affects the viability of all facilities.

---

497. Conversation with author.

To operate economically, nuclear facilities must run continuously (except during refueling). They cannot cycle on and off depending on market price. They have to accept whatever the market offers. Nuclear builders, operators and investors no longer choose to depend on daily markets to provide the stream of revenues needed to cover high-fixed costs, future waste treatment and unscheduled stoppages. They want contracts or regulatory backing that assures them an adequate cash flow. The government has returned to the scene, either as guarantor or as regulator. The free market will not support nuclear power.

Without a steady, assured stream of payments in prospect, nobody will build the nuclear facilities. To attract builders, the government has to insure risks and subsidize financing (thereby dumping the risk onto the government or the public) or provide a regulatory framework that protects nuclear investment. Governments and consumers, already subsidize other no-carbon sources. So why not nuclear?

The giant nuclear project can go forward, then, with direct government backing, subsidy, insurance or ownership or if undertaken by a large regulated utility that builds and operates the project, knowing that it can pass on costs to its franchise customers. No ordinary commercial organization has the scale needed to tackle nuclear risks unaided. Risk matters because it affects cost of capital which accounts for perhaps half the cost of nuclear electricity. Transparency, though, might deter nuclear development. Right now, nobody spells out the risks to the public.

Nuclear power always was a government project, one way or another, always too expensive or too risky for competitive markets.

Lesson 8: Beware the giant project

Giant projects take years to complete, require their multiple owners and builders to coordinate efforts, need government approvals, drop a big slug of new supply into a market and call for significant capital investment. A. J. Merrett and Allen Sykes, who developed the notion that giant projects were different, had in mind a massive hydro-electric project in the Canadian north or a huge mining venture in a remote underdeveloped country that would clearly tax the logistic, engineering, managerial, financial and risk assessment competence of the builders.[498] Mismanaged or failed giant projects litter the landscape.

Kashagan typifies the giant project gone awry. Discovered in 2000 in the northern end of the Caspian Sea, "the world's biggest oil find in three decades"[499] involved the joint efforts of three experienced oil majors and Kazakhstan's state oil company, was supposed to cost $13 billion to develop, cost $43 billion (and still counting) by 2014, came in eight years late (with more delays possible), had to halt early production because poisonous gases

---

498. A. J. Merrett, and Allen Sykes, *The Finance and Analysis of Capital Projects* (London: Longmans, 1982.)
499. "Cash all gone," *The Economist*, Oct. 11, 2014, p. 73.

corroded the pipeline (possibly another $5 billion to correct) and came to market just as oil prices collapsed. A journalist blamed lack of "clear leadership" and "government meddling" for the debacle.[500] But this vast endeavor, set in a harsh environment, with multiple partners, reporting to the government of a former Soviet republic had all the characteristics of a giant project case study from day one.

Building a nuclear power station is a giant project. Nuclear pioneers thought of it as like building a coal-fired plant, just bigger. They underestimated engineering complications and operating risks, misunderstood environmental opposition and could not keep up with changing safety rules. Nuclear projects came in years late and hugely over budget. Fred Schweppe, legendary MIT engineer, when asked whether giant power facilities like nuclear plants were dinosaurs, replied that they were not if the builders knew the facility's cost, time to construct and market for its output.[501] They thought they did, but they did not.

Planning big requires thorough risk assessment, enormous self-confidence or colossal ignorance. Getting it right may require divine intervention. That is why smart managers hedge their bets.

Lesson 9: Breaking up costs money

In Adam Smith's world, specialists learned from repetition, profited from economies of scale, and as a result produced at the lowest cost. It follows, then, a firm that wants to offer a competitive product should assemble its product from parts bought in the market from those low-cost specialists. Consumers could do the same, dispensing with the assembler. Which led economist Ronald Coase to observe:

> … if production is regulated by price movements, production could be carried out without any organization at all, well might we ask why is there any organization?[502]

Coase concluded that:

> The main reason why it is profitable to establish a firm would seem to be that there is a cost of using the price mechanism...[503]

> ... the operation of the market costs something... [504]

---

500. *Ibid.*
501. Conversation with the author.
502. Ronald Coase, "The Nature of the Firm," in Williams and Winter, *op. cit.*, p. 10.
503. _____, *op. cit.*, p. 21.
504. _____, *op. cit.*, p. 22.

That is, finding, bargaining for and assembling all those parts costs too much money. Avoiding all those transaction costs may justify assembling the product in-house.

Splitting the electricity industry into component parts sharpened the operating skills and cost controls of the new companies. They had to rely on their own line of business for profits and bonuses – no cross subsidies from other divisions in the company. But the split-up also created new costs: to operate the market, pay traders, assemble portfolios of power purchases, insure for risk, advertise, run supply businesses and pay for additional corporate overhead.

Apple and Toyota successfully assemble components from separate suppliers. But they do not shop around seeking prices from all and sundry. They specify what they want. They control every aspect of the design and production as well as the price of the final product. Their suppliers might as well be their subsidiaries. The electricity retailers, in contrast, must purchase bulk power in the market, insure against price risk, cannot choose their sources, and must use a delivery mechanisms over which they have no control. Nobody is in charge of shaping, pricing and delivering the final product. Who takes the rap when the lights go out?

"Maybe there was a reason that the companies were vertically integrated," is the way consultant Stephen Peck put it, with characteristic English understatement.[505]

Lesson 10: You can shift the risk, but you can't make it disappear

The old utilities assumed that they knew what the customer needed and that they could deliver the goods on time and on budget. If they erred on any count, they could passed on the costs to hapless customers (or taxpayers), who, ultimately, took all risks. The utility ran a low-risk operation with low capital costs because the consumer or taxpayer insured it.

Reformers wanted to assure that nobody other than the decision makers should bear the risks of those decisions. No more would local consumers get stuck with the costs of the incomplete, unworkable or overly expensive project. The project owner would foot the bill, which shifted risk but did not make it go away. Project owners, considering that any one project could fail commercially, will seek a higher return on all projects to compensate for that additional risk, and that extra compensation could translate into higher prices overall (unless offset by more efficient operations). In effect, consumers no longer pay directly for the costs of bad projects, but they eventually pay to cover the new risk taken by all generators.

---

505. Stephen C. Peck, in personal discussion with the author. See, also, Stephen C. Peck, "Coasian Insights in Electricity Industry Structural Reform," Flèche, May 2003.

Admittedly, power plant builders have made spectacularly stupid investments which will never earn a compensatory return, but disasters make future investors more cautious, tighten supply and eventually raise prices. The Darwinian nature of capitalism selects out the eternally stupid.

Lesson 11: What is not in the price is not in the market

Free market types like to believe that the market will solve all problems, put a price on all needs (new capacity, clean air, reliability, national security) and produce lowest cost solutions for the economy. And it would if buyers and sellers valued all the benefits and took into account all costs (including those borne by society). Economists call the deal between buyers and sellers that includes all that information a "complete" contract.

Unfortunately, electricity markets do not offer complete contracts. Nobody voluntarily pays more to get fuel diversity, clean air, low-carbon emissions or capacity to assure reliable service. In the old days, the monopoly utility, after consultation with regulator or government, assembled a set of resources consciously designed to meet societal needs at the lowest possible cost. Market participants, today, if they want to survive commercially, seek the cheapest short-term price and leave the consequences of supporting societal needs to others if they can get away with it. So the government has to set standards and then force all participants to support these common services and public goods, or no one will. The government has to step in and reassert its role as regulator in order to correct the market defect.

The market participants would pay ready-to-operate reliability units when they need them but investors no longer will build those units based on the hope that they can sell high-priced power at times of stress often enough to warrant the investment. [506] To assure the availability of those units, the government has to force all users to either finance them or reduce demand during peak periods. [507] Alternatively, suppliers could contract with consumers to provide a specified level of reliability and pay penalties to the consumers when failing to meet specifications, making reliability an insurance issue for individual

---

506. Consumers could pay less for lower quality service and accept cut-offs during periods of network stress, as is the case for interruptible service sold to industrial customers, who, despite the name of their service, often became upset by the interruptions. Or, the reliability supervisor could pay the customer to allow the interruption. Either way is an alternative to putting peaking capacity in place. These demand-side alternatives to generating capacity threaten the value of existing peaking capacity, though if developed after the peaking capacity goes on line, endangers the value of that capacity.

507. Reducing consumer demand when it comes at times inconvenient to the network has a decades long history and it probably spares the network from expensive and avoidable capital expenditures, and thus saves consumers money in the end. But there is something perverse in a situation in which a business, rather than trying to give customers what they want (service at peak) tries to convince them not to take it. What about, as an alternative, trying to figure out how to provide that need economically. Shouldn't businesses serve customer needs rather than vice versa?

consumers rather than an engineering issue for the entire grid, a financial rather than engineering solution.

To assure that high-cost-no-carbon energy sells, the government forces wholesalers to buy it. They pay whatever it takes to get the green power and pass the costs on to consumers, scarcely a procedure that discovers the lowest cost means to reduce greenhouse gas emissions. As an alternative, the government could put a heavy tax on carbon emissions and let the market sort out the best solutions on a price basis.

Capital makes up a large part of the cost of infrequently operated or low-carbon plant. Most builders – whether regulated or not – go to the same small group of engineers, financiers and equipment manufacturers to get their projects built and financed. Most similar, standardized projects, then, should have similar costs, and no firm should have a significant advantage, except for cost of capital. Governments trying to solve emissions or reliability problems need to minimize risks and cost of capital, if possible, and regulated entities tend to have lower cost of capital. Maybe they could eliminate the complications and just do deals on a regulated basis.

Infrastructure investors can raise huge amounts of low-cost capital and they accept low returns for a low risk, regulated long term contract, whether or not the investment has moving parts. Policy makers should seek the most economical – not regulated or unregulated – solutions.

The market has provided solutions to reliability, fuel diversity and national security problems when there were no problems. Once problems emerged, the government had to step in.

Lesson 12: RPI-X inevitably morphs into rate of return regulation in disguise

RPI-X first focused on price even though return on investment played a role in its formulation. The utility tried to cut its costs faster than price dropped. After the easy cost reductions, cost of capital and capital spending become more important in the price formula, as demonstrated by the battles between regulator and company over the size of the capital program. When regulators offer returns above cost of capital, utilities fights hard to spend as much as possible, because that increases regulated asset value (rate base). Then UK regulation starts to look like rate of return regulation on steroids.

When the regulator combines RPI-X with a performance check list, the utility focuses on items the regulator measures, keeps capital expenditures within prescribed bounds, avoids innovations that provide no benefit to it (because nobody thought of them in advance of the rate plan) and emphasizes cost cutting (the only strategy it can employ to beat the regulatory formula). Originality does not pay unless the regulator measures it. OFGEM's RIIO plan, to its credit, at least included an innovation stimulus package.

Multi-year rate of return regulation with an inflation adjustment and periodic profit adjustments is still rate of return regulation, with bells and whistles attached, just well camouflaged.

Lesson 13: Sometimes it pays to experiment, or at least do some research first

Politicians often ignore information that points to unwelcome conclusions. They rely instead on bedrock beliefs. They applied those beliefs to the UK's electricity sector without seeking evidence to validate them. Every change constituted a nationwide experiment, a leap of faith. The centralized country UK seemingly lacks a place to test ideas on a small scale somewhere before instituting them everywhere.

Economists used to claim that they could not test their ideas. In lieu of experimentation, they quantified their work with simulations based on how they thought people would act, which meant like economists, not necessarily like most people. Now they do perform experiments to test how people might behave under different circumstances and whether incentives produce desired results. They attempt to replicate market conditions in a computer laboratory, pay winners money in order to make conditions as realistic as possible and even scan their subjects' brains during the decision-making process. Admittedly, experiments cannot duplicate reality. Experimental subjects may act differently in a laboratory setting than outside. But experiments do provide insights and could guide expectations.

The British made decisions based on beliefs and went straight from theory to practice. Research and experiment costs far less than taking the plunge on a nationwide basis and then discovering the mistakes.

Lesson 14: When they get tired of competing they merge

In economics texts, competing firms cut costs and lower prices to sell more of their products. Some fail. The survivors, facing less competition, raise prices, which encourages newcomers to enter the market and the process goes on. In the real world, competitors may merge with each other instead, claiming that size permits them to attain economies of scale which will lower their costs, with the implication that cost reductions might benefit consumers. But when only a few firms remain in business (an oligopoly), the survivors set prices to preserve or raise profits. Eventually this oligopoly breaks down, but eventually can take a long time.

The UK's electricity industry coalesced into a few firms, evidence that size and vertical integration offered advantages or companies would not have merged. If so, UK regulators could have promptly appropriated some of the advantages for consumers, as do American regulators, stopped obsessing about market structure and focused on getting the price right.

A handful of global firms dominate low margin commodities markets. For years, OPEC set petroleum prices (not always successfully). Few markets have more than four large cell phone companies. In the United States, the successors to the Bell System (broken up by an anti-trust judgment) merged and reintegrated. American railroads, which had competed vigorously and drove each other into bankruptcy repeatedly, merged into four big players, and in 2011 an executive of Union Pacific declared, "In our 150 year history, we're just getting close to earning our cost of capital..."[508] Airlines throughout the world have merged, to lower costs, lessen competition and raise prices.

More competition should lead to lower prices. Sometimes it leads to fewer competitors.

Lesson 15: A country not wishing to maintain a domestically owned energy industry will have no difficulty selling it

State-controlled firms that dominate the energy sector outside the USA and UK operate on a scale unmatched by most British or American firms. They may expand abroad for national reasons (such as advancing nuclear ambitions or securing fuel supplies) and accept below-market returns in order to achieve their aims. In addition, huge private firms, sovereign wealth funds and state pension funds seeking steady returns from infrastructure investments and able to raise low cost money have made large acquisitions in the energy arena.

But why worry? To the British, electric companies are purely commercial entities, tradable in global capital markets. Ownership does not matter as long as the lights stay on. British politicians fussed more when Americans took over chocolate manufacturer Cadbury – a Quaker-founded firm with deep roots in the British psyche – than when foreigners bought out almost the entire British electric industry. When Cadbury sold out, the British felt the loss. The electric companies did not generate that feeling.

"The British do not have explicit national champions... They would consider it vulgar" wrote a business columnist.[509] Nationality of ownership, however, matters when decision-making and choice jobs move abroad, and developments in the headquarters country trump other needs. German firms responded to the nuclear setback at home

---

508. Robert Wright, "Union Pacific back on the rails," *Financial Times*, Aug. 23, 2011, p. 13.

509. Jonathan Guthrie, "It's time for Britain's national champions to raise their game," *Financial Times*, 8 December 2014, p. 18.

by withdrawing from the UK nuclear effort. Russian politics may determine Gazprom's sales policy more than commercial needs. The Cameron government, on the other hand, seemed unconcerned about – and even encouraging of – handing the entire nuclear industry to foreigners, including a Chinese state company. The May government that followed, hesitated, put in a few restrictions, and then marched forward. As long as the locals get some jobs, why worry?

In a perfectly competitive world managers would make decisions on an economic – not nationalistic – basis, and consumers would come out ahead. But in a world in which other nations take an active interest in their energy companies, indifference to the role of ownership looks like unilateral disarmament.

Mercantilism and nationalism are alive and well in the energy sector.

Lesson 16: The regulator is the customer

Competitive firms design products to attract and satisfy customers. Management consultants Michael Treacy and Fred Weirsema, categorized successful firms by one of "three distinct value disciplines," each of which "produces a different kind of customer value." [510] They offer "operational excellence" (economically produced, middle-of-the-road, mass marketed products), "product leadership" (best, most innovative products) or "customer intimacy" (products tailored to a particular customer's needs).

Electric companies contemplating a value discipline would have to explain it to regulators who weigh each offering for the appearance of fairness and try to allocate its costs and benefits. Considering the trouble involved, the regulators would most likely choose a one-size-fits-all service and tariff as their easiest option. They would continue the tradition of standardization that made the physical provision of electricity economical, even though modern communications and computation might permit electric companies to offer different solutions and tailor services to individual customer needs.

The old product works so why change it? Regulators and Bell System managers opposed digital switching because analog switches worked perfectly well. They saw no need for the new product and they might have delayed one of the twentieth century's technological revolutions if they had prevailed. [511] With the regulator as gatekeeper, customers might not get the opportunity to approve or reject the product.

---

510. Michael Treacy and Fred Weirsema, *The Discipline of Market Leaders* (Reading, MA: Addison-Wesley, 1995), introduction p. 2.
511. The regulator may inhibit the consumer's adoption of a product innovation in a different way, by setting tariffs for the interaction between the new product and the existing grid that overcharges for the connection and thereby make the new product seem too expensive to consumers. Regulators that do not like a product can delay its reaching the mass market. Do they act to protect a befuddled consumer from making the wrong decision or to protect the incumbent utility that they regulate?

The regulator acts as surrogate for the consumer and the electric company prospers by offering only products on the regulator's shopping list. Even the "unregulated" UK electricity seller, after the 2012 reforms, must make the offering fit a template. Offering something different requires an exception, and one exception leads to another, and then the whole scheme unravels.

The concept of regulator-as-customer channels the route of innovation through one consumer, discourages innovators from promoting ideas that the regulator might not like and fosters a culture of getting along by going along.

Lesson 17: Confuse the customer to make a bigger profit

Electricity customers seeking low prices pick the high priced offering? Make the price list complicated enough, offer too many choices, and people do that. Consider travelers buying airline tickets, faced with price packages that include or exclude taxes, charges for carry-on bags, for checked luggage (how many bags?), for extra legroom, for early boarding, not to mention for food on the plane and maybe even for a pillow. How many people have the time or energy to digest all that information? They could buy the airplane for less effort.

The United States' complicated drug plan for seniors forces them to correctly predict next year's prescriptions in order to pick the best offering. Consumers Union, a consumer protection organization, complained, "When consumers can't assess their choices, health insurance markets don't operate efficiently. Health plans can exploit that confusion..."[512]

Unscrupulous firms target vulnerable consumers. In a notorious case (one of the few in America that sent a CEO to jail), Lincoln Savings told its marketers, "Always remember, the weak, meek and ignorant are always good sales targets."[513] Americans, at least, say it directly.

To complicate matters, electricity customers do not consume electricity. They consume services or products provided by electrical devices. They care about electricity only when it ceases. They may prefer an uncomplicated process to buy a product that they will ignore except in its absence. They are, in a sense, disengaged customers, participating in a process devised by economists, systems analysts and policymakers rather than product designers and marketers. Businesses can take advantage of customers who do not pay attention.

Companies peddling the same product as everyone else naturally steer customers to the high-priced variants of their offerings. When the government steps in to regulate com-

---

512. Consumers Union, *Health Policy Brief*, January 2012, "What's Behind the Door: Consumers' Difficulties Selecting Health Plans," p. 7.

513. Scott McCartney, "Lincoln Savings Memo: 'The Weak, Meek and Ignorant Are Good Targets' ", September 9,1990, AP News Archive (Beta).

plexity in order to protect consumers, sellers will have to accept lower profits because customers can now figure out the most economical price schedule, or raise prices on the low-end offering, or find another line of business. But until then, why give up a good profit-maximizing strategy, especially when everyone else does it?

Customer choice raises profit margins.

Lesson 18: Innovation disrupts long term planning and renders investments obsolete

Electric utilities make regulator-sanctioned investments. The smart grid and meter became industry favorites in the early 2000s, thanks to government backing and a smart sales push by potential suppliers. No point spending on projects that do not pay off in terms of regulatory favor or quick return. One broker mistakenly wrote, "In a world where utility ... growth is hard to find, material capex [capital expenditure] cuts ... are very attractive."[514] Not so when regulators favor that particular capex, because it goes into rate base (regulated asset value) and becomes a money maker. Innovations can augment regulated asset base, better employ existing assets or destroy the value of those assets.

Smart grids and meters provide two-way communications and real time monitoring. System operators use their sensors and control devices to improve asset utilization, anticipate and prevent problems and accelerate recovery from outages. They provide real-time information that allows consumers to tailor consumption to price, and that could reduce demand at peak periods. But benefits must exceed metering costs, and some consultants implementing the investments voiced doubts.[515] While the smart grid should help electric companies operate more efficiently and reliably and might help customers make smart choices if they want to make them, it will also harm the prospects of generators that sell peak period power.

The electric car can reinforce existing industry structure. Recharging the car during low demand periods will increase electricity sales from existing power plants. Recharging them when the wind blows could reduce carbon emissions from transportation even more.

Smart grids and electric cars can help utilities get more out of existing assets.

---

514. Bobby Chada, *et. al.*, "Utilities: The Ripple Effect," Morgan Stanley, February 6, 2013, p. 8.
515. One consultant told the editor of a trade journal, "The value proposition for the end user is variable at this point in time," consultant talk for "Who knows if it will pay off?" and even worse, another said "There's no business case for smart metering if it's just to shift demand, at least not in the near term." Even if the technology turned out better than expected, those comments indicate that the client utilities at the time may have decided to latch on to something they saw more as a rate base builder than an economical service. (Michael T. Burr, "Smart Grid at a Crossroads," *Public Utilities Fortnightly*, January 2013, pp. 29 and 27.)

Large scale, efficient electric storage devices, charging up during periods of low demand and discharging when demand rises, could replace peaker power plants. They could improve the economics of renewable energy by using it to recharge when the grid neither needs the energy nor will pay for it and to discharge it into the grid when the grid needs power. Storage would add value to renewables and allows the existing grid to serve more customers.

Microgrids are local electric networks that can operate independently of the grid. Small distributed generators, locally sited renewables, cogeneration facilities and energy storage devices can provide power when the microgrid disconnects from the utility network. Microgrids could make local service more reliable by reducing dependency on the national network and taking pressure off the grid when it is stressed, certainly a plus for consumers.

Those innovations, however, could easily undermine any government's forty year plan. With a smart grid in place, transmission operators know network conditions on a real time basis, require a smaller margin of safety and could run the grid with less capacity. Build more or operate better? Microgrids that drop off the network or storage devices that discharge power at peak times reduce the need for conventional generating capacity and for utility transmission and distribution facilities previously thought of as low-risk investments. Electric cars that charge their batteries with off-peak renewable energy might look like a safer decarbonization strategy than nuclear power plants. Economical storage devices and the electric car used as a traveling battery could ease the way for microgrids to operate with minimal recourse to grid power, making the electricity network the legacy supplier, like the old airlines, encumbered by high costs, fighting against the low-priced upstarts. The government has imposed a host of new costs on the electricity network, which will raise its prices and in doing so, make it more vulnerable to disruption by lower-cost outsiders.

If in-house innovations can disturb plans, imagine the impact of innovation from firms that have no industry investment to defend. Apple owned no telecom assets before it marketed the smart phone and Amazon no bookstores before it started to sell books and Google had no experience in home energy before it bought a thermostat firm for an eye-popping sum. Huge firms with vast cash resources on the prowl for opportunities could enter the market, selling electricity customers a badly needed new idea. Outsiders really can mess up the best laid plans.

Hedge your bets. Don't make forty year plans. Even the Soviets stopped at five years.

Lesson 19: Read Adam Smith – the other book

Adam Smith explored psychology and motivation in *The Theory of Moral Sentiments*, the book that made his reputation. Smith theorized about real people, not the *homo economicus* who inhabits microeconomics classrooms. People, he said, do the right thing because of professional pride or moral duty or patriotism, not just for money. Those who ran the old electric industry worked cooperatively toward the goal of providing economical and reliable electricity to consumers because that was their job and they believed in those quaint goals. Deregulation and separation of functions into individual business units changed that picture, making the new motivation to maximize the profit of each of the new businesses in every legal way possible.

Policy makers assumed that only profit can motivate market players. They dismissed incentives that did not fit into their neoliberal ideological model. They broke the nexus between industry decision making and the motivations encompassed in Smith's idea of moral sentiments because they just knew that the old industry executives, lacking obvious economic incentives, could not possibly run efficient enterprises that produced the desired product at the right price, and they assumed that whatever other problems arose in the new system could be handled best by the market or the residual regulatory process.

Policy makers changed the rules and expected that a band of competitive profit maximizers operating within the same network, with a sharp regulator riding herd over them, would give the public more reliable service at lower prices than the old crew. They did not consider whether the sidelining of the old values might have detrimental consequences.[516]

Economists, lately, have discovered that Adam Smith, the original capitalist, was right. People really do things for more than money and sometimes when offered money they do what you least expect.

Lesson 20: Not enough competition may work better than not enough regulation

Market forces take time to work. In the meantime sellers can take advantage of consumers, but only for so long before annoyed consumers move their business to less greedy suppliers or to substitute products once they appear on the scene. After privatization, competitive unregulated suppliers and generators pulled out all stops to extract what they could from gullible consumers and poorly designed wholesale markets but on the whole they earned no more than an average profit for their efforts. The competitive market worked badly, slowly and unfairly, but in the end it worked.

The regulated electricity business consistently earned more than needed to pay for its services, despite all the bluster of the regulators. The "tough" price orders permitted companies to earn returns high enough to attract capital and then some. If the past is an

---

516. As pointed out by Bowles (*op. cit.*), creating a monetary incentive may cancel rather than augment a non-monetary (cultural or professional) incentive.

indicator, customers will continue to pay up until technology changes in a way that lets them get off the regulated network. [517]

"Reasonable" regulators make the electricity business much more profitable than real customers.

Technology and Fixes

What made the United Kingdom different? It had a competitive edge. It started the electric revolution, produced the greatest electricity theorists, invented the machinery and had easy access to Continental manufacturers and innovators. The British practically invented the public utility concept when they chose to regulate gas light. They pioneered sophisticated pricing concepts. With London the undisputed financial capital of the world, they could raise money for any promising enterprise, and their industrialists knew how to organize and run businesses.

What made the United Kingdom different, outweighing all advantages, was a central government that habitually interfered in markets and technology. Everything had to go through Parliament. Government committees vetted innovation. The government could delay technological progress in order to favor a political goal.

Thomas P. Hughes, the historian, described technology as a "complex system of interrelated factors with a large technical component." [518] "Technical" referred to the tools. "Technology" included economic, political, cultural and other factors. "Technology usually has the structure of a goal-seeking, problem solving, open system." [519]

So, what was the goal in the early days: to develop a new industry that would provide the most reliable and economic service to as many consumers as possible, enlarge the country's worldwide business opportunities, protect vested interests, or prop up a powerful ideological movement? Policy makers ignored technological developments abroad, played down innovations that might lead the industry in the "wrong" direction and changed the goal from solving problems for customers to preventing problems for competitors.

The government took the opposite approach when it launched National Grid, embracing a new technology in order to solve a problem for consumers. Utilities and investors

---

517. American markets, incidentally, show the same pattern: poor returns in the competitive sector and high returns in the regulated sector.

518. Thomas P. Hughes, "Technology, History and Technical Problems," in Chauncey Starr and Philip C. Ritterbush, eds., *Science, Technology and the Human Prospect* (NY: Pergamon Press, 1980), p. 142.

519. *Ibid.*

could cooperate or not, but most did because benefits were obvious, and customers came out ahead.

Nationalization neither unleashed the wonders of scientific management nor changed the technological path set by National Grid. It did avert diseconomies of small-scale distribution threatened by the expiration of franchises, although distribution was the industry sector that least benefited from scale economies and Labor could have dealt with franchise expiration without nationalizing the entire industry. Nationalization failed egregiously in one respect: the owner lost sight of the industry's original goals by using it to support coal mining, nuclear power and trade union employment.

Privatization built on the CEGB/National Grid foundation: regional distributors, nationwide transmission, selecting the lowest cost generators for operation. Culturally, privatization replaced public service with short-term profit maximization as the norm for decision making. The privatized industry's chief successes came not from embracing new technology or from reorganizing the market but rather from making common sense commercial decisions (such as getting out of high-priced coal contracts and cutting unnecessary staff) that moved it closer to its primary goal of serving customers economically and reliably.

Nationalizers had to build a system to service the postwar recovery. They met those primarily engineering challenges, perhaps conservatively and inefficiently, but they met them. The privatizers reworked an overstaffed industry with excess capacity and bad contracts. They had to meet one technical challenge, devising a computer system for the Pool, but dealt mainly with financial and commercial issues rather than how to keep the power on. They assembled a package that may have benefited industry players more than consumers and lacked an effective market mechanism to attract capital for long term investment.

The 2012 reforms refocused the industry on providing reliable, low-carbon electricity. Despite government assurances based on computer models that extended to 2050, the reforms looked like an expensive path to the chosen goals. F.A. Hayek, the intellectual icon of Thatcherite conservatism, argued in *The Road to Serfdom*, that society should "employ foresight and systematic thinking in planning our common affairs."[520] Central government should create "conditions under which the knowledge and initiatives of individuals are given the best scope" for them to plan, rather than have the government plan it all, using a "consciously constructed 'blueprint.'"[521] The Coalition dumped Hayek in favor of central planning.

---

520. Hayek, *loc. cit.*
521. *Ibid.*

To employ a market mechanism, the government must declare: "We will reward you to reduce greenhouse gases in the atmosphere in order to prevent global warming. If you figure out how to do it best, you will make the biggest profit. "To employ central planning, the government says, "In order to reduce carbon emissions we will ban certain activities and pay you to engage in others that we choose, to reduce carbon emissions." The former strategy encourages innovators to search for solutions. The latter limits the role of innovation to improving the chosen solutions. [522] The world's most entrepreneurial genetic scientist put money on developing an organism that will eat the carbon and turn it into fuel. The UK government put its money on windmills. Policy makers dislike betting on innovation in the abstract. They cannot envision the answer that innovation will produce. But they do know what a windmill looks like.

Nuclear, wind or solar power are means to reduce or eliminate greenhouse gas emissions from energy production. The goal, however, is to mitigate or halt global climate change in the most efficient manner, not to promote select electric generators. The nation might reach the goal more efficiently by installing energy saving windows and phasing out incandescent light bulbs than by erecting nuclear power stations. Investors, though, will put money where incentives will produce the fattest, surest and fastest profits, not on fixes that produce the biggest emission reduction for money spent. The government's chosen solutions could crowd better ones. The nation needs a mix of tools to fix the problem. Should a government minister or the market should determine the mix?

Network, competition and capital

Electricity supply operates on a network, requires heavy capital investment and provides a service fundamental to modern society. Power generation stresses the environment, requires huge quantities of water, emits air pollutants and produces nuclear wastes. Transmission lines, giant windmills and massive power stations mar the landscape. Until new technologies change this picture society will oversee the business, demanding that it provide a nearly universal, reliable and environmentally benign service at a low price.

For almost a century, the industry provided increasingly reliable service to more consumers. It easily raised capital for expansion despite limitations on earnings, regularly reduced costs, and lowered prices while earnings rose. Integration and economies of scale paid off for everyone.

---

522. Proponents of protection for specific energy sources argue that the market for those resources would not develop without that protection. With an assured market, they can raise money, reduce costs and gain economies of scale. That may be so, but the policy may reduce the prospects for other resources that might offer equal or better means of producing power or saving energy or reducing carbon emissions.

The industry ran as a regulated monopoly, a condition justified by economies of scale: one large firm could serve the market at a lower unit cost of output than many smaller operations. To assure that one big, low-cost firm would supply the market, the state prohibited the entry of competing firms. To assure that the sole firm in the market passed on its low costs in the form of low prices, the state regulated prices. Costs declined as scale increased for over 80 years.

That picture changed in the 1960s and 1970s. Coal-fired power stations reached maximum efficiency, so building bigger ones no longer produced savings. The leap to even larger nuclear power stations raised costs more. Finally, manufacturers developed gas-fired generators both smaller and more efficient that the big coal-fired stations. Economies of scale no longer obtained. No need for a monopolist anymore. The small competitors could produce low-cost power, and if they competed with each other, the competition would drive down the price.

Those new cost curves emerged during a period of ideological upheaval. The Carter administration had unleashed competition on formerly regulated sectors. Reagan embraced those reforms and extended them. The Pinochet regime in Chile bought into a Chicago School competitive economic model. Mrs. Thatcher wanted to reduce governmental influence on the economy. And then capitalism triumphed. The Berlin Wall fell. Communism collapsed.

Power plants, the free marketeers argued, are like oil refineries. They do not require regulation. Let them compete. Let customers choose their retail suppliers, too. If consumers can figure out how to buy gas for their cars, they can figure out how to buy electricity. The United Kingdom, Chile, parts of the United States and the European Union reorganized electricity networks, broke them apart and established central markets for generation. The new electric industry reduced operating costs while still keeping the lights on but without the spectacular savings or new products seen after other deregulations.

The reformers did not foresee the potential for higher capital and transaction costs or for the market manipulation that ensued after they replaced a cooperative culture with a purely economic set of incentives. Nor did they anticipate that reformed markets might not easily attract capital for long term investments needed to lower greenhouse gas emissions or to maintain reliability. At the time of reforms, few worried about reliability or greenhouse issues, and if they did, they had confidence that the market could solve the problems. In the United States, in the absence of market signals, regulators have had to offer bonuses to attract capital for transmission lines, states to mandate purchase of renewable resources, and regional transmission organizations to institute special contracts to secure generation capacity. The United Kingdom government, dealing with similar issues, had to impose solutions, too.

Policymakers and regulators had a static view of the electricity market, focusing on cost reductions rather than new products. If they had actually controlled the lighting market for the past two centuries, astoundingly cost-effective candles would now surely light the world. Edison did not perfect the candle. He created a new product. The cellular telephone's progress tells a similar story. For almost three decades, telecommunications providers and equipment manufacturers relentlessly reduced the size of the phone until it fit into the palm of the hand. It took an outsider, Apple, to create a completely new product, the smart phone.

Richard Schuler, a regulator, economist and engineer argued that "the real benefits that might evolve from market-based rather than regulated supply industries" come from "the pace and type of change that markets induce – many times not foreseen." [523] Deregulation and privatization sped improvements, such as gas-fired generators, smart grid and meters, and financial innovations to control risk. American experience shows that both regulated and government-owned utilities adopted these new technologies. Of course they were not disruptive in the sense that both the old utilities and their restructured brethren adopted the new technologies without substantial changes in their business models. Maybe the real innovation has to come from outsiders unhindered by legacy assets, able to offer their products directly to consumers, after the breakup of the utility monopoly.

The deregulators focused on what economist John Maurice Clark called the "normative ideal" of "pure and perfect competition." [524] Clark argued that departures from that norm "are not only inseparable from progress but necessary to it." [525] Progress means more than the same thing at a lower price. In competitive markets of the textbook variety, competitors grind each other down, reduce costs and prices, but do not produce new products. Regulated utilities, too, produce few new products. Regulators inhibit innovation because they not only have to approve new ideas but also have so many ways to limit the profits derived from the innovation. So why rock the boat for a questionable reward? Yet deregulated electricity suppliers have offered few innovations. The big ones developed a simple business model: hold down costs and offer prices close enough to competitors to neither alienate customers nor attract government scrutiny. The small ones have a hard time staying in business for long.

Policy makers saw a lazy regulated industry. Economies of scale had played out, leaving no rationale for a monopolistic structure. So they dismembered the industry and introduced competition. The old electric companies, though, subsequently reassembled in order to regain lost economies, achieve the right balance of risk and reward and reestab-

---

523. Leonard S. Hyman and Richard E. Schuler, "Electricity Restructuring, Consumer Prices and the Cost of Capital: Lessons for the Modeling of Future Policy," presented at HICSS-47, Jan. 2014, p. 7.

524. John Maurice Clark, *Competition as a Dynamic Process* (Washington, DC: The Brookings Institution, 1961), p. ix.

525. *Ibid.*

lish pricing power. The unregulated sector had liberated itself from the old obligation to serve the public, and could withdraw assets from public service as means to manipulate price. Regulated utilities, in contrast, still had to serve and invest or lose their franchises, but they had the advantage of knowledge over an understaffed regulator and an ever changing ministry.

After deregulation, policy makers had to reimpose regulation indirectly by determining the means with which the industry would meet environmental, societal and economic goals since the market did not. It may looked like regulation with a competitive twist (lowest bidder gets to execute the mandate) but government choices of and specifications for projects could have affected electric bills more than differences between bidders. No doubt, when conditions warrant, policy makers will change the market framework again.

Planners project that society will consume increasing volumes of clean electricity and the incumbent electric industry will supposedly make the massive investments to meet those needs, even though it has no obligation to do so. And it might not. Outsiders, alternate suppliers, individuals with on site generation and storage, can take advantage of declining costs and new technology to take market share. Energy saving trends could wreck the projections altogether. Government planners obsess over directives to a fading electricity industry that seeks profit opportunities more from following government edicts than by creating competitive products for customers. Technology, rather than government policy, will open the door to real change, perhaps to competition aimed at the industry itself, rather than a game played by industry insiders.

Conclusion

One hundred forty years of UK industrial history boiled down to six take-aways reads like this:

- Government interference focusing on industry structure and ownership did not help consumers – with one exception.
- Nationalization was unnecessary, but its worst aspects resulted not from market organization but rather from overstaffing, bad coal contracts and nuclear mismanagement.
- Privatization's benefits came largely from common sense decisions such as laying off unnecessary staff and getting rid of bad coal contracts, rather than from market structure.
- Unregulated firms, despite consumer-unfriendly activities, earned only average profits, and at least in that sense competitive markets worked.

- Regulated firms made extraordinary profits, far more than needed to attract capital, and the regulatory framework evolved into rate-of-return-on-steroids disguised as something else.

- Competitive electricity markets, as structured, cannot attract long-term capital and the government's favored investments (nuclear, renewables, capacity facilities) may crowd out competitive generators.

Nobody will put the industry back together again. The neo-liberal reformers who made the structural changes in the 1980s and 1990s destroyed the public service culture that had existed before. Rejoining the parts will not re-create that culture. Furthermore, a retreat from the facade of market-based solutions would be too wrenching. And, why bother, anyway, now that the government is on its way to controlling at least half the generation through contracts and mandates of various types? But the best reason not to spin wheels on still another restructuring is that change will come anyway.

Technology and capital, more than regulatory framework or ownership, set most of the costs of providing electricity. The managers of the old utilities worked hard (although not always brilliantly) to run reliable and affordable operations because that was their job as public servants. Deregulation or restructuring or privatization could not drastically reduce costs – beyond dispensing with the politically imposed ones – or create new products because they did not change the industry's technology. Deregulators, in a sense, cleaned the barnacles and weeds off the ship's hull. That sped up the ship but did not change her shape.

When policy makers tire of deregulation and decide to try something else, they will make just as little difference without accompanying technological change. If the industry could find or develop a clean, cheap, quickly-built, large scale, universally acceptable and reliable source of power, the pendulum could swing back to network-based, regulated, central station monopolies. The rapid decline in the costs of network-based renewables should assure the network of a place in the market, but a smaller one than before, given the capital and intellectual effort pouring into advances in small scale and on-site technologies.

The real trick is to develop policies that maintain the electricity infrastructure while opening the way for change. Nobody knows whether the successful new technology will involve solar power windows and curtains and small batteries or giant, solar power satellites beaming energy into a central grid. The nation needs to support options not likely to be offered by the market.

In the end, though, the outsiders – firms with huge resources – will have the greatest incentive to implement the most effective and disruptive technologies. They will force the next round of changes and consumers, by their choice of technologies, will make the final pronouncement on the direction of the electricity industry.

Appendix: Technics and Metrics

For centuries the UK's baffling monetary system required translation for the decimally-inclined. The electric industry uses engineering terminology and financial concepts that puzzle the innumerate, industry reformers and maybe even energy secretaries who arrive anew on the scene almost annually. This appendix provides simplified explanations. A more thorough review is available elsewhere.[526]

Currency

Before the UK switched to the metric system in 1971, its currency, the **pound sterling** (£), was divided into 20 **shillings** (s) and the shilling into 12 **pence** (d), meaning that the pound consisted of 240 pence.

After 1971, the UK abolished the shilling and divided the pound into 100 pence (symbol "p"). Tables in this book generally decimalize currency references in order to make tables consistent.

Fiscal Years

The UK government **fiscal year** (FY) begins on the first of April and ends the following thirty-first of March. (FY 2014/2015 ending on the 31st of March 2015 is called FY 2015.) British utilities operating as government agencies reported on a fiscal year basis. Many changed subsequently to calendar years. Tables in this book that compare fiscal year to calendar year data equate fiscal year to previous calendar year. That is, fiscal 1980/1981 data for the UK would be compared to calendar year 1980 data for the USA because 9 of the 12 months were in calendar 1980.

---

526. Leonard S. Hyman, Andrew S. Hyman and Robert C. Hyman, *America's Electric Utilities: Past, Present and Future* (Vienna, VA: Public Utilities Reports, 2005).

Electrical Prefixes and Measures

Electrical terminology affixes Greek prefixes that indicate quantity to units of measurement named after dead scientists and inventors:

Kilo –	Thousand (10^3) –	abbreviated k or K
Mega –	Million (10^6) –	abbreviated M
Giga –	Billion (10^9) –	abbreviated G
Tera –	Trillion (10^{12}) –	abbreviated T

The **Watt** measures work produced in one second. The **kiloWatt**, one thousand times larger, is the commonest electrical unit of work. A plant with a **capacity** of 10 kiloWatts (kW) can produce 10 kiloWatts simultaneously in one second. The **kiloWatt-hour (kWh)** measures how many kiloWatts a facility produces in one hour, or, to put it another way, the output of a facility with a capacity of one kiloWatt that operates continuously for one hour. Look at it this way. The local bakery has an oven that holds 10 pies at one time. It has a capacity of 10 pies. Each pie takes one hour to bake. Thus the bakery can produce 10 pies per hour.

Voltage, also called electromotive force, is akin to pressure that pushes electricity through the network. Its basic unit is the **Volt** but the **kiloVolt** (kV) most often used as a measure. Generators measure size (capacity) in kW or MW, while transmission lines are rated in kV (the higher the voltage the bigger the line).

Industry sources vary their use of capitals in electrical terms. A KWH is a kWh is a kwh.

Capacity, Reserves and Efficiency

Demand (in kW) measures total generating capacity required by customers at a given moment. Because the system has a limited ability to store electricity, at time of highest demand (**peak**) it must have generators available to meet that demand. In order to assure service, the system maintains a **reserve** (measured in kW), whose adequacy is judged by the **reserve margin** or **capacity margin** (spare capacity as a percentage of capacity or of peak load). The network with too low a reserve margin risks blackouts while too high a margin burdens customers who have pay to keep those extra units available.

The peak period lasts for only part of the day (see Figure A). Generators dedicated to meeting peak operate for only a few hours. If customers could reduce usage during peak hours and move some of it to other parts of the day, the network could reduce the generating capacity it maintains and use the remaining generators more, as shown in Figure B.

In Figure A, the electric system keeps 14 kW of capacity available, with a reserve of three kW, in order to meet peak demand of 11kW at hour 14. If customers shift their demand to an off peak period, as in Figure B, the electric company sells as much electricity as before, cuts peak demand by three kW, keeps the same reserve of three kW, but now needs only 11kW of capacity to serve customers. It has to support less plant that sits idle most of the day.

Figure A. Normal Peak

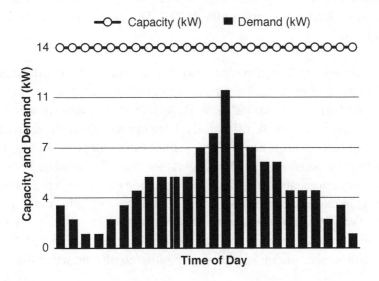

Figure B. Reduced Peak, Reduced Capacity, Same Sales

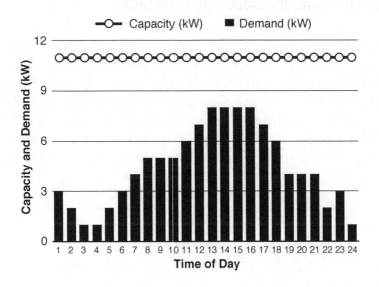

Capacity factor measures average load as a percentage of capacity and **load factor** average load as a percentage of peak load. The higher either factor, the better the utility makes use of its plant. Too high a capacity or load factor, though, might signal that the utility runs plant beyond optimal levels, which could lead to equipment breakdowns.

Generators burn huge amounts of fuel but one-to-two-thirds of its heat content goes up the smokestack rather than producing electricity. Ideally, the generator should burn as little fuel as possible to produce a kWh of electricity. **Heat rate** measures generator efficiency by the amount of heat (caloric content of fuel) needed to generated one kiloWatt-hour of electricity, in **British thermal units** (BTUs). The lower the heat rate, the more efficient the generator.

Base load generators are big, low-cost units that run most of the time. **Intermediate** (or mid merit) units have higher costs. They run when needed and when the price is high enough to make their operation profitable. **Peaking** units are usually smaller, have high operating costs, and can turn on or off quickly. They operate during peak periods and emergencies when no other units are available. Those three categories of plant produce the energy required by consumers. Other generators provide **capacity.** They are kept ready to maintain system reliability, rather than to provide energy on a regular basis. **Renewable** energy generators produce only when conditions are right, that is, when the wind blows or the sun shines. Their output is **intermittent**. As a result, the network operator cannot rely upon them in the same way as base load generators and the network has to carry an extra reserve of quick starting generators to turn on when the renewable goes off line.

Economies of Scale and the Declining Cost Curve

Typically, businesses have U-or saucer–shaped cost curves. (See Figure C.) At low levels of output, their cost per unit is high. They cannot get volume discounts from suppliers, they pay staff not fully occupied and their machinery is not fully utilized. As sales rise, the firm makes better use of assets, the staff works more for the same pay as before, and the purchasing manager can secure volume discounts. After a while, as volume soars, the organization becomes bureaucratic, people get in each other's way, the boss does not know what is going on, the machinery runs without time off for maintenance, customers get annoyed about delays, and costs per unit produced rise.

Figure C. Cost per Unit of Output: the Normal Cost Curve

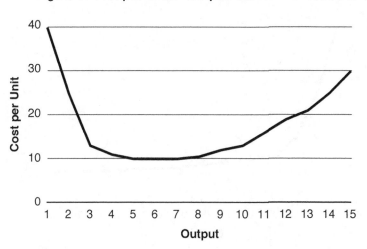

In Figure C, cost per unit of output bottoms out at five units. Thus, in a market for 15 units a year, three firms producing five units each could supply the market at the lowest cost. Consumers would benefit from competition between three firms each operating at optimal size.

Contrast that picture with the cost curve of the average utility during the good old days of economies of scale (Figure D). As the utility grew in size, costs per unit produced declined. If the market were 15 units per year, one firm producing 15 units would have a lower cost of production than, say, three firms producing five units each. One big firm could serve consumers better than several smaller, less efficient competing firms. A **natural monopoly** exists. In order to make sure that only one firm, operating at the lowest cost, serves the market, the state prevents competitors from entering the market. In order to prevent the natural monopolist from taking advantage of its position, the state regulates its prices or its profits.

Figure D. Declining Cost Curve, Economies of Scale

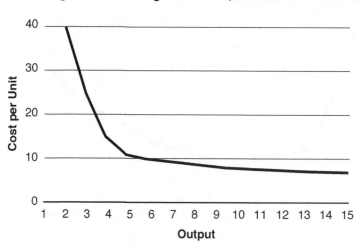

Financial Concepts and Ratios

Cost of capital is return required to convince investors to put money (capital) into a business. Risk determines cost of capital. The higher the risk, the greater the return required. If an investment costs £100 and has a cost of capital of 10%, it must earn at least £10 per year in order to attract investors. They will not invest to earn only £8. If earnings subsequently rise to £12, more than cost of capital, the investment's market value will rise above £100. Regulators permit a utility to earn cost of capital on investment dedicated to public service. When the utility earns less, the market value of its traded securities falls below that of the dedicated investment and when it earns more, market value trades above that investment. James Bonbright, legendary authority on utility regulation and finance, argued that utilities should earn cost of capital as a minimum, not as a maximum. The regulator has to allow a return sufficient to attract capital and provide a buffer against misfortune, but not to rob consumers.

Return on invested capital (ROIC), also called return on capital, (ROC) or return on capital employed (ROCE) is a key measure of the profitability of total investment. To calculate it, first determine operating income, the sum remaining from revenues after subtracting all operating expenses, including taxes and depreciation of assets. From operating income, the corporation pays interest on debt and dividends on stock, and anything left over constitutes earnings reinvested in the business. Some analysts prefer to use pretax operating income, which is calculated before subtracting tax payments.

Return on invested capital is calculated by dividing operating income by invested capital (money invested in the business by shareholders plus profits retained plus borrowed money) and then showing that number as a percentage. Companies that borrow to raise part of their capital have to divide the operating income into two components, **interest** paid on the debt and **net profit** available to stockholders. A company raises £1,000 by selling £500 of debt with an interest cost of 5% (£25) and £500 of common stock equity. It has an operating income of £100. After paying £25 of interest, it has £75 left as a net profit for the stockholders. The company earns a 10% return on capital (£100 /£1,000), pays 5% on its debt and the stockholders' **return on equity** is 15% (£75/£500).

To get around dodgy accounting practices employed to make things look better than they are, financial analysts look at two other indicators. **Earnings before interest and taxes** (EBIT) measures the earning power of the enterprise. EBIT is also called pretax operating income (POI). **Earnings before interest, taxes, depreciation and amortization** (EBITDA) is cash profit derived from operations, the cash coming in the door that the tax collectors and creditors can rely on for payments. People that buy whole businesses rely on these measures. Analysts often value a company by its **enterprise value** (EV), the market value of its stocks and bonds outstanding, as a multiple of its EBITDA.

The c**apitalization ratio** shows the percentages of invested capital contributed by borrowings and shareholder money. The greater the percentage from lenders (debt), the riskier the company. Too much **leverage** or **gearing** (high debt levels) raises the risk for investors. The **debt ratio** is the percentage of capitalization contributed by debt, and the **equity ratio** the percentage made up by stockholder investment and retained earnings. Bond investors and analysts also rely on various measures of **coverage,** such as how many times over the company earns the interest that it pays. (the higher the better) and the ratio of outstanding debt as a multiple of EBITDA (the lower the better).

Dividing the company's net income and dividends paid by the number of shares (units of ownership in the company) outstanding produces **earnings per share** (EPS) and **dividend per share** (DPS) respectively. Investors measure a stock's price by the ratio of price to earnings per share, the **price/earnings** or **P/E ratio**. A stock with an EPS of £10 and a price of £100 has a price/earnings (P/E) ratio of 10x. Some investors reverse the order and calculate an **earnings/price** or **E/P ratio** (also called "earnings yield") of 0.10 or 10%. Fast growing stocks sport high P/E (and low E/P) ratios. Most established companies pay part of their earnings to stockholders in the form of dividends. The **dividend yield** is the dividend as a percentage of the stock price. A stock selling for £100 that pays a £5 dividend has a is 5% dividend yield. Investors profit on their share investment from the dividend and from price increase in the stock (**capital gain**). The combination of dividend plus capital gain is **total return**.

The financial statement on the company's books will look like this:

	Revenues
less	All operating expenses (including depreciation but not income taxes)
equals	Operating income (pretax)
less	Interest expense
equals	Income before income taxes
less	Income tax expense
equals	Net income

For analytical purposes, the following format makes ratio calculation easier:

Financial Ratios: A Simple Example

Line	Item	Amount (£)	Comment (Lines)
	Income statement		
1	Revenue	1,000	
2	Operating expenses		
3	Depreciation and amortization	100	
4	Income taxes	150	
5	Other operating expenses	350	
6	Total operating expenses	600	3 + 4 + 5
7	Operating income	400	1 - 6
8	Interest expense	100	5% on £2,000
9	Net income	300	7 - 8
10	Dividend paid	150	
	Balance sheet		
11	Assets		
12	Current assets	100	Receivable in one year
13	Net plant	4,000	Plant less depreciation
14	Total assets	4,100	12+13
15	Liabilities and capital		
16	Current liabilities	100	Payable in one year
17	Capital		
18	Debt	2,000	Payable after one year
19	Equity	2,000	Stockholder investment
20	Total capital	4,000	18 + 19
21	Total liabilities and capital	4,100	16 + 20
	Other information		
22	Number of shares	100	
23	Price of stock year end (£)	60	
24	Price of stock beginning of year (£)	55	

How to Calculate Financial Ratios (line numbers from "Financial Ratios: A Simple Example," above, are italicized when inserted in formulas, all other numbers are £except where noted otherwise)

25. Earnings before interest and taxes (EBIT): *1-3-5* = 1,000 - 100 -350 = 550
26. Earnings before interest, taxes, depreciation and amortization (EBITDA): *1 -5* = 650
27. Pretax operating income (POI): *1-3-5* = 1,000 - 100 - 350 = 550
28. Market value of common stock (MV): *22* x *23* = 100 x 60 = 6,000
29. Enterprise value (EV): *18* + MV = 2,000 + 6,000 = 8.000
30. Return on capital (ROC): *7/20* = 400/4,000 = 0.10 = 10.0 %
31. Pretax return on capital (PROC): POI/*20* = 550/4,000 = 0.1375 = 13.75%
32. Return on equity (ROE): *9/19* = 300/2,000 = 0.15 =15.0%
33. Debt ratio (DR): *18/20* = 2,000/4,000 = 0.50 = 50.0%
34. Equity ratio (ER): *19/20* = 2,000/4,000 = 0.50 = 50.0%
35. EV/EBITDA: EV/EBITDA = 8,000/650 = 12.31x
36. Pretax interest coverage (PIC): POI/*8* = 550/100 = 5.5x
37. Debt/EBITDA: *18*/EBITDA = 2,000/650 = 3.08x
38. Earnings per share (EPS): *9/22* = 300/100 = 3.00
39. Dividend per share (DPS): *10/22* = 150/100 = 1.50
40. Dividend payout ratio (PO): *10/9* = 150/300 = 0.50 = 50.0%
41. Price /earnings ratio (P/E): *23*/EPS = 60/3.00 = 20x
42. Earnings /price ratio (E/P): EPS/*23* = 3.00/60 = 0.05 = 5.00 %
43. Dividend yield (DY): DPS/*23* = 1.50/60 = 0.025 = 2.5%
44. Total return (TR): [(*23* -*24*) + DPS]/*24* = [(60 -55) + 1.50]/55 = 6.50 /55 = 0.1182 = 11.82%

Current vs Historical and Real vs Nominal

Inflation raises asset values and decreases purchasing power, distorts inter-period comparisons and undermines the value of accounting data.

Businesses tend to keep books on a **historical cost** basis except in countries with high inflation, where they used an inflation adjusted or **current cost** basis.[527] Choice of cost changes the look of the books without changing anything tangible. The switch from historical to a current cost accounting made the British electricity industry look less profitable without affecting its cash income. Say that Utility bought its assets for £1,000 years ago, collects £200 per year selling electricity and depreciates plant over 20 years, thus writing off £50 as an expense every year. It lays out £50 every year for the fuel and other operating expenses. Thus, Utility earns a profit of £100 and a return on plant of 10%.

Utility pretends that it paid current prices for its assets, for accounting purposes, or £2,000 and depreciating those assets over 20 years would require a depreciation expense of £100 per year. However, Utility still sells the same amount of electricity for £200 and has the same operating expenses of £50. Using the current cost accounts, however, the profit falls to £50, which produces only a 2.5% return on the current cost investment of

---

527. The calculation of current costs is a process open to all sorts of problems and judgement calls. Should the accountants simply write up assets using a cost of living index? Should they assume that a builder today would reproduce an old asset using old methods at current costs or should they put in the cost of a modern assets that fulfills the same functions? Current cost accounting just fudges the numbers differently than historical cost accounting. The cash flows show the real numbers and they should be identical using both procedures.

£2,000. From a cash basis, though, Utility laid out £1,000, not the current cost, and it still collects the same £200 as before from customers and shells out the same £50 for operating expenses. Nothing has changed other than that the business look less profitable.

	Historical Cost	Current Cost
Revenue	£200	£200
Operating expense	50	50
Cash flow from operations	150	150
Depreciation	50	100
Net profit	100	50
Assets	1,000	2,000
Return (net profit/assets)	10.0%	2.5%

Inflation distorts comparisons. Worker earns £1,000 in year 1 and £2,000 in year 10. Salary doubles. Worker is twice as well off? The cost of a package of household goods that Worker buys rises from £900 in year 1 to £1,800 in year 10. A **cost of living index** with the first year as a base would rise from 100 to 200 (cost of living doubles). Salary doubles but so does the **cost of living**, and Worker is no better off in **real terms** than ten years ago, even though the **current** or **nominal** numbers reported at the time show Worker making twice as much as before. That is, the pound's purchasing power halves in the ten years. To show real numbers, calculate all numbers using purchasing power of currency in a particular year, usually the first or last year of the analysis as a base for the analysis.

	Year 1	Year 10
Nominal wages	£1,000	£2,000
Cost of living index (year 1 = 100)	100	200
Real wages (in year 1£)	£1,000	£1,000
Cost of living index (year 10 = 100)	50	100
Real wages (in year 10£)	£2,000	£2,000

For simple calculation, divide the nominal number by a price index converted from 100 to 1.00 for the base year. To determine change in the value of the currency, use a **cost of living index** or an index that takes into account all the price changes in the economy, such as the **GDP deflator**.

British regulators set prices using estimates of expenditures to be made over a multi- year period. Prices rise or fall over time. Regulators cannot accurately predict the current costs in each of the upcoming years of and they do not try. They work with **real prices**, that is, they calculate as if prices for goods and services will not change from year to year. Subsequently, in each year, they adjust the cost charged to customers for the year's inflation. In the United States, regulators tend to work with nominal numbers and do not make automatic adjustments for inflation. Thus, in inflationary periods, American utilities have to ask for price changes on a frequent basis, and the rate case involves much more than an automatic price adjustment. The regulator likes to look at everything, starting from scratch.

Calculating the Cost of Capital the CAPM Way

OFGEM uses the Capital Asset Pricing Model (CAPM) to calculate cost of capital. CAPM says that return should rise as risk increases. Beta (β), the variability of the security's return versus that of the market, measures risk. (To simplify, a stock that rises or falls 20% when the market rises or falls 10% has a beta of 2. If it rises or falls 5% compared to a market increase or decrease of 10% it has a beta of 0.5.) To determine the cost of equity capital, CAPM starts with return on risk-free capital (government bonds), and then adds on an appropriate premium for the risk of investing in the stock of the company. For those who like formulas:

$$R \ = \ \text{cost of equity capital for a specific security}$$
$$\beta \ = \ \text{beta of that security}$$
$$Rf \ = \ \text{return on risk free capital (the government bond)}$$
$$Rm \ = \ \text{average return earned by all stocks (market rate of return)}$$
$$\text{and:}$$
$$R \ = \ Rf + \beta \ (Rm - Rf)$$

For instance, the electric company stock has a β of 0.8 and risk free bonds yield 3%. In the past, the average stock earned a 7% total return while the average risk-free bond yielded 4% (meaning that the equity risk premium was 3% for owning a stock over a bond). So the cost of equity capital for the electric company is:

$$R = 3\% + 0.8 \ (7-4\%) = 5.4\%.$$

CAPM can produce biased results because regulators and hired experts can choose data sets to produce the desired conclusions. Beta and the equity premium change over time, too, so the past may not predict the future. The formula, furthermore, does not work equally well for all types of stocks (producing unrealistically low predicted return for stocks with low price-to-book ratios and low betas, categories that include utility stocks). Which is why many US regulatory agencies use multiple methods to determine cost of capital.

According to Fischer Black, who helped to invent CAPM, "Estimating expected return is hard … We need decades of data for accurate estimates …"[528] Should regulators, then, have relied on data from the few years after privatization to draw conclusions? Next, within a few years of privatization, most of the companies in the electric industry ceased to trade in the market. Thus the CAPM model calculation had to rely on a narrow sample or use substitutes or assume that all betas eventually converge to the market average of 1.0, or just rely on guesses for beta. Harry M. Markowitz, who won a Nobel Prize for his contribution to the development of CAPM, wrote:

> The capital asset pricing model (CAPM) is an elegant theory … Both the original CAPM … and the alternate CAPM … imply that the expected return of a stock depends in a simple (linear) way on its beta … This conclusion has been used for estimating expected returns, but it has lost favor … because of poor predictive results …the conclusion that expected returns are linear functions of beta does not hold when real-world limits … are introduced …[529]

Fortunately, UK regulators use CAPM studies that produce wide ranges for cost of capital so they have room to exercise judgment. A one percentage point change in return, in the late 1990s, would have produced roughly a 3% change in required revenue, so rate of return matters, but possibly less than other factors in the price setting proceedings. Despite the specificity of the numbers used, setting a rate of return is not rocket science, and possibly no form of science, but rather judgment buttressed by well chosen numbers.

---

528. Fischer Black, "Estimating Expected Return," Financial Analysts Journal, January-February 1995, p. 168.
529. Harry M. Markowitz, "Market Efficiency: A Theoretical Distinction and So What," Financial Analysts Journal, September/October 2005, pp. 17-18.

Appendix: Statistical Tables

Table A. Population (millions) and Real Gross Domestic Product (1990 $ billions)

Year	UK Real GDP ($ billions)	US Real GD ($ billions)	UK Population (millions)	US Population (millions)	UK Real GDP per Capita ($)	US Real GDP per Capita ($)
	[1]	[2]	[3]	[4]	[5]	[6]
1882	122.5	177.2	32.9	53.0	3,725	3,341
1887	129.8	200.5	34.2	59.4	3,797	3,372
1892	140.0	246.0	35.6	65.9	3,933	3,732
1897	162.8	273.5	37.3	72.5	4,360	3,774
1902	181.0	351.7	39.1	79.5	4,628	4,426
1907	195.4	442.9	40.8	87.3	4,784	5,071
1912	206.5	498.3	42.4	95.7	4,868	5,207
1917	241.3	545.4	43.5	103.8	5,544	5,254
1922	196.4	612.8	44.4	110.5	4,427	5,546
1926	213.2	779.2	45.2	117.9	4,713	6,610
1927	230.4	786.8	45.4	119.5	5,075	6,584
1932	227.8	616.4	46.3	125.4	4,916	4,914
1937	280.7	833.4	47.3	129.5	5,937	6,438
1942	353.0	1,320.4	48.4	135.4	7,294	9,753
1946	317.0	1,306.9	49.2	141.9	6,440	9,207
1947	312.3	1,287.2	49.5	144.7	6,306	8,896
1950	344.9	1,457.6	50.2	152.3	6,847	9,573
1960	448.9	2,022.2	52.4	180.7	8,571	11,193
1970	594.9	3,045.8	55.6	205.1	10,694	14,854
1980	719.5	4,161.0	56.3	227.8	12,777	18,270
1989	932.3	5,419.2	57.1	248.8	16,288	21,783
1990	935.9	5,464.8	57.2	249.9	16,302	21,866
2000	1,264.4	7,664.6	58.9	281.4	21,467	27,237
2010	1,487.7	9,021.8	62.3	308.7	23,880	29,225
2015	1,578.2	10,466.8	65.1	321.8	24,166	32,526

Notes for Table A (all columns):
1882-1990 data in Geary-Khamis Dollars.
All post 1990 data in $1990 as estimated by author from standard economic sources.
UK v. USA for post 1990 comparison based on OECD purchasing power parity analyses.
Sources for Table A (all columns):
1. 1882-1990 data from Angus Maddison, *Monitoring the World Economy* ("Maddison")
   (Paris: Organisation for Economic Cooperation and Development, 1995).
2. Post 1990 data from US Dept. of Agriculture, Economic Research Service, "Real Historic Gross
   Domestic Product (GDP) and growth Rate of GDP for Baseline countries/regions (in billions of 2010
   dollars) 1969-2014, Updated 12/18/2014." World Bank Development Indicators, International Finance
   Statistics of the IMF, 2017, adjusted by author to most recent data and put on 1990 base.

Table B. Electricity Generation by Ownership (million kiloWatt hours)

Year	UK Utility and Wholesale Generators	UK Isolated and Industry	UK Total	US Utility and Wholesale Generators	US Isolated and Industry	Total US
	[1]	[2]	[[3]	[4]	[5]	[6]
1882	0	0	0	0	1	1
1887	15	10	25	175	325	500
1892	30	20	50	300	800	1,100
1897	95	55	150	800	1,700	2,500
1902	451	263	714	2,507	3,462	5,969
1907	1,275	802	2,077	5,862	8,259	14,121
1912	2,172	1,751	3,923	11,569	13,183	24,752
1917	4,198	1,260	5,458	23,438	17,991	43,429
1922	4,541	2,238	6,779	43,632	17,572	61,204
1926	6,992	2,741	9,733	69,353	24,869	94,222
1927	8,452	3,164	11,616	75,418	25,972	101,390
1932	12,248	3,382	15,630	79,393	19,966	99,359
1937	22,905	6,000	28,905	118,913	27,563	146,476
1942	35,654	10,000	45,654	185,979	47,167	233,146
1946	41,233	22,697	63,950	223,178	46,431	269,609
1947	42,580	22,000	64,580	255,739	51,661	307,400
1950	54,521	18,000	72,521	329,141	59,533	388,674
1960	122,748	13,565	136,313	755,374	88,814	844,188
1970	232,378	15,674	248,052	1,531,609	108,162	1,639,771
1980	269,945	14,132	284,077	2,293,990	60,369	2,354,359
1989	297,890	15,747	313,637	2,891,498	93,677	2,985,175
1990	302,936	16,824	319,760	2,907,158	130,830	3,037,988
2000	341,783	33,934	375,717	3,645,432	156,673	3,802,105
2010	347,580	30,976	378,556	3,980,970	144,082	4,125,060
2015	307,596	28,760	336,756	3,919,294	158,307	4,077,601

Notes for Table B

1. Columns 1-3 for 1882-1892. Author's estimates based on lamp data.
2. Columns 1-3 for 1895-1917. Byatt adjusted data in *Garcke's Manual,* which appears in Hannah, was adjusted by author for 25% estimated system losses plus self generation by industrial firms, also estimated by author (partly from data in Byatt) to calculate total electric generation.
3. Columns 1-3 for 1992-2012. Author's estimates for missing self-generation data.
4. Columns 2 and 5. Generation on customer premises and industrial self generation.
5. Columns 5-6 for 1882-1897. Industrial load estimated at 2-2.5 times that purchased from utility. Losses of 25% assumed from generation to sales.

Sources for Table B.

1. Columns 1-3. for 1895-1917. I.C.R. Byatt, *The British Electrical Industry* ("Byatt") (Oxford: Clarendon Press, 1979), pp. 74-76, 95, 107. *Manual of Electrical Undertakings* ("Garcke's *Manual*"), (London:

Electrical Press, various dates). Leslie Hannah, *Electricity Before Nationalisation* ("Hannah, *Electricity*") (Baltimore: Johns Hopkins Press, 1979), pp. 422-428.

2. Columns 1-3 for 1922-2012. Department of Energy & Climate Change (DECC), "Electricity generated and supplied 1920-2012." DECC, *Digest of UK Energy Statistics* (DUKES), 2013. Ministry of Energy, *Digest of Energy Statistics*, 1970.

3. Columns 4-6 for 1882-1897. Elfun Society, *The Edison Era 1876-1892* (Schenectady, NY: Elfun Society, 1976). *Electrical World*, Vol. 80, no. 11, September 9, 1922, p. 546.

4. Columns 4-6 for 1902-1970. Department of Commerce, *Historical Statistics of the United States: Colonial Times to 1970* (Washington, DC: U.S. Government Printing Office, 1975), p. 821 ("Historical Statistics 1970"). Energy Information Administration (EIA), *Annual Energy Review*, various issues. EIA, *Monthly Energy Review*, various issues. Edison Electric Institute (EEI), *Statistical Yearbook of the Electric Utility Industry*, ("EEI, *Statistical Yearbook*") various issues. EEI, *Pocketbook of Electric Utility Industry Statistics*, various issues.

5. All columns after 1970 from DECC and EIA.

Table C. Electric Generation per Capita (kWh) and per Dollar of Real GDP
(kWh per $1990 thousand)

Year	UK kWh per capita	US kWh per capita	UK kWh per $ real GDP (kWh per $ 1,000)	US kWh per $ real GDP (kWh per $ 1,000)
	[1]	[2]	[3]	[4]
1882	0.0	0.0	0.0	0.0
1887	0.7	8.4	0.19	2.49
1892	1.4	16.7	0.36	4.47
1897	4.0	34.5	0.92	9.14
1902	18.3	68.4	3.95	16.97
1907	50.9	161.8	10.63	31.88
1912	92.5	258.6	19.00	49.67
1917	125.5	418.4	22.62	79.63
1922	152.9	553.9	34.52	99.88
1926	221.2	799.2	45.65	120.92
1927	255.9	848.5	50.42	128.86
1932	337.6	793.8	68.61	161.19
1937	611.1	1,131.1	102.97	175.76
1942	943.3	1,721.9	129.33	176.57
1946	1,299.8	1,798.6	201.74	206.30
1947	1,304.6	2,121.4	206.79	238.81
1950	1,444.6	2,552.0	210.27	266.65
1960	2,184.5	4,671.8	303.66	417.46
1970	4,461.4	7,995.0	416.96	538.37
1980	5,045.8	10,335.2	394.83	565.82
1989	5,492.8	11,998.3	336.41	550.85
1990	5,590.2	12,156.8	341.66	555.91
2000	6,378.9	13,511.4	297.15	496.06
2010	6,076.3	13,362.8	254.53	457.22
2015	5,166.7	12,671.2	213.13	389.57

Notes and Sources for Table C:

See Tables A and B.

Table D. Electric Utility Sales to Ultimate Customers (million kWh), per Capita (kWh) and per Dollar of Real Gross Domestic Product (kWh per $1990 thousand)

Year	UK Electric Utility Sales to Ultimate Customers (million kWh)	US Electric Utility Sales to Ultimate Customers (million kWh)	UK Sales per Capita (kWh)	US Sales per Capita (kWh)	UK Sales per $1000 Real GDP	US Sales per $1000 Real GDP
	[1]	[2]	[3]	[4]	[5]	[6]
1882	0	1	0.0	0.0	0.00	0.01
1887	5	131	0.1	2.2	0.04	0.65
1892	10	225	0.3	3.4	0.07	0.91
1897	72	550	1.9	7.6	0.44	2.01
1902	339	2,200	8.7	27.7	1.87	6.26
1907	956	5,160	23.4	59.1	4.89	11.65
1912	1,629	9,833	38.4	102.7	7.89	19.73
1917	3,149	21,490	72.4	207.0	13.05	39.24
1922	3,759	35,883	84.7	271.3	19.14	58.56
1926	5,817	56,089	128.7	475.7	26.66	71.98
1927	6,981	61,251	153.8	512.6	30.30	77.85
1932	10,176	63,711	219.8	508.1	44.67	103.36
1937	19,169	99,359	334.5	767.3	68.29	118.51
1942	30,386	159,408	627.8	1,177.3	86.08	120.73
1946	34,798	190,795	707.3	1,342.3	109.77	145.99
1947	35,858	217,581	727.4	1,503.7	114.82	169.03
1950	45,474	280,539	905.9	1,842.0	131.85	192.47
1960	102,363	683,199	1,953.5	3,780.8	228.03	341.57
1970	192,907	1,391,359	3,469.6	6,783.8	324.27	456.81
1980	224,250	2,126,094	3,983.1	9,333.2	311.67	510.96
1989	270,400	2,646,810	4,805.6	10,638.3	290.04	488.41
1990	274,430	2,683,976	4,797.7	10,740.2	293.22	490.78
2000	330,090	3,421,414	5,604.2	12,158.5	261.06	446.19
2010	328,829	3,754,841	5,278.2	12,163.3	221.09	416.20
2015	303,157	3,758,922	4,657.0	11,681.0	192.09	359.13

Notes for Table D:

See Tables A-C.

Sources for Table D:

1. See Tables A-C.
2. Columns 1-3. Hannah, *Electricity*, pp. 430, 431. Leslie Hannah, *Engineers, Managers, and Politicians* ("Hannah, *Engineers*") (Baltimore: Johns Hopkins University Press, 1982), p. 292. DECC, DUKES, various issues, Tables 5. 2, 5.3.
3. Columns 4-6. Department of Commerce, *Statistical Abstract of the United States*, various issues. Department of Commerce, *Historical Statistics of the United States 1789-1945* (Washington, D.C.: U.S. Government Printing Office, 1949), p. 159 ("Historical Statistics 1945"). EEI, *Historical Statistics of the Electric Utility Industry through 1970* (New York: EEI, no date) p. 60. EEI, *Statistical Yearbook*, 1992. EIA, *Electric Power Monthly*, various issues.

Table E. UK Utility Sales to Ultimate Customers by Customer Category (%)

	Lighting	Power	Traction/ Railway	Domestic	Commercial	Industrial	Traction/ Railway	Other (Public Lighting)
	[1]	[2]	[3]	[4]	[5]	[6]	[7]	[8]
1882	100.0	0.0	0.0					
1887	100.0	0.0	0.0					
1892	100.0	0.0	0.0					
1897	98.6	1.4	0.0					
1902	72.0	8.2	19.8					
1907	43.5	33.1	23.4					
1912	31.1	51.8	17.1	12.6	16.6	51.8	17.0	2.0
1917				9.2	12.1	65.9	11.4	1.4
1922				9.8	12.2	65.3	11.0	1.7
1926				13.0	14.0	61.8	9.7	1.5
1927				13.2	13.2	62.7	9.3	1.6
1932				19.9	15.5	54.2	8.4	2.0
1937				25.4	15.4	52.3	6.2	0.7
1942				22.1	10.7	63.0	3.8	0.4
1946				33.5	11.1	50.7	3.9	0.8
1947				35.5	11.1	49.0	3.8	0.6
1950				32.8	12.7	50.4	3.1	1.0
1960				34.5	14.2	48.8	1.6	0.9
1970				40.7	16.8	40.4	1.2	0.9
1980				39.1	20.8	38.0	1.2	0.9
1989				34.1	28.2	35.8	1.0	0.9
1990				34.2	28.4	35.7	1.0	0.8
2000				33.9	30.7	34.0	0.7	0.7
2010				36.1	30.4	31.7	1.2	0.6
2015				35.7	31.8	30.4	1.5	0.6

Notes for Table E:
1. Columns 1-3. Author's estimates based on data from Byatt.
2. Columns 4-6. 1912 and 1917 partially estimated by author.

Sources for Table E:
1. Columns 1-3. Byatt, *op. cit.*, p. 98.
2. Columns 4-6. 1912 and 1917 partially estimated by author.
3. Columns 1-6. Data allocated by author to maintain consistency in time series.

Table F. US Utility Sales to Ultimate Customers by Customer Category (%)

Year	Lighting	Power	Railway	Residential	Commercial (Small Light and Power)	Industrial (Large Light and Power)	Railway	Other
	[1]	[2]	[3]	[4]	[5]	[6]	[7]	[8]
1882	100.0	0.0	0.0					
1887	91.6	8.4	0.0					
1892	84.4	13.3	2.3					
1897	88.2	9.1	2.3					
1902	60.5	35.7	3.8					
1907	41.3	33.1	25.6					
1912	30.4	36.0	33.6					
1917	27.8	47.6	24.6					
1922	26.7	57.6	15.7					
1926	26.3	61.7	12.0	13.3	16.9	57.0	9.6	3.2
1927				13.7	17.6	56.4	9.0	3.3
1932				20.5	19.0	48.6	7.2	4.7
1937				19.5	18.2	51.7	5.7	4.9
1942				18.7	17.1	55.4	4.2	4.6
1946				22.5	17.3	51.8	3.7	4.7
1947				22.9	17.6	52.2	3.3	4.0
1950				26.5	18.0	49.6	2.1	3.8
1960				29.7	16.9	49.8	0.7	2.9
1970				32.2	22.5	41.2	0.3	3.8
1980				34.5	24.6	37.3	0.2	2.4
1989				34.3	27.3	34.8	0.2	3.4
1990				34.1	27.5	34.7	0.2	3.5
2000				35.3	30.7	30.8	0.2	3.0
2010				38.5	32.4	25.9	0.2	3.0
2015				37.4	33.5	26.3	0.2	3.0

Notes for Table F.
Columns 4-8. Author's estimates for 2010, 2015 adjusted by author to previous classifications.

Sources for Table F.
1. Columns 1-3. See Table D. *Electrical World*, Vol. 80, no. 11, September 9, 1922, p. 546; *Electrical World*, Vol. 91, no. 1, January 7, 1928, p. 18.
2. Columns 4-8. See Table D. EEI, *Statistical Yearbook*, various issues. EIA, *Electric Power Monthly* and *Annual Electric Review*, various issues.

Table G. Nominal Average Electricity Prices (p/kWh and ¢/kWh)

Year	UK All Electric (p/kWh)	UK Industrial (p/kWh)	UK Lighting/ Residential (p/kWh)	US All Electric (¢/kWh)	US Industrial (¢/kWh)	US Lighting/ Residential (¢/kWh)	UK All Electric (¢/kWh)	UK Industrial (¢/kWh)	UK Lighting/ Residential (¢/kWh)
	[1]	[2]	[3]	[4]	[5]	[6]	[7]	[8]	[9]
1882	–	–	3.96	–	–	25.00	–	–	19.29
1887	–	–	3.25	–	–	23.50	–	–	15.76
1892	–	–	2.95	–	–	22.00	–	–	14.37
1897	–	–	2.19	–	–	19.80	–	–	10.64
1902	1.40	0.60	1.69	3.93	1.40	16.20	6.88	2.93	8.25
1907	0.95	0.50	1.38	3.40	1.10	10.50	4.63	2.43	6.72
1912	0.75	0.42	1.51	2.98	1.00	9.10	3.65	2.05	7.35
1917	0.62	0.40	1.30	2.93	1.20	7.52	2.95	1.90	6.19
1922	0.86	0.54	2.20	2.77	1.80	7.38	3.81	2.39	9.75
1926	0.73	0.47	1.50	2.71	1.49	7.00	3.55	2.28	7.29
1927	0.65	0.39	1.41	2.71	1.46	6.82	3.16	1.90	6.85
1932	0.55	0.32	1.00	2.85	1.53	5.60	1.93	1.12	3.50
1937	0.44	0.27	0.71	2.17	1.14	4.30	2.17	1.33	3.51
1942	0.42	0.30	0.68	1.79	0.94	3.67	1.70	1.21	2.75
1946	0.47	0.37	0.59	1.81	0.98	3.22	1.89	1.49	2.38
1947	0.47	0.38	0.60	1.77	0.97	3.09	1.89	1.53	2.42
1950	0.50	0.39	0.56	1.81	1.02	2.88	1.39	1.10	1.57
1960	0.62	0.52	0.67	1.69	0.97	2.47	1.73	1.47	1.89
1970	0.79	0.65	0.84	1.59	0.95	2.10	1.88	1.56	2.01
1980	3.33	2.74	3.84	4.49	3.44	5.12	7.77	6.38	8.96
1989	5.16	3.99	6.29	6.41	4.83	7.95	8.46	6.54	10.31
1990	5.36	3.96	6.64	6.57	4.75	7.85	9.54	7.05	11.81
2000	4.80	3.00	6.68	6.81	4.64	8.24	7.30	4.55	10.16
2010	9.17	6.06	11.85	9.83	6.77	11.54	13.43	9.39	18.37
2015	11.75	7.79	15.46	10.41	6.91	12.65	17.96	11.91	23.63

Notes for Table G.

1. Columns 1-6. Average price of electricity sold.
2. Columns 1-3. 1882-1912, author's estimates based on Hannah, *Electricity* and Byatt, *op. cit.* 1894-1914, author's estimates based on Sparks, *op. cit.* 1902-1907 industrial prices based on Byatt, *op. cit.*, p. 129, less distribution costs. 1917 estimated by author.
3. Columns 4-6. 1882-1907 based on data from Byatt and U.S Department of Commerce. 1902-1912 industrial prices estimated by author.
4. Columns 7-9. UK prices translated into US currency at average exchange rate for year.

Sources for Table G.
1. Columns 1-6. 1882-1907, Byatt, *op. cit.*, pp. 24, 129.
2. Columns 1-3. Hannah, *Electricity*, pp. 429-430. 1912-1947, Hannah, *Electricity*, pp. 429-430., Garcke's *Manual* (various issues). 1948-1987, Electricity Council, *Handbook of Electricity Supply Statistics 1987* (London: Electricity Council, no date). 1970-2015, DECC, DUKES, various issues.
3. Columns 4-6. See Table B.
4. Columns 7-9. Lawrence H. Officer, "Dollar-Pound Exchange Rate from 1791," Measuring Worth, 2015.

Table H. Real Price of Electricity (1990 p/kWh and 1990 ¢ /kWh)

Year	UK All Electric (p/kWh)	UK Industrial Electric (p/kWh)	UK Lighting/ Residential Electric (p/kWh)	US All Electric (¢ /kWh)	US Industrial Electric ¢/kWh	US Lighting/ Residential Electric (¢/kWh)
	[1]	[2]	[3]	[4]	[5]	[6]
1882	–	–	153.90	–	–	320.10
1887	–	–	146.99	–	–	345.59
1892	–	–	129.96	–	–	319.30
1897	–	–	102.15	–	–	311.32
1902	51.66	26.42	74.42	53.98	19.23	222.53
1907	39.58	20.86	57.57	41.77	13.51	128.99
1912	29.74	16.65	51.55	33.56	11.26	102.48
1917	13.89	8.96	29.12	22.66	9.28	58.16
1922	18.63	11.70	47.65	19.49	12.67	51.94
1926	16.83	10.83	34.59	18.42	10.13	47.59
1927	15.44	9.26	33.48	18.87	10.17	47.49
1932	14.43	8.39	26.23	25.75	13.82	50.59
1937	11.41	6.95	18.28	17.64	9.27	34.96
1942	7.51	5.37	12.17	13.10	6.88	26.87
1946	7.02	5.52	8.81	10.68	5.79	19.01
1947	6.51	5.27	8.32	9.12	5.00	15.93
1950	6.19	4.92	7.01	8.83	4.98	14.06
1960	5.98	5.06	6.52	6.61	3.81	9.67
1970	5.23	4.34	5.58	4.69	2.80	6.20
1980	6.06	4.98	5.51	6.63	5.08	7.56
1989	5.50	4.25	6.70	6.67	5.03	8.28
1990	5.36	3.96	6.64	6.57	4.75	7.85
2000	3.87	2.41	5.38	5.57	3.70	6.74
2010	5.77	3.82	7.46	6.43	4.43	7.56
2015	6.86	4.55	9.02	6.27	4.16	7.62

Notes for Table H.
Columns 1-6. Data in Table H deflated by GDP deflator for each country with 1990=100. See Table L for deflators.
Sources for Table H.
Columns 1-6. See Table G for sources of prices and Table **M** for sources of GDP deflators.

Table I. Gas Consumption (million therms)

Year	UK Total Gas Consumption	UK Consumption for Electric Generation	UK Consumption for Non-generation Purposes	US Total Gas Consumption	US Consumption for Electric Generation	US Consumption for Non-generation Purposes
	[1]	[2]	[3]	[4]	[5]	[6]
1882	0.31	0	0.31	0.32	0	0.32
1887	0.40	0	0.40	2.09	0	2.09
1892	0.45	0	0.45	2.15	0	2.15
1897	0.54	0	0.54	2.25	0	2.25
1902	0.66	0	0.66	3.47	0	3.47
1907	0.78	0	0.78	4.73	0	4.73
1912	0.89	0	0.89	6.71	0	6.71
1917	1.00	0	1.00	9.27	0.15	9.12
1922	1.11	0	1.11	9.38	0.28	9.10
1926	1.32	0	1.32	15.31	0.54	14.77
1927	1.34	0	1.34	16.73	0.64	16.09
1932	1.37	0	1.37	17.14	1.09	16.05
1937	1.50	0	1.50	25.81	1.72	24.09
1942	1.64	0	1.64	32.65	2.40	30.25
1946	2.00	0	2.00	42.32	3.12	39.20
1947	2.08	0	2.08	49.10	3.85	45.25
1950	2.34	0	2.34	65.00	6.50	58.50
1960	2.71	0	2.71	129.30	17.85	111.45
1970	5.85	0.06	5.79	217.95	40.15	177.80
1980	17.36	0.14	17.22	202.35	38.10	164.25
1989	19.81	0.23	19.58	196.02	35.55	160.47
1990	20.37	0.22	20.15	196.03	33.32	162.71
2000	37.72	10.73	26.99	238.24	53.15	185.09
2010	36.62	12.75	23.87	246.17	75.50	170.67
2015	26.83	7.27	19.56	281.80	99.80	181.99

Notes for Table I.
1. Columns 1-6. All figures are in therms. The therm is equivalent to 100,00 British thermal units (BTUs). There are 10.32 therms per thousand cubic feet (MCF) of natural gas. Manufactured gas had roughly half the energy content of natural gas. Thus comparisons made on the basis of cubic feet may be misleading. All data adjusted by author from cubic feet to therms where necessary. One therm = 29.31 kWh.
2. Columns 1-3. UK data for 1882-1960 consists entirely of manufactured gas.
3. Columns 4-6. Natural gas played a significant role in US gas supply by the late 1880s. Calculations based entirely on manufactured gas consumption miss the transformation of the American gas industry. Industry-wide data for the period before 1900 is sketchy and appears to be based, largely, on information collected by one trade journal.

Sources for Table I.
1. Columns 1-3. Philip Chantler, *The British Gas Industry: An Economic Study* (Manchester: Manchester University Press, 1936) pp. 6-7. DECC, "Historical gas data: gas production and consumption (1920-2012)," Energy Trends Table 4.1 Gas (June 2015). London and Cambridge Economic Service, *The British Economy: Key Statistics 1900-1964* (London: Times Publishing, no date), Table B. B.R. Mitchell, with Phyllis Dean, *Abstract of British Historical Statistics* (Cambridge: Cambridge University Press, 1962), p. 269. Political and Economic Planning (PEP), *Report on the Gas Industry in Great Britain* (London: PEP, March 1939), pp. 42, 43, 202.
2. Columns 4-6. American Gas Association, *Gas Data Book,* various issues. Department of Commerce, *Statistical Abstract,* various issues. Department of Commerce, *Historical Statistics of the United States 1789-1945. Electrical World,* Vol. 91, no. 1. EIA, "Historical Energy Consumption 1775-2009." EIA, *Monthly Energy Review,* various issues. Environmental Protection Agency, *Survey of Town Gas and By-product Production and Locations in the U.S. (1880-1950),* Feb. 1985, p. 21. Robert C. Lofftness, *Energy Handbook* (NY: Van Nostrand Reinhold, 1978), p. 149.

Table J. Gas Consumption per $ of Real Gross Domestic Product
(therms per $1990 thousands)

Year	UK Total Gas Consumed	US Total Gas Consumed	UK Gas Consumed Less for Electric Generation	US Gas Consumed Less for Electric Generation
	[1]	[2]	[3]	[4]
1882	2.53	1.810	2.53	1.81
1887	3.08	10.43	3.08	10.43
1892	3.21	8.74	3.21	8.74
1897	3.32	8.23	3.32	8.23
1902	3.65	9.87	3.65	9.87
1907	3.99	10.68	3.99	10.68
1912	4.31	13.06	4.31	13.06
1917	4.14	17.00	4.14	16.72
1922	5.66	15.31	5.66	14.85
1926	6.18	19.65	6.18	18.96
1927	5.82	21.26	5.82	20.45
1932	6.01	27.81	6.01	26.04
1937	5.34	32.05	5.34	28.91
1942	4.36	24.72	4.36	22.91
1946	6.31	32.38	6.31	29.99
1947	6.56	38.14	6.56	35.15
1950	6.78	44.59	6.78	40.13
1960	6.04	63.94	6.04	55.11
1970	9.83	71.56	9.73	58.38
1980	24.13	48.63	23.93	39.47
1989	21.25	36.17	20.99	29.61
1990	21.77	35.87	21.53	29.77
2000	31.38	30.73	22.79	24.24
2010	24.94	25.64	17.60	19.17
2015	17.00	26.92	12.39	17.38

Notes and Sources for Table J.
See Tables A and I.

Table K. Gas Consumption per Capita (therms)

Year	UK Total Gas Consumption	US Total Gas Consumption	UK Gas Consumption Less for Electric Generation	US Gas Consumption Less for Electric Generation
	[1]	[2]	[3]	[4]
1882	9.42	5.98	9.42	5.98
1887	11.7	35.28	11.70	35.28
1892	12.64	32.63	12.64	32.63
1897	14.48	31.03	14.48	31.03
1902	16.88	43.65	16.88	43.65
1907	19.12	54.18	19.12	54.18
1912	20.99	68.03	20.99	68.03
1917	22.99	89.31	22.99	87.86
1922	25.00	84.88	25.00	82.35
1926	29.21	129.86	29.21	125.28
1927	29.52	140.00	29.52	134.64
1932	29.59	136.68	29.59	127.99
1937	31.71	199.31	31.71	186.02
1942	33.88	241.13	33.88	223.41
1946	40.65	298.24	40.65	276.25
1947	42.02	339.92	42.02	312.72
1950	46.61	426.78	46.61	384.11
1960	51.72	715.55	51.72	616.77
1970	105.21	1062.65	104.13	866.89
1980	308.35	888.28	305.86	800.83
1989	346.93	787.80	342.91	644.98
1990	356.12	784.43	352.27	651.10
2000	640.41	846.62	458.23	657.75
2010	587.80	797.44	383.15	552.87
2015	412.14	875.70	300.46	565.54

Sources and Notes for Table K.

See Tables A and I.

Table L. Nominal (p/therm and ¢/therm) and Real
(1990 p/therm and 1990 ¢/therm) Gas Prices

Year	UK Price (p/therm)	US Price (¢/therm)	UK Price (¢/therm)	UK Real Price (1990 p/therm)	US Real Price (1990 ¢/therm)
	[1]	[2]	[3]	[4]	[5]
1882	3.78	3.28	18.40	1.47	0.42
1887	3.67	2.70	17.80	1.69	0.40
1892	3.67	2.40	17.87	1.65	0.35
1897	3.67	2.70	17.84	1.71	0.43
1902	3.67	2.96	17.91	1.62	0.41
1907	3.67	3.33	17.87	1.53	0.41
1912	3.98	3.98	19.38	1.58	0.45
1917	8.40	3.91	39.98	1.88	0.30
1922	6.31	6.60	27.95	1.37	0.46
1926	5.40	5.77	26.24	1.24	0.39
1927	5.34	5.52	25.95	1.27	0.38
1932	4.59	6.92	16.06	1.20	0.63
1937	4.66	5.08	23.02	1.20	0.41
1942	4.60	4.77	18.58	0.82	0.26
1946	4.55	4.60	18.33	0.68	0.27
1947	4.64	4.67	18.70	0.64	0.24
1950	5.46	4.63	15.29	0.68	0.23
1960	8.96	6.05	25.18	0.87	0.24
1970	8.70	6.41	20.88	0.58	0.19
1980	19.91	31.30	46.39	0.36	0.46
1989	33.72	45.00	55.30	0.36	0.47
1990	35.54	45.90	63.26	0.36	0.46
2000	19.73	50.40	29.99	0.16	0.41
2010	50.66	67.50	78.52	0.32	0.44
2015	87.70	59.50	134.04	0.51	0.36

Notes for Table L.
1. Columns 1-2. 1882-1902 data based on Byatt, with interpolations from dates with data to cited years, based on cost and price indices. Prices translated from cubic foot to therm basis by author when needed.
2. Column 1. 1907-1917 and 1937-1947 estimated by author based on government price indices.
3. Column 2. 1882-1902 author's estimates based on data from Byatt and various price indices, taking into account approximate mix of manufactured and natural gas. 1902-1927 estimated by author based on weighted prices of natural and manufactured gas. Weighted average cost to end use customers.
4. Column 3. Average exchange rate for year used to convert from UK to US currency.
5. Columns 4-5. Calculated in real terms using GDP deflator, with 1990=100.

Sources for Table L.
1. Columns 1-2. 1882-1902, Byatt, pp. 24,26. Also fragmentary data from various contemporary sources.

2. Column 1. 1920-1936, PEP, *op. cit.*, p. 206., and *Report by the Joint Committee of the House of Lords and the House of Commons on Gas Prices*, 1937. 1949-1987, Electricity Council, *op. cit.*, pp. 148, 149. 1970-2012, DUKES, various issues.
3. Column 2. 1902-1927, 2009, *Statistical Abstract. 1932-1960*, American Gas Association*, 1966 Gas Data Book*. 1970-2015, American Gas Association, *Gas Facts* (various issues) and EIA, "Natural Gas Prices."
4. Columns 3-5. See table M for sources of indices used.

Table M. GDP Deflator and Foreign Exchange Rate

Year	Exchange Rate ($ per £)	UK GDP Deflator	US GDP Deflator
	[1]	[2]	]3]
1882	4.87	2.57	7.81
1887	4.85	2.18	6.80
1892	4.87	2.27	6.89
1897	4.86	2.14	6.36
1902	4.88	2.27	7.28
1907	4.87	2.40	8.14
1912	4.87	2.52	8.88
1917	4.76	4.46	12.93
1922	4.43	4.62	14.21
1926	4.86	4.34	14.71
1927	4.86	4.21	14.36
1932	3.50	3.81	11.07
1937	4.94	3.88	12.30
1942	4.04	5.59	13.66
1946	4.03	6.70	16.94
1947	4.03	7.21	19.40
1950	2.80	8.01	20.49
1960	2.81	10.31	25.55
1970	2.40	15.00	33.88
1980	2.35	55.07	67.76
1989	1.64	93.82	96.03
1990	1.78	100.00	100.00
2000	1.52	124.21	122.26
2010	1.55	158.87	152.73
2015	1.53	171.39	165.98

Notes for Table M
1. Columns 1-3. Annual averages.
2. Columns 1-3: According to calculations by economic consultants Smithers & Co., the US/UK real exchange rate was relatively stable from the early 1900s through the end of World War II, with the

exception of two brief periods in the late teens and the early thirties. After the war, the rate moved far off the trend for a brief period in the early 1950s and the mid 1980s. Overall, from 1899 though 2013, the real exchange rate barely budged. This indicates that price comparisons between the UK and the USA may not reflect purchasing power differences in some years, but on average they do. (Andrew Smithers, "The theory that shows that sterling is overvalued," *Financial Times*, April 23, 2014, p. 20.)

3. Column 2. 1882-1900, author's estimates based on cited sources.
4. Column 3. 1882-1887, author's estimates based on N BER-Kendrick and BLS data. 1892-1929, NBER-Kendrick Index, 1929-2015, Bureau of Economic Analysis (BEA) data.

Sources

1. Column 1. Lawrence H. Officer, *op. cit.*
2. Column 2. 1882-1900, Mitchell with Deane, *op. cit.*, pp. 367-372. 1900-1964, London and Cambridge Economic Service, *The British Economy: Key Statistics 1900-1964* (London: The Times Publishing Co., no date), Tables A,B. 1964-22, David and Gareth Butler, *British Political Facts 1900-1985* NY: St. Martin's Press, 1986), pp. 380-382, UK Office for National Statistics, "GDP Deflator Table," Federal Reserve Bank of Saint Louis (FRED), "GDP Implicit Price Deflator in United Kingdom" (from OECD). Samuel H. Williamson, "What Was the U.K. GDP Then?," *MeasuringWorth* 2015.
3. Column 3. Department of Commerce, *Long Term Economic Growth 1860-1970* (Washington, DC: U.S. Government Printing Office, 1973), pp. 222-223. Bureau of Economic Analysis reports. Louis Johnston and Samuel H. Williamson, "What Was the U.S. GDP Then?," *MeasuringWorth* 2015.

Table N: Operating Expenses (¢/kWh), Average Size of Power Plants (kW) and Annual Sales per kW of Year End Capacity (kWh)

Year	UK Operating Costs (¢/kWh sold)	US Operating Costs (¢/kWh sold)	UK Average Size Power Station (kW)	US Average Size Power Station (kW)	UK Annual Sales per kW of Capacity (kWh)	US Sales per Annual kW of Capacity (kWh)
	[1]	[2]	[3]	[4]	[5]	[6]
1892	–	–	–	–	500	900
1897	–	–	–	–	582	1,201
1902	3.66	2.19	300	539	685	1,830
1907	3.17	1.86	600	847	996	1,905
1912	2.33	1.51	2,937	1,467	1,159	1.904
1917	2.28	1.32	4,687	2,061	1,575	2,389
1922	3.68	1.43	4,870	3,813	1,238	2,292
1926	2.62	1.29	9,713	6,736	1,242	2,399
1927	1.94	1.31	10,731	6,765	1,328	2,442
1932	1.05	1.28	16,197	8,539	1,381	1,853
1937	1.43	1.01	23,151	9,091	2,151	2,789
1942	1.49	0.84	31,997	11,550	2,602	3,538
1946	1.67	0.98	38,963	13,056	2,774	3,792
1947	–	–	–	–	2,825	4,158
1950	–	–	–	–	3,044	4,071
1960	–	–	–	–	3,392	4,053
1970	–	–	–	–	3,369	4,079
1980	–	–	–	–	3,266	3,464
1989	–	–	–	–	3,856	3,586
1990	–	–	–	–	3,970	3,651

Notes for Table N

1. Columns 1,2: Direct cash operating expenses for all operations.
2. Column 3: 1902, 1907 estimated by author.
3. Column 5: new series 1960-1990.
4. Columns 5, 6: 1892 estimated by author. Calculations by author based on sales to ultimate customers and year end capacity.

Sources for Table N

1. Column 1: Garcke's *Manual* (various issues), Byatt, p. 129. Hannah, *Electricity,* pp. 427-433.
2. Column 2: *Statistical Abstract* ,1924, p. 340, 1940, p. 405, 1948, p. 497. *Electrical World*, Jan. 3, 1931, p. 26.
3. Columns 3, 5: Garcke's *Manual*, various editions.
4. Columns 4,6: *Historical Statistics 1945*, pp. 157-159. *Historical Statistics 1970*, pp. 822-828.

Table O. Reserve Margin (%), Load Factor (%) and Heat Rate (BTU per kWh)

Year	UK Reserve Margin (%)	US Reserve Margin (%)	UK Load Factor (%)	US Load Factor (%)	UK Heat Rate (BTU)	US Heat Rate (BTU)
	[1]	[2]	[3]	[4]	[5]	[6]
1922	52.7	38.5	28.0	53.0	31,250	29,600
1926	44.6	50.0	30.0	52.2	26,700	23,600
1927	39.4	52.1	30.0	53.6	24,050	22,600
1932	43.3	88.8	33.9	50.3	19,400	18,450
1937	21.0	47.9	35.9	56.0	16,000	17,850
1942	19.7	25.1	42.1	63.1	15,800	16,100
1946	11.3	11.4	41.9	65.5	16,000	15,700
1947	10.8	6.1	42.4	57.9	16,016	15,600
1950	12.4	10.3	50.4	59.4	14,475	14,030
1960	17.7	31.5	48.4	65.5	12,835	10,701
1970	39.9	19.0	52.0	63.9	11,929	10,508
1980	45.0	30.7	58.0	61.0	10.946	10,489
1989	31.3	28.6	61.0	62.2	10,686	10,312
1990	29.8	25.6	62.2	60.4	10,565	10,366
2000	23.5	10.8	67.8	61.2	8,686	10,201
2010	37.0	27.1	64.7	63.5	8,714	9,756
2015	36.2	NA	68.2	NA	8,445	9,189

Notes:
1. Column 1: 1922 estimated by author.
2. Column 2: 2010 estimated by author.
3. Column 3: Based on several time series. 1989 estimated by author.
4. Column 4: 1922, 2010 estimated by author.
5. Column 5: All data estimated by author based on coal and oil equivalents calculated by UK government.
6. Column 6: 1922 estimated by author.

Sources:
1. Columns 1,3,5: 1922-1932, *Garcke's Manual* (various issues), 1937-1950, Ministry of Fuel and Power publications. 1922-1947, Hannah, *Electricity*, pp. 432-433. 1950-2015, DUKES and related publications.
2. Columns 2,4,6: EEI *Yearbook* (various issues), EIA *Monthly Energy Review, Electric Power Monthly, Electric Power Annual* (various issues).

Table P. International Comparisons: UK, USA, France-Germany-Italy

	1960	1970	1980	1990	2000	2010	2015
	[1]	[2]	[3]	[4]	[5]	[6]	[7}
GDP per capita ($1990)							
UK	8,571	10,694	12,777	16,302	21,467	23,880	24,166
USA	11,193	14,854	18,270	21,866	27,237	28,225	32,526
France-Germany-Italy	6,813	10,669	14,087	17,094	22,177	23,336	24,796
Electricity production per capita (kWh)							
UK	2,185	4,461	5,046	5,590	6,379	6,076	5,167
USA	4,672	7,995	10,335	12,157	13,511	13,363	12,671
France-Germany-Italy	1,651	3,152	4,755	6,173	7,032	7,223	7,279
Electricity production per $1990 GDP (kWh)							
UK	0.304	0.417	0.395	0.342	0.297	0.254	0.213
USA	0.417	0.538	0.566	0.556	0.496	0.457	0.390
France-Germany-Italy	0.242	0.296	0.337	0.360	0.317	0.310	0.294
Industrial production index (1990=100)							
UK	55.1	73.4	81.4	100.0	114.0	97.6	100.0
USA	40.6	65.5	87.9	100.0	149.3	144.9	156.9
France-Germany-Italy	38.7	66.7	85.3	100.0	112.0	110.3	113.5
Average price of electricity (¢/kWh)							
UK	1.7	1.9	7.8	9.5	7.3	13.4	18.0
USA	1.7	1.6	4.5	6.6	6.8	9.8	10.4
France-Germany-Italy	–	–	7.9	11.5	8.1	12.6	21.7
Population (millions)							
UK	52.4	55.6	56.3	57.2	58.9	62.3	64.1
USA	180.7	205.1	227.8	249.9	281.4	308.7	3218
France-Germany-Italy	168.4	182.3	188.6	192.3	198.1	204.9	208.6
GDP ($1990 billions)							
UK	448.9	594.9	719.5	935.9	1264.4	1487.7	1585.0
USA	2022.2	3045.8	4161.0	5464.8	7664.6	9021.8	9437.0
France-Germany-Italy	1147.3	1944.9	2656.9	3287.7	4393.2	4781.5	4708.5
Electricity production (TWh)							
UK	136.3	248.1	284.1	319.8	375.7	378.6	338.9
USA	844.2	1639.8	2354.8	3038.0	3802.1	4125.1	4092.9
France-Germany-Italy	278.1	574.7	896.7	1187.0	1393.0	1480.0	1487.0

Notes:

1. For UK and USA see Appendix tables – Real GDP, Table A. Electricity production per capita, Table C. Electric production per $ real GDP, Table C. Average price electricity, Table G. Population, table A. GDP, Table A. Electric production, Table B.
2. Data for France-Germany-Italy have been aggregated for all calculations. Industrial production index weighted by GDP.
3. Gross domestic product (GDP) in US $ 1990. Purchasing power parity translations of all GDPs to dollars.
4. Data may differ from numbers shown in other tables in order to present consistent time series.

Sources:

1. GDP and population from International Monetary Fund, *World Economic Outlook Data Base,* April 2011, National Accounts, People., national population estimates, and U.S. Department of Agriculture, *op. cit.* See Appendix Table A. World Bank and other sources for 2015.
2. GDP deflator from U.S. Bureau of Economic Analysis (BEA).
3. Industrial Production Index (IPI) data for France-Germany-Italy from OECD *StatExtracts.*, for UK from Office for National Statistics, for USA from Federal Reserve.
4. Electricity data for USA from EIA, for France-Germany-Italy from Eurostat, for UK from UK National Statistics.

Table Q. REC Financial Statements
(Fiscal Years 1989/1990-1996/1997) (billion £)

Fiscal Year Ended 31 March	1990	1991	1992	1993	1994	1995	1996	1997
	[1]	[2]	[3]	[4]	[5]	[6]	[7]	[8]
Income statement								
Revenues	13.31	13.63	14.88	15.47	15.26	15.42	14.75	16.04
Operating income								
Pretax (PT)	0.75	1.36	1.67	1.65	1.81	2.01	1.79	2.23
Aftertax (AT)	0.46	1.00	1.28	1.26	1.56	1.68	1.70	2.12
Net income								
Pretax	0.93	1.23	1.47	1.67	1.77	2.00	1.67	1.75
Aftertax	0.64	0.87	1.08	1.28	1.52	1.67	1.58	1.64
Book capitalization								
Equity	6.06	5.52	6.16	6.77	7.72	7.32	6.53	6.94
All debt	0.00	1.89	1.79	1.97	1.33	2.21	5.57	5.30
Total	6.06	7.41	7.95	8.74	9.05	9.53	12.1	12.24
Offering price capitalization								
Equity	–	5.37	6.01	6.62	7.57	7.17	6.38	6.79
All debt	–	1.89	1.79	1.97	1.33	2.21	5.57	5.30
Total	–	7.26	7.80	8.59	8.90	9.38	11.95	12.09
First day premium								
Capitalization Equity	–	6.44	7.08	7.69	8.64	8.24	7.45	7.86
All debt	–	1.89	1.79	1.97	1.33	2.21	5.57	5.30
Total	–	8.33	8.97	9.66	9.97	10.45	13.02	13.16
National Grid consolidated								
Equity (book)	–	5.52	6.37	7.24	8.48	8.35	6.53	6.94
Net income (AT)	–	1.05	1.30	1.53	1.81	1.94	1.83	1.64

Notes:

Adjusted wherever possible to exclude extraordinary income.

Debt and capitalization as defined by rating agencies.

Estimates made for missing data including that of acquired companies.

Sources:

S&P Global Credit Review, Nov.. 1996, Oct. 1997.

Moody's Electric Utility Sourcebook, Oct. 1995, 1996, 1997.

Moody's Power Company Sourcebook, Oct. 1999.

Moody's International Manual, various issues.

Major UK Companies Handbook, various issues.

Privatization prospectuses, company and brokerage reports.

Table R. REC Statement Analysis
(Fiscal Years 1989/1990-1996/1997) (%)

FY	1990	1991	1992	1993	1994	1995	1996	1997	Average 1991-1997
	[1]	[2]	[3]	[4]	[5]	[6]	[7]	[8]	[9]
Reported (book)									
Operating margin	5.6	10.0	11.2	10.7	11.6	13.0	12.1	13.9	11.8
Return on capital									
Pretax	12.4	18.4	21.0	18.9	20.0	21.1	14.8	18.2	18.9
Aftertax	7.6	13.5	16.4	14.4	17.2	17.6	14.0	17.3	15.8
Return on equity	10.6	15.8	17.5	18.9	19.7	22.8	24.2	23.6	20.4
Equity ratio	100.0	73.0	77.1	77.5	85.3	76.8	54.0	56.7	71.5
Flotation value									
Return on capital									
Pretax	–	18.7	21.4	19.2	20.3	21.4	15.0	18.4	19.2
Aftertax	–	13.8	16.4	14.7	17.5	17.9	14.2	17.5	16.0
Return on equity	–	16.2	18.0	19.3	20.1	23.3	24.8	24.2	20.8
Equity ratio	–	74.0	77.1	77.1	85.1	76.4	53.3	56.2	71.3
First day premium value									
Return on capital									
Pretax	–	16.3	18.6	17.1	18.2	19.2	13.7	16.9	17.1
Aftertax	–	12.0	14.3	13.0	15.6	16.1	13.1	16.1	14.3
Return on equity	–	13.5	13.7	16.6	17.6	20.3	21.2	20.9	17.9
Equity ratio	–	77.3	78.9	79.6	86.7	78.9	57.2	59.7	74.0
National Grid consolidated									
Return on equity									
Book	–	19.0	21.1	22.6	23.4	26.5	28.0	23.5	23.5
Flotation	–	19.6	21.6	23.1	23.9	27.1	28.7	24.0	24.0
First day	–	16.3	18.4	19.9	20.9	23.5	24.8	20.6	20.6

Notes:

Operating margin = [(Pre-tax operating income)/(Revenue)] x 100.

Pretax return on capital= [(Pre-tax operating income)/(Capitalization)] x 100.

Aftertax return on capital= [After-tax operating income)/(Capitalization)] x 100.

Return on equity= [(After-tax net income)/(Common equity)] x 100.

Equity ratio = [(Common equity)/(Capitalization)] x 100.

Sources:

See Appendix Table Q.

Table S. Generator Financial Statements (£ billions)
(Fiscal Years 1989/1990-1996/1997)

FY	1990	1991	1992	1993	1994	1995	1996	1997
	[1]	[2]	[3]	[4]	[5]	[6]	[7]	[8]
Income Statement								
Revenues	6.61	7.02	7.71	7.54	6.57	6.84	6.88	6.39
Operating income								
Pre-tax	0.90	0.69	0.85	1.05	1.21	1.24	1.40	1.34
After-tax	0.57	0.46	0.59	0.75	0.83	0.96	0.91	0.99
Net income								
Pre-tax	0.87	0.78	0.87	1.01	1.20	1.25	1.50	1.31
After-tax	0.54	0.55	0.61	0.71	0.82	0.97	1.01	0.94
Capitalization (book)								
Equity	3.61	3.07	3.49	3.98	4.56	4.43	4.84	4.41
All debt	0.00	0.67	0.84	0.96	1.47	1.11	1.48	3.06
Total	3.61	3.74	4.33	4.94	6.03	5.54	6.30	7.47
Capitalization (flotation)								
Equity	–	3.60	4.02	4.51	5.09	4.96	5.37	4.94
All debt	–	0.67	0.84	0.96	1.47	1.11	1.46	3.06
Total	–	4.27	4.86	5.47	6.56	6.07	6.87	8.00
Capitalization (first day)								
Equity	–	4.36	4.78	5.27	5.85	5.72	6.13	5.70
All debt	–	0.67	0.84	0.96	1.47	1.11	1.46	3.06
Total		5.03	5.62	6.23	7.32	6.83	7.59	8.78

Notes:

1. National Power and PowerGen.

See Tables Q and R for procedures followed.

Sources:

See Tables Q and R.

Table T. Generator Statement Analysis
(Fiscal Years 1989/1990-1996/1997) (%)

FY Ended 31 March	1990	1991	1992	1993	1994	1995	1996	1997	Average
	[1]	[2]	[3]	[4]	[5]	[6]	[7]	[8]	[9]
Reported (book)									
Oper. margin (PT)	13.6	9.8	11.0	13.9	18.4	18.1	20.3	21.0	16.1
Return on capital									
Pre-tax	24.1	18.4	19.6	21.2	20.1	22.4	22.2	17.9	20.3
After-tax	15.8	10.6	18.6	15.2	13.8	17.3	14.4	13.3	14.0
Return on equity	15.0	17.9	17.5	17.8	18.0	21.9	20.9	21.3	19.3
Equity ratio	100.0	82.0	80.6	80.6	75.6	80.0	76.8	59.0	76.4
Flotation									
Return on capital									
Pre-tax	–	16.2	17.5	19.2	18.4	20.4	20.3	16.8	18.4
After-tax	–	10.8	12.1	13.7	12.7	15.8	13.2	12.4	13.0
Return on equity	–	15.3	15.2	15.7	16.1	19.6	18.8	19.0	17.1
Equity ratio	–	84.3	82.7	82.4	77.6	81.7	78.2	61.8	78.4
First day									
Return on capital									
Pretax	–	13.8	15.1	16.9	16.5	18.2	18.4	15.3	16.3
After tax	–	9.1	10.5	12.0	11.3	14.1	12.0	11.3	11.5
Return on equity	–	12.6	12.8	13.5	14.0	17.0	16.5	16.5	14.7
Equity ratio	–	86.7	85.1	84.6	79.9	83.7	82.9	64.9	81.1

Notes and Sources:

See Tables Q and R.

Table U. National Grid Financial Statements
(Fiscal Years 1989/1990-1996/1997) (£ billions)

FY	1990	1991	1992	1993	1994	1995	1996	1997
	[1]	[2]	[3]	[4]	[5]	[6]	[7]	[8]
Income Statement								
Revenues	1.07	1.14	1.32	1.39	1.43	1.43	1.49	1.46
Operating income								
Pretax	0.43	0.47	0.56	0.59	0.63	0.65	0.66	0.66
After tax	0.29	0.34	0.40	0.44	0.49	0.48	0.47	0.48
Net income								
Pretax	0.43	0.39	0.50	0.53	0.58	0.61	0.61	0.60
After tax	0.29	0.26	0.34	0.38	0.44	0.44	0.42	0.42
Capitalization (book)								
Equity	1.68	0.94	1.15	1.41	1.70	1.97	1.10	1.39
All debt	0.00	0.90	0.75	0.94	0.81	0.36	1.32	1.04
Total	1.68	1.84	1.90	2.35	2.51	2.33	2.42	2.43
Capitalization (Flotation)								
Equity	—	0.91	1.12	1.38	1.67	1.94	1.07	1.36
Debt	—	0.90	0.75	0.94	0.81	0.36	1.32	1.04
Total	—	1.81	1.87	2.32	2.48	2.30	2.39	2.40
Capitalization (first day)								
Equity	—	1.18	1.39	1.65	1.94	2.21	1.34	1.63
Debt	—	0.90	0.75	0.94	0.81	0.36	1.32	1.04
Total	—	2.08	2.14	2.59	2.75	2.57	2.66	2.67

Sources:

See Tables Q and R.

Note:

Assumes that NG shares were offered and traded at same premium as RECs.

Table V. National Grid Statement Analysis
(Fiscal Years 1989/1990-1996/1997) (%)

FY	1990	1991	1992	1993	1994	1995	1996	1997	Average 1991-1997
	[1]	[2]	[3]	[4]	[5]	[6]	[7]	[8]	[9]
Reported (book)									
Op. margin (PT)	40.2	41.2	42.4	42.4	44.1	45.4	44.3	45.2	43.6
Return on capital									
Pretax	25.6	25.5	29.5	25.1	25.1	27.9	27.3	27.2	26.8
After-tax	17.3	18.5	21.1	18.7	19.5	20.6	19.4	19.8	19.7
Return on equity	17.3	27.7	29.6	27.0	25.9	22.3	38.2	30.2	28.7
Equity ratio	100.0	51.1	60.5	60.0	67.7	84.5	45.5	57.2	60.9
Flotation									
Return on capital									
Pretax	—	26.0	29.9	25.4	25.6	28.3	25.5	27.5	26.9
Aftertax	—	18.8	21.4	19.0	19.8	20.9	19.7	20.0	19.9
Return on equity	—	28.6	30.4	27.5	26.3	22.6	39.2	30.9	29.4
Equity ratio	—	50.2	59.9	59.5	67.3	84.3	44.8	56.7	60.4
First day									
Return on capital									
Pretax	—	22.6	23.4	17.0	21.1	23.7	22.9	22.5	21.9
Aftertax	—	16.3	18.7	17.0	17.8	18.7	17.7	18.0	17.7
Return on equity	—	22.0	24.5	23.0	22.7	19.9	31.3	25.8	24.2
Equity ratio	—	56.7	65.0	63.7	70.5	86.0	50.4	61.0	64.8

Sources:

See Tables Q and R.

Notes:

1. For calculations, see Table R.
2. All ratios for National Grid are calculated as if National Grid were sold to the public simultaneously with the RECs and its shares traded at the same premium or discount as the RECs at the time of initial offering, based on NG being a key part of investor valuation of the RECs. In all likelihood, NG would have commanded an even higher valuation because of its profitability.

Table W. Scotland Financial Statements (Fiscal Years 1989/1990-1996/1997)
(£ billions)

FY	1990	1991	1992	1993	1994	1995	1996	1997
	[1]	]2]	[3]	[4]	[5]	[6]	[7]	[8]
Income Statement								
Revenues	1.59	1.81	2.06	2.21	2.36	2.55	3.16	3.89
Operating income								
Pretax	0.35	0.38	0.43	0.49	0.53	0.56	0.61	0.89
Aftertax	0.35	0.38	0.37	0.38	0.45	0.43	0.47	0.70
Net income								
Pretax	0.22	0.25	0.36	0.44	0.50	0.53	0.58	0.78
Aftertax	0.22	0.25	0.30	0.33	0.42	0.40	0.44	0.59
Capitalization (book)								
Equity	0.49	0.66	1.26	1.44	1.66	1.88	2.07	2.40
All debt	1.09	1.08	0.64	0.28	0.26	0.32	0.88	2.30
Total	1.58	1.74	1.90	1.72	1.92	2.20	2.95	4.70
Flotation								
Equity	–	2.45	3.05	3.23	3.45	3.68	3.87	4.20
All debt	–	1.08	0.64	0.28	0.26	0.32	0.88	2.30
Total	–	3.53	3.69	3.51	3.71	4.00	4.75	6.50
First day								
Equity	–	2.69	3.29	3.47	3.69	3.92	4.11	4.44
All debt	–	1.08	0.64	0.28	0.26	0.32	0.88	2.30
Total	–	3.77	3.93	3.75	3.95	4.24	4.99	6.74

Sources:

See Tables Q and R.

Notes:

Floatation and first day valuations for Scottish shares for fiscal 1991 as if shares sold at end of fiscal year at same premiums as actually prevailed at time of sale in May 1991.

Table X. Scotland Statement Analysis. (Fiscal Years 1989/1990-1986/1987) (%)

FY	1990	1991	1992	1993	1994	1995	1996	1997	Average
	[1]	[2]	[3]	[4]	[5]	[6]	[7]	[8]	[9]
Reported (book)									
Pretax op. margin	22.0	21.0	20.9	22.2	22.5	22.0	19.3	22.9	21.6
Return on capital									
Pretax	22.2	21.8	22.6	28.5	27.6	25.5	20.7	18.9	24.0
Aftertax	22.2	21.8	19.5	22.1	23.4	19.5	15.9	14.9	19.2
Return on equity	44.9	37.9	23.8	22.9	25.3	21.3	21.3	24.6	23.2
Equity ratio	31.0	37.9	66.3	88.7	86.5	85.5	70.2	51.1	73.9
Flotation									
Return on capital									
Pretax	–	10.8	11.7	14.0	14.3	14.0	12.8	13.7	13.4
Aftertax	–	10.8	10.0	10.8	12.1	10.8	9.9	10.8	10.7
Return on equity	–	10.2	9.8	10.2	12.2	10.9	11.4	14.0	11.4
Equity ratio	–	69.4	82.3	92.0	93.0	92.0	81.5	64.6	82.1
First day									
Return on capital									
Pretax	–	10.1	10.9	13.1	13.4	13.2	12.2	13.2	12.3
Aftertax	–	10.1	9.4	10.1	11.4	10.1	9.4	10.4	10.1
Return on equity	–	9.3	9.1	9.5	11.4	10.2	10.7	13.3	10.5
Equity ratio	–	71.4	83.7	92.5	93.4	92.5	82.4	65.9	83.1

Sources:
See Table W.

Notes:
Average for 1991-1997 for "Reported" data and 1992-1997 for "Adjusted for Offering Price" data.

Table Y. Fuel Mix for UK Electricity Generation (1902-2015) (%)

Year	Coal [1]	Hydro [2]	Oil [3]	Gas [4]	Nuclear [5]	Other [6]
1902	95.9	0.1	1.0	0.0	0.0	3.0
1907	95.9	0.1	1.0	0.0	0.0	3.0
1912	95.9	0.1	1.0	0.0	0.0	3.0
1917	95.9	0.2	1.0	0.0	0.0	3.0
1922	95.8	0.2	1.0	0.0	0.0	3.0
1926	95.0	0.2	2.7	0.0	0.0	2.4
1927	96.7	0.4	0.7	0.0	0.0	2.2
1932	93.4	3.1	0.6	0.0	0.0	2.9
1937	93.7	3.4	0.2	0.0	0.0	2.7
1942	97.6	3.3	0.1	0.0	0.0	1.7
1946	97.5	2.9	0.2	0.0	0.0	1.6
1947	98.1	2.8	0.4	0.0	0.0	0.9
1950	96.1	2.0	0.4	0.0	0.0	3.0
1960	86.6	1.9	9.5	0.0	1.8	2.2
1970	74.3	1.8	14.0	0.0	10.5	0.5
1980	83.3	1.3	3.1	0.0	12.8	0.5
1989	69.1	1.4	5.5	0.0	22.8	2.2
1990	69.0	1.5	6.5	0.1	20.6	2.3
2000	30.9	1.3	1.6	39.0	24.0	3.2
2010	28.4	0.8	1.3	46.7	16.9	5.9
2015	22.5	1.9	0.6	29.7	20.9	24.4

Sources:
DECC:, DUKES, Table 5.1, "Electricity Supplied by Fuel Source," "Fuel Input for Electricity Generation."

Notes:
Percentage of kWh produced by each fuel. No pre -1920 data from DECC, series post-1920 are inconsistent and lack data on non-utility generation for much of the period. Fortunately the UK depended largely on coal for electric generation between 1902 and 1950, so the only question is what number in the 90s best represents coal's percentage of the fuel mix. All numbers from 1902 to 1917 are author's estimates. Other is largely renewables.

Table Z. Fuel Mix for USA Electric Generation (1902-2015) (%)

Year	Coal [1]	Hydro [2]	Oil [3]	Gas [4]	Nuclear [5]	Other [6]
1902	48.5	51.5	0.0	0.0	0.0	0.0
1907	60.2	39.8	0.0	0.0	0.0	0.0
1912	69.5	29.8	0.7	0.0	0.0	0.0
1917	63.5	32.1	3.2	1.2	0.0	0.0
1922	57.1	34.7	6.2	2.0	0.0	0.0
1926	60.0	32.2	4.0	3.8	0.0	0.0
1927	60.5	32.5	2.7	4.3	0.0	0.0
1932	51.1	36.3	3.8	8.8	0.0	0.0
1937	53.4	32.4	4.7	9.5	0.0	0.0
1942	58.1	29.6	3.5	8.8	0.0	0.0
1946	53.0	30.6	6.8	9.6	0.0	0.0
1947	55.2	27.0	7.5	10.3	0.0	0.0
1950	47.4	26.0	11.4	15.2	0.0	0.0
1960	52.7	17.7	6.6	22.9	0.1	0.0
1970	44.9	15.3	12.6	25.9	1.3	0.0
1980	50.7	12.2	10.7	15.1	11.0	0.3
1989						
1990	52.5	9.6	4.1	12.2	19.0	2.6
2000	51.7	7.2	2.9	15.8	19.8	2.6
2010	44.8	6.3	0.9	24.0	19.6	4.4
2015	32.9	6.1	0.6	32.7	19.4	8.3

Sources:

U.S. Dept. of Commerce, *Historical Statistics ... to 1945*, pp. 155-158, *Historical Statistics ... to 1970*, pp. 820-825.

Edison Electric Institute, *EEI Pocketbook of Electric Utility Industry Statistics*, 1978.

EIA, *Electric Power Monthly*, Tables 7.2a, 7.2b.

Notes:

All producers of electricity, kWh produced by each fuel.

1902-1907 estimated by author based on Department of Commerce data.

Bibliography

Books and pamphlets

Allen, Stuart and Selina O'Connor, eds., *The Guide to World Equity Markets 1992* (London: Euromoney Publications, 1992).

Bartlett, C. J., *A History of Postwar Britain 1945-1974* (London: Longman, 1977).

Bonbright, James C., *Principles of Public Utility Rates* (NY: Columbia University Press, 1961).

Bowers, Brian, *Electricity in Britain* (Manchester: Greater Manchester Museum of Science and Industry, no date).

Bowles, Samuel, *The Moral Economy* (New Haven: Yale University Press, 2016).

Brown, C.N.N., *JW Swan and the Invention of the Incandescent Electric Lamp* (London: Science Museum, 1978).

Bussing, Irvin, *Public Utility Regulation and the So-Called Sliding Scale* (NY: Columbia University Press, 1936).

Byatt, I.C.R., *The British Electrical Industry 1875-1914* (Oxford: University Press, 1979).

Chantler, Philip, *The British Gas Industry* (Manchester: Manchester University Press, 1938).

Chick, Martin, "Nationalization, privatization and regulation," in Kirby, Maurice W. and Mary B. Rose, eds., *Business Enterprise in Modern Britain: From the Eighteenth to the Twentieth Century* (London: Routledge, 1994).

Clark, John Maurice, *Competition as a Dynamic Process* (Washington, DC: The Brookings Institution, 1961).

_____, J. Maurice, Studies in the Economics of Overhead Costs (Chicago: University of Chicago Press, 1962).

Clegg, H.A. and T.E. Chester, *The Future of Nationalization* (Oxford: Basil Blackwell, 1953).

Coase, Ronald "1991 Nobel Lecture: The Institutional Structure of Production," "The Nature of the Firm," in Oliver G. Williams and Sidney G. Winter, eds., *The Nature of the Firm: Origins, Evaluation and Development* (NY: Oxford U. Press, 1993).

Dampier, Sir William Cecil, *A History of Science* (Cambridge University Press, 1948).

Dimock, Marshall E., *British Public Utilities and National Development* (London: George Allen and Unwin, 1933).

Drucker, Peter F., *Managing for Results* (NY: Harper & Row, 1964).

Dodds, John W., *The Age of Paradox: a Biography of England 1841-1851* (NY: Rinehart, 1952).

Elfun Society, *The Edison Era 1876-1892* (Schnectady, NY: Elfun Society, 1976).

Foster, C.D., *Privatization, Public Ownership and the Regulation of Natural Monopoly* (Oxford: Blackwell Publishers, 1992).

Friedel, Robert and Paul Israel, with Bernard S.Finn, *Edison's Electric Light* (New Brunswick: Rutgers University Press, 1986).

Gordon, Bob, *Early Electrical Appliances* .(Aylesbury: Shire Publications, 1984).

Gaitskell, Hugh, *The Diary of Hugh Gaitskell 1945-1956*, edited by Philip M. Williams (London: Jonathan Cape, 1983.

Greig, James, *John Hopkinson: Electrical Engineer* (London: HMSO, 1970).

Hadfield, Charles, *The Canal Age* (NY: Frederick A. Praeger, 1969).

Hannah, Leslie, *Electricity before Nationalisation* (Baltimore: Johns Hopkins University Press, 1979).

_____, Engineers, Managers and Politicians (Baltimore: Johns Hopkins University Press, 1982).

Havighurst, Alfred, *Britain in Transition* (Chicago: University of Chicago Press, 1985).

Hayek, F.A., *The Road to Serfdom* (Chicago: University of Chicago Press, 2007).

Helm, Dieter, *Energy, the State and the Market* (Oxford: Oxford University Press, 2003).

Hennessey, R.A.S., *The Electric Revolution* (London: Scientific Book Club, 1972).

Henney, Alex, *A Study of the Privatisation of the Electric Supply Industry in England & Wales* (London: EEE Ltd, 1994).

Horrocks, Sally and Thomas Lean, *An Oral History of the Electricity Supply Industry: Scoping Study for proposed National Life Stories Project* (London: British Library, November 2011).

Hughes, Thomas P. , *Networks of Power* (Baltimore: Johns Hopkins University Press, 1983).

_____, "Technology, History and Technical Problems," in Chauncey Starr and Philip C. Ritterbush, eds., *Science, Technology and the Human Prospect* (NY: Pergamon Press, 1980).

Hunt, Sally, *Making Competition Work in Electricity* (NY: John Wiley, 2002).

Hyman, Leonard S., ed., *The Privatization of Public Utilities* (Vienna, VA: Public Utilities Reports, 1995).

Hyman, Leonard S., Andrew S. Hyman and Robert C. Hyman, *America's Electric Utilities: Past, Present and Future* (Vienna, VA: Public Utilities Reports, 2005).

Jehl, Francis, *Menlo Park Reminiscences* (Volume One), NY: Dover, 1990).

Josephson, Matthew, *Edison* (NY: McGraw-Hill, 1959).

Joskow, Paul L. and Richard Schmalansee, *Markets for Power* (Cambridge, MA: The MIT Press, 1983).

Kahn, Alfred E., *The Economics of Regulation* (NY: John Wiley & Sons, 1970, 1971).

_____, *Letting Go* (East Lansing: Institute of Public Utilities and Network Industries, Michigan State University, 1998).

_____, *The Passing of the Public Utility Concept: A Reprise* (NY: National Economic Research Associates, May 1983).

_____, "Competition in the Electric Industry is Inevitable and Desirable," in N.Y. State Energy Research and Development Authority, *The Electric Industry in Transition* (Arlington, VA: Public Utilities Reports, 1994), pp. 26, 27.

Kirby, Maurice W. and Mary B. Rose, *Business Enterprise in Modern Britain* (London: Routledge, 1994).

Kirby, Richard Shelton, Sidney Withington, Arthur Burr Darling and Frederick Gridley Kilgour, *Engineering in History* (NY: Dover Books, 1990).

Landes, David S., *The Unbound Prometheus: Technological Change and Industrial Development in Western Europe from 1750 to the Present* (Cambridge: Cambridge University Press, 1969).

Merrett, A.J. and Allen Sykes, *The Finance and Analysis of Capital Markets* (London: Longmans, 1982).

Mitchell, Sally, ed., *Victorian Britain* (NY: Garland, 1988).

Neuberger, Julia, ed., *Privatisation... Fair Shares for All or Selling the Family Silver* (London: PAPERMAC, 1987).

Parkinson, C. Northcote, *Left Luggage* (Boston: Houghton Mifflin, 1967).

Payne, Peter L., *The Hydro* (Aberdeen: Aberdeen University Press, 1988).

Posner, Michael V., *Fuel Policy* (London: Macmillan, 1973).

Rostas, L., *Comparative Productivity in British and American Industry* (London: Routledge, 1999).

Routledge, Robert, *Discoveries and Inventions of the Nineteenth Century* (NY: Crescent Books, 1989).

Schweppe, Fred C., Michael C. Caramanis, Richard D. Tabors and Roger E. Bohn, *Spot Pricing of Electricity* (Boston: Kluwer Academic, 1988).

Shaw, George Bernard, *The Intelligent Woman's Guide to Socialism and Capitalism* (NY: Brentano's, 1928).

Smart, John E., *The Deptford Letter-books: An Insight on SZ de Ferranti's Deptford Power Station* (London: Science Museum, 1976).

Smith, Adam, *An Inquiry into the Nature and Causes of the Wealth of Nations* (NY: The Modern Library, 1937).

_____, *The Theory of Moral Sentiments* (Amherst, NY: Prometheus Books, 2000).

Smith, Ken, *A Civil War Without Guns – 20 Years On* (London: Socialist Publications, 2004).

Starr, Chauncey and Philip C. Ritterbush, eds., *Science, Technology and the Human Prospect* (NY: Pergamon Press, 1980).

Treacy, Michael, and Fred Wiersema, *The Discipline of Market Leaders* (Reading, MA: Addison-Wesley, 1995).

Turvey, Ralph, *Optimal Pricing and Investment in Electricity Supply* (London: George Allen and Unwin, 1968).

Walker, Michael A., *Privatization Tactics and Techniques* (Vancouver: The Fraser Institute, 1988).

Statistical Sources

American Gas Association, *Gas Data Book* (Arlington, VA: AGA, various years).

_____, *Gas Facts* (Arlington, VA: AGA, various years).

Bank of England, "Series 1 to 2." (Treasury stock yields).

Butler, David and Gareth Butler, *British Political Facts 1900-1985* (NY: St. Martin's Press, 1986).

Central Intelligence Agency, *Handbook of Economic Statistics* (Washington, DC: US Government Printing Office, various dates).

Economic History Services, "What Was the UK GDP Then?"

Edison Electric Institute, *Historical Statistics of the Electric Utility Industry* (NY: EEI, 1961).

_____, *Statistical Yearbook of the Electric Utility Industry* (Washington, DC: EEI, various years).

_____, *EEI Pocketbook of Electric Utility Industry Statistics* (Washington, DC: EEI, various years).

Electricity Council, *Handbook of Electricity Supply Statistics 1987* (London: Electricity Council, no date).

Eurogas, "Natural Gas Consumption in EU 27 in 2007," 13 March 2008.

European Commission, Eurostat, "Electricity prices by type of user – Euro per kWh., 1998-2006.

_____, Eurostat, "Electricity prices for EU households and industrial consumers on 1 January 2006."

_____, Eurostat, "Electricity prices for EU households and industrial consumers on January 2007. "

_____, Eurostat, "Gas and Electricity Market Statistics Data 1990-2006. "

_____, Eurostat, "Competition indicators in the electricity market," news release, 26 September 2002.

Federal Reserve Bank of St. Louis, *FRED*.

H.M Treasury, GDP deflators at market and money prices," September 2006.

IDS Statistics, "Retail Price Index, – All Items, 1986-1994."

Infoplease, "Economic Statistics by Country, 2004."

International Energy Agency, *Energy Balances of OECD Countries 1980-1989*.

_____, *Energy Balances of OECD Countries 1995-1996*.

_____, *Energy Statistics of OECD Countries 1970-1979, Vol. 2* (Paris: OECD), 1997).

_____, *Electricity Information 2001 with 2000 data*.

International Monetary Fund,, *International Financial Statistics Yearbook*, various years.

_____, *World Economic Outlook Data Base,* April 2011.

KEMA Consulting GmbH, "Review of European Electricity Prices On behalf of Union of the Electricity Industry – EURELECTRIC, Final Report," November 2005.

Loftness, Robert L., *Energy Handbook* (NY: Van Nostrand Reinhold, 1978).

London and Cambridge Economic Service, *The British Economy: Key Statistics 1900-1964*
(London: Times Publishing, 1965).

Maddison, Angus, *Monitoring the World Economy 1820-1992* (Paris: Development Centre, Organisation for Economic Cooperation and Development, 1995).

Measuring Worth, "What Was the U.K. GDP Then," 2010, 2015.

_____. "What Was the U.S. GDP Then," 2010.

Mitchell, B.R., *European Historical Statistics 1750-1970* (NY: Columbia University Press, 1975).

Mitchell, B.R., with Phyllis Deane, *Abstract of British Historical Statistics* (Cambridge:At University Press, 1962).

National Archives, "Chartered Gas Light and Coke Company."

Officer, Lawrence H., "Dollar-Pound Exchange Rate from 1791," *Measuring Worth*, 2013.

Organisation for Economic Cooperation and Development, "GDP Implicit Price Deflator in United Kingdom."

_____, *StatExtracts*. "Gross domestic product 2000-2008."

Spadaro, Joseph V., Lucille Langlois and Bruce Hamilton, "Greenhouse Gas Emissions of Electricity Generation Chains: Assessing the Difference," *IAEA Bulletin*, 42/2/2000. p. 19.

Union of the Electricity Industry (EURELECTRIC), "Electricity Markets: Getting the Picture Straight and Boosting Market Integration," (Updated version: 28 March 2006.)

U.K. Department of Energy & Climate Change (and predecessor agencies), *Digest of United Kingdom Energy Statistics* (and predecessor publications) (London: The Stationery Office, various years).

_____, "Quarterly Energy Prices," March 2011.

_____, *Energy Trends*, June 2002, March 2011, Table 5.1, "Fuel used in electric generation and electricity supplied."

_____, *UK Energy in Brief* (London: The Stationery Office, various dates).

_____, "Electricity generated and supplied 1920-2012"

_____, Statistical Release, "UK Climate Change Sustainable Development Indicator: 2010 Greenhouse. Gas Emissions, Provisional Figures and 2009 Greenhouse Gas Emissions, Final Figures by Fuel Type and End- User," 25th March 2011, Tables 7 and 8.

_____, Statistical release, "UK Climate Change Sustainable Development Indicator: 2012 UK Greenhouse Gas Emissions, Provisional Figures and 2011 Greenhouse Gas Emissions Final Figures by Fuel Type and End Uses," 28th March 2013.

_____, "Historical Coal Data: Availability and Consumption 1853 to 2012."

_____, "Gas Since 1882, Historical Data, Numbers."

UK Individual Shareholders Society, "UK Stock Market Statistics," August 14, 2014, sharesssoc.org.

U.K. Office of National Statistics (and predecessor agencies), *Annual Abstract of Statistics* (including predecessor *Statistical Abstract of the United Kingdom)*, (London: HMSO, various years).

_____, "Percent of total equity: Individuals Not seasonally adjusted Updated on 9/12/2009," Series DEY 1: SRS.

_____, "GDP Deflator Table."

_____, "Real time GDP database."

_____, "Retail Prices Index: monthly index numbers of retail prices 1948-2006 (RPI)(RPIX). "

_____, "Gross domestic product preliminary estimate," *Statistical Bulletin*, various dates.

U.K. Department of Trade and Industry, *Energy Statistics 2001* (London: HMSO, 2001).

_____, "Quarterly Energy Prices," June 2007.

U.K. Parliament, "Daily Hansard Written Answers, TRADE AND INDUSTRY, 17 October 2005: Column 786W."

United Nations, *1990 Energy Statistics Yearbook* (NY: UN, 1992).

U.S. Department of Commerce, Bureau of the Census, *Historical Statistics of the United States 1789-1945* (Washington, DC: U.S. Government Printing Office, 1949).

_____, *Historical Statistics of the United States: Colonial Times to 1970, Part 2* (Washington, DC: U.S. Government Printing Office, 1975).

_____, *Statistical Abstract of the United States* (Washington, DC: U.S. Government Printing Office, various years).

U.S. Department of Commerce, Bureau of Economic Analysis, "Real Gross Domestic Product, First Quarter 2011," May 26, 2011, Table 2.

U.S. Department of Commerce, Social and Economic Statistics Administration, *Long Term Economic Growth 1860-1970* (Washington, DC: U.S. Government Printing Office, June 1973).

U.S. Department of Energy, Energy Information Administration, *Annual Electric Review* (Washington, DC: U.S. Government Printing Office, various dates).

_____, *Annual Energy Review* (Washington, DC: U. S. Government Printing Office, various dates).

_____, *Electric Power Monthly* (Washington, DC: U.S. Government Printing Office, various dates).

_____, *Emissions of Greenhouse Gases in the United States 2009,* March 2011, Tables 6, 7, 12.

_____, *Monthly Energy Review* (Washington, DC: U.S. Government Printing Office, various dates).

_____, "Historical Energy Consumption 1775-2009."

_____, "Natural Gas Prices."

_____, "Survey of Town Gas and By-product Production and Locations in the U.S. (1880-1950)," February 1985.

U.S. Environmental Protection Agency, *Inventory of U.S. Greenhouse Gas Emissions and Sinks: 1990-2008,* Washington, DC, April 15, 2010 (EPA 430-R-10-006), Table ES-2.

World Bank, World Development Indicators database, Gross domestic product, 2009, 2010, PPP. "

Trade and Industry Publications

Butterworth, Andy, "Derivatives Under NETA," *Power Finance & Risk,* August 21, 2000, p. 5.

Carbon Sequestration Leadership Forum, "An Energy Summary of the United Kingdom," 2006.

"EDF says reported strike price of £165/MWh is 'Wrong'; DECC silent on upper limit," i-Nuclear, July 19, 2012, i-nuclear.com.

Electric Utility Week, June 27, 1994, "U.K.'s Nuclear Electric Says It's Ready Now for Privatization via Flotation,," p. 14, ""U.K.'s Open-Access Policy Leads 13,500 Users to Switch Utilities Since April," p. 15.

Electrical World, January 4, 1908; January 4, 1913; January 5, 1918; January 7, 1922; September 9, 1922; January 6, 1923; January 1, 1927; January 7, 1928; January 5, 1929; January 3, 1931; January 7, 1933; January 15, 1938; June 1, 1974.

Electrical World and Engineer, January 2, 1904.

Electricity Association, Policy Research, "Electric companies in the United Kingdom – a brief chronology," 21st October 2002.

Ellis, Walter,, "Is the U.K. gassing up?," *Global Energy Business,* July/August 1999, p. 29.

"Stelzer Tells Britain To Consider U.S.-Style Regulation," *Energy Daily,* August 1, 1997, p. 3.

Electronic Design, February 15, 1976, "Getting Electricity to Work for Man," hbci.com.

EnergyLinx, "Electricity distribution Network Operators," 2002.

Energy Networks Association, "gas factsheet 01," no date,

_____, Policy Research, "Who owns whom in the UK electricity industry," 21st October 2002.

_____, "Introduction to the UK Electricity Industry, The structure of the electricity industry," 2002.

_____, "Introduction to the UK Electricity Market, Competition in Supply," 2002.

EnergyQuote, "Market Developments/Electricity Market Timeline," 2008.

Fells, Ian, "The paradoxes of UK energy policy," *Energy Economist,* Special Edition – 2000 Collection, p. 3.

Fishlock, David, "Milord," *Nuclear Industry,* Second Quarter 1989, p. 41.

Garcke's Manual 1951-1952, Frederick C. Garrett, ed. (London: Electrical Press, 1952).

Howles, L.R., "Nuclear Station Achievement: 1984 Annual Review," *Nuclear Engineering International,* May 1985.

Lambert, Jeremiah D., "Privatizing Electricity in Britain: The Role of the National Grid," *Public Utilities Fortnightly,* March 30, 1989, p. 14.

Manual of Electrical Undertakings 1896, Emil Garcke, ed., (London: P. S. King, 1896).

Manual of Electrical Undertakings and Directory of Officials 1908, Emil Garcke, ed. (London: Electrical Press Ltd., 1906).

Manual of Electrical Undertakings and Directory of Officials 1911, Emil Garcke, ed., (London: Electrical Press Ltd., 1911).

Manual of Electrical Undertakings and Directory of Officials 1925-1926, 1935-1936, 1940-41, 1946-1947, 1947-1948, Frederick C. Garrett, ed. (London: Electrical Press Ltd, 1926, 1936, 1941, 1947, 1948).

Megawatt Daily's Market Report, "British regulator modifies distribution rate cuts," October 11, 1999, p. 2.

Nicholson, Jim, "U.K. Pool licks its wounds," *Platt's*, July/August 1999, p. 21.

Platts, "UK carbon tax frozen at GBP18/mt ($f30.6/mt) of CO2 from 2016-2020," 19 March 2014, platts.com.

Platts Electric Power Daily, "Mirant Considers Divesting Its 49% Share of U.K.-Based Western Power distribution," July 16, 2002, p. 3.

Platts Insight, December 2010.

Power Finance & Risk, "The Prospects for NETA," December 4, 2000, p. 5.

Power in Europe, UK privatisation: Electricity efficiency: who cares?," Financial Times Business Information, 25 May 1989, p. 10.

Power in Europe, "UK electricity privatisation," Financial Times Business Information, 17 August 1989, p. 3.

Power Line, June 1989, July 1989, August 1989.

Practical Law Company, "Trafalgar House/Northern Electric" [sic], Volume VI (1995): March.

Privatisation International, January 1992, March 1992, May 1992, January 1993, May 1993, February 2001.

Privatisation Yearbook 1994, Rodney Lord, ed., November-December 1993.

World Nuclear News, "Europe lists concerns over Hinkley deal," 03 February 2014, world-nuclear-news.org.

_____, "Comment hints at Hinkley Point C approval," 23 September 2014, world-nuclear-news.org.

_____, "European Commissioners approve Hinkley Point project," 8 October 2014, world-nuclear-news,org.

_____, "Hinkley Point C contract terms," 08 October 2014, world-nuclear-news.org.

Academic Research, Professional Papers, Consulting Reports and Presentations

Aiken, Maxwell, "An accounting history of capital maintenance: legal precedents for managerial autonomy in United Kingdom," *Accounting Historians Journal*, June 1, 2005.

Averch, Harvey and Leland L. Johnson, "The Behavior of the Firm Under Regulatory Constraint," *American Economic Review* 52(5): 1052-1069.

Awerbuch, Shimon, "Restructuring our electricity networks to promote decarbonisation," Tyndall Centre for Climate Change Research, Working Paper 49, March 2004.

Baker, John, "British Privatisation: A rationale for the US...," presented to Edison Electric Institute Financial Conference, 16th October 1989.

Black, Fisher, "Estimating Expected Return," *Financial Analysts Journal*, January-February 1995.

Borenstein, Severin and James Bushnell, "The U.S. electricity Industry after 20 Years of Restructuring," Energy Institute at Haas, September 2014.

Borin, Seth, Todd Levin and Valerie M. Thomas, "Estimates of the Cost of New Electricity Generation in the South," Working paper # 54, Georgia Tech School of Public Policy, March 26, 2010.

Bower, John, "Why Ofgem?," Oxford Institute of Energy Studies, August 2003.

Brattle Group, "Electricity Markets: The Price of Power," *ENERGY*, 2000, no. 1.

Buckland, Roger and Patricia Fraser, "The scale and patterns of abnormal returns to equity investment in UK electricity distribution," *Global Financial Journal*, 13 (2002).

Burr, Michael T., "Smart Grid at a Crossroads," *Public Utilities Fortnightly*, January 2013, p. 29.

Butler, "UK Electricity Networks," Imperial College and Parliamentary Office of Science and Technology, September 2001.

Chang, Max, David White, Ezra Hausman, Nicole Hughers and Bruce Biewald, "Big Risks, Better Alternatives," Synapse Energy and Economics, October 6, 2011.

Chick, Martin, "Nationalization, privatization and regulation," in Maurice W. Kirby and Mary B. Rose, *op. cit.*, p. 315.

Consumers Union, "What's Behind the Door: Consumers' Difficulties Selecting Health Plans," *Health Policy Brief*, January 2012.

Cornwall, Ian and Alex Henney, "Regulation in England and Wales" (London: EEE Ltd and Cornwall Consulting Ltd, no date).

Cornwall Energy, "Gas and Electricity Costs to Consumers," Paper for the National Right to Fuel Campaign, January 2008.

Day, Christopher J. and Derek W. Bunn, "Generation Asset Divestment in the England and Wales Electricity Market: A Computational Approach to Analysing Market Power," London Business School, April 1999.

Del Bo, Chiara and Massimo Florio, "Electricity Investment: An Evolution of the New British Energy Policy and Its Implications for the European Union," March 20, 2012.

Dimson, Elroy, Paul Marsh and Mike Staunton, "Irrational Optimism," *Financial Analysts Journal*, January/ February 2004.

Domah, Preetum and Michael G. Pollitt, "The Restructuring and Privatisation of the Electric Distribution and Supply Business in England and Wales: A Social Cost Benefit Analysis," July 2000.

EEE Ltd, "The New Electricity Trading Arrangements in England and Wales,", no date.

EEE Ltd, Cameron McKenna LLP and Steven Stoff, "A Proposal for a Study of the Independent Transmission Company Plus (ITC-PLUS)," January 2003.

EEE Ltd and Cornwall Consulting Ltd, "Regulating Transmission and Distribution Charges, "no date.

Energy Links Consultancy, "Gas Moratorium, Battle for Power, Executive Summary," January 1998.

Evans, Joanne and Richard Green, "Why did British electricity prices fall after 1998?," Research Memorandum 35-2003, Business School, University of Hull, 2003.

Foreman-Peck, James S. and Christopher J. Hammond, "Variable Costs and the Visible Hand: the Re-Regulation of Electricity Supply, 1932-1937," *Economica*, February 1997.

Galal, Ahmed, Leroy Jones, Pankaj Tandon and Ingo Vogelsang, "Synthesis of Cases and Policy Summary," "Questions and Approaches to Answers," World Bank Conference on the Welfare Consequences of Selling Public Enterprises, June 11-12, 1992.

Giuletti, Monica, Luigi Grossi and Michael Waterson, "Price transmission in the UK electricity market: was NETA beneficial?," draft, December 2008, University of Warwick.

Gorini de Oliveira, Ricardo, and Mauricio Tiomno Tolmasquim, "Regulatory performance analysis case study: Britain's electricity industry," *Energy Policy*, 332 (2004) 1261.

Green, Richard, "England and Wales – A Competitive Electricity Market?," Program on Workable Energy Regulation, University of California, PWP-60, September 1998.

_____ and David M. Newbery, "Competition in the Electricity Industry in England and Wales," *Oxford Review of Economic Policy*, Vol. 13, No. 1 (1997).

_____ and Tanga McDaniel, "Competition in Electricity Supply: will'1998' be worth it?," University of California Energy Institute, May 1998.

Hanson, A.H., "Electricity Reviewed: The Herbert Report," *Public Administration*, Vol. 34, June 1956. Hattori, Toru, Tooraj Jamasb and Michael G. Pollitt, "A comparison of UK and Japanese electricity distribution performance 1985-1998: lessons for incentive regulation," Cambridge-MIT Institute, CMI Working Paper 03, October 2003.

Helm, Dieter and Andrew Powell, "Pool Prices, Contracts and Regulation in the British Electricity Supply Industry," *Fiscal Studies*, 13, No.1, 1992.

Henney, Alex, 'The illusory politics and imaginary economics of Neta," *Power UK*, 85, March 2001.

_____, "The Collapse of the government's Electric Generation Policies," 11 May 2012.

_____, Alex, *The Electricity Supply Industry of Eleven West European Countries*, (London: EEE Ltd, Spring 1992).

Holmes, Andrew, John Chesshire and Steve Thomas, "Power on the Market: Strategies for Privatising the UK Electricity Industry," A Power Europe Report, 1987, Financial Times Business Information.

Hrab, Roy, "Privatization: Experience and Prospects," Panel on the Role of Government, University of Toronto, Research Paper 22, February 2004.

Hyman, Leonard S., "Privatization: The Hows and Whys," *Public Utilities Fortnightly*, February 1, 1993.

_____, "Restructuring Electricity Policy and Financial Models," *Energy Economics*, Vol. 32, Issue 4.

_____ and Richard E. Schuler, "Electricity Restructuring, Consumer Prices and the Cost of Capital," HICSS-47 conference, January 2014.

Institute of Civil Engineers, "Discussion of Electrical Energy," *Proceedings of the Institute of Civil Engineers*, (London: The INST. CE, VOL. CVL, 1891).

Jenkinson, Tim, "Regulation and Cost of Capital," in Michael Crew and David Parker, eds., *International Handbook on Economic Regulation* (Cheltenham: Edward Elgar, 2006).

Jewell, Kevin, "Manipulated, Misled, Ignored, Abused: Residential Consumer Experience with Electric Deregulation in the United Kingdom," Consumers Union and Public Services International Research Unit, University of Greenwich, Fall 2003.

Johnson, Christopher, "The Economics of Britain's Privatisation," World Electricity Conference, 16 & 17 November 1987, Financial Times Conferences.

Johnson, Stanley, "Environmental Problems Facing the Electricity Industry," World Electricity Conference, 16 & 17 November 1987, Financial times Conferences.

Jones, Sir Philip, "Opening Address," World Electricity Conference, 16 & 17 November 1987, Financial Times Conferences.

Joskow, Paul L., "Incentive Regulation in Theory and Practice: Electricity Distribution and Transmission Networks," MIT, January 21, 2006.

Keay, Malcolm, "UK Energy Review – still in search of an energy policy?" *Oxford Energy Comment*, July 2006.

Kellogg, Susan, "British Energy Bailout Continues the Nuclear Bouncing ball," Scientech Issue Alert, September 17, 2002.

Kleindorfer, Paul R., D.-J. Wu and Citru S. Fernando, "Strategic gaming in electric power markets," *European Journal of Operational Research*, 130 (2001).

Littlechild, Stephen C., "Privatisation, Competition and Regulation," 29th Wincott lecture, 1999, London Institute of Economic Affairs, 2000.

Lowe, Peter, "Public Utility Regulation in Britain: Some Lessons from the Pre-1945 Experience," *Economic Issues*, Vol.1, Part 1, March 1996.

Lowe, Peter, "The Reform of Utility Regulation in Britain: Some Current Issues in Historical Perspective," *Journal of Economic Issues*, Vol. XXXII, No.1, March 1998.

MacKerron, Gordon, "Cost and Benefits of 100% Electricity Marketing Opening," NERA, APRIL 2001.

Markowitz, "Market Efficiency: A Theoretical Distinction and So What?," *Financial Analysts Journal*, September/October 2005.

Miller, Donald, "Power Generation to 2000 – Nuclear and Private Sector," World Electricity Conference, 16 & 17 November 1987, Financial Times Conferences.

Mon, Gonzalo, "History of the Manufactured Gas Business in the United States'" International Symposium and Trade Fair on the Clean-Up of Manufactured Gas Plants, Prague, Czech Republic, September 1995.

NERA, "Energy Supply Margin Update," December 2010.

Newbery, David M., "Privatisation and liberalisation of network utilities," *European Economic Review* 41(1997).

_____, "Pool Reform and Competition in Electricity," November 1997, IEA/LBS Lectures on Regulation Series VII 1997.

_____, "The Regulator's Review of the English Electricity Pool," Department of Applied Economics, University of Cambridge, 28 August 1998.

_____, "Privatising Network Industries," CESifo conference on Privatisation Experiences in the EU, 1-2 November 2003.

_____, "Electricity Liberalisation in Britain: the quest for a satisfactory market design," The Cambridge-MIT Institute Electricity Project, 14 July 2004.

Otero, Jesús and Catherine Waddam Price, "Price Discrimination, Regulation and Entry in the UK

Practical Law Company, "Trafalgar House/Northern [sic] Electric," March 1995.

Residential Electricity Market," *Bulletin of Economic Research*, 53:3, 2001.

Primeaux, Walter J., Jr., "Deregulation of Electric Utility Firms: An Assessment of the Cost Effects of Complete Deregulation vs Deregulation of Generation Only," University of Illinois, September 1982.

Peck, Stephen C., "Coasian Insights in Electricity Industry Structural Reforms," Flèche, May 2003.

Robinson, Terry and Andrzej Baniaki, "The volatility of prices in the English and Welsh electricity pool," *Applied Economics*, 34, 2002.

Ross, Duncan A., "Electricity Privatisation – Changing the Balance – The Future for the Area Boards – New Opportunities," European Study Conference, 24 June 1988.

Ruff, Larry, "New Electricity Trading Arrangements for England & Wales," *Energy Regulation Brief.*, NERA, August 1999.

Sparks, Charles C., "Inaugural Address," *The Journal of the Institution of Electrical Engineers*, Vol. 54, No. 252, December 1915.

Sweeting, Andrew, "Market Power in the England and Wales Wholesale Electricity Markets 1995-2000," MIT Center for Energy and Environmental Policy Research, August 2004.

Thomas, Steve, "The Wholesale Electricity Market in Britain – 1990-2001," August 2001, PSIRU, University of Greenwich.

_____, "The Impact of Privatisation on Electricity Prices in Britain," Presentation to the IDEC Seminar on Public Utilities, São Paulo, August 6-8, 2002.

Timera Energy, "Timera take on the 1st UK capacity auction," January 12, 2015, timera-energy.com.

TB&A inforum, "UK Restructuring Creates Competitive Climate," Vol.I, Issue 1, November-December 1993, p. 1.

Vogelsang, Ingo, "The United Kingdom, Vol.1: Background; British Telecom," World Bank Conference on the Welfare Consequences of Selling Public Enterprises, June 11-12, 1992.

"Walter Citrine," Spartacus Educational, spartacus.schoolnet.co.uk, February 29, 2012.

Warren, Andrew, "Regulation of a Privatised Electricity Industry," World Electricity Conference, 16 & 17 November 1987, Financial times Conferences.

Willis, Amy, "NETA and Power Supply Security in the UK," October 30, 2003.

Wilson, Chris M. and Catherine Waddams Price, "Do Consumers Switch to the Best Suppliers?," University of East Anglia, May 2006.

Wise, George, "Swan's way: a study in style," *IEEE Spectrum*, April 1982.

Wolfram, Catherine D., "Electricity Markets: Should the Rest of the World Adopt the United Kingdom's Reforms?," *Regulation*, Vol. 22, no. 4, 1999.

Government and Regulatory Documents

British Gas, "Proposed Merger of National Grid Group plc and Lattice Group plc to create National Grid Transco plc, Response from British Gas," May 2002.

Commons Hansard – Written Answers, 10 July 1997.

Competition Commission, *PowerGen Plc and Midlands Electricity Plc: A report on the proposed merger,* April 1996.

_____, *Northern Ireland Electricity Plc: A report on reference under Article 15 of the Electricity (Northern Ireland) Order 1992*, April 1997.

_____, *AES and British Energy: A report on references made under section 12 of the Electricity Act 1989,* 2001.

_____, *National Power Plc and Southern Electric Plc: A report on the proposed merger,* April 2006.

Department for Business Enterprise & Regulatory Reform, Energy – Its Impact on the Environment and Society," Annex 2C, 2006. www.berr.gov.uk/files/file20324.

_____, *Meeting the Energy Challenge: A White Paper on Nuclear Power,* January 2008.

Department of Energy & Climate Changer *2050 Pathways Analysis,* July 2010.

_____, *Electricity Market Reform Consultation Document* (London: The Stationery Office, December 2010.)

_____, "Energy Plans Go Before Parliament," press release, 23 June 2011.

_____, *National Policy Statement for Nuclear Power Generation* (EN -6), Vol. II, June 2011.

_____, *Overarching National Policy Statement for Energy,* Version for Approval (EN-1), June 2011.

_____, *Ofgem Review Final Report,* July 2011.

_____, *UK Renewable Energy Roadmap,* July 2011.

_____, *Planning our electric future: a White Paper for secure, affordable and low-carbon electricity,* July 2011.

_____, *The Carbon Plan: delivering our low carbon futures.* Presented to Parliament pursuant to sections 12 and 14 of the Climate Change Act of 2008, December 2011.

_____, "An Energy Bill to Power Low-Carbon Economic Growth, Protect Consumers and Keep the Lights On." Press Notice 2012/151, 29 November 2012.

_____, "Supplementary Memorandum to Delegated Powers and Regulatory Reform Committee on Part 2 (electricity market reform) of the Energy Bill, 17 June 2013.

_____, Press release, "Initial agreement reached on new nuclear power station at Hinkley," 21 October 2013.

_____, "Smart meter roll-out for the domestic and small and medium non-domestic sectors,(GB)," 30/1/2014.

Department of Trade and Industry, *The Social Effects of Energy Liberalisation, The UK Experience,* June 2000.

_____, Press Release, P/99/812.

Electricity Act 1989. (London: HMSO, 1989).

Electricity Consumers' Council, *Annual Report 1986/87.*

Energy Act 2013, 18th December 2013.

Energy Bill (HC 135).

Energy Bill (HC Bill 100).

European Commission, "State aid: Commission concludes modified UK measures for Hinkley Point nuclear plant are compatible with EU rules," press release, 8 October 214.

Gore, Donna and Grahame Danby, Grahame Allen and Patsy Roberts, "Utilities Bill," House of Commons Library Research Paper 00/7, 26 January 2000.

House of Commons, *Report from the Select Committee on Lighting by Electricity together with the Proceedings of the Committee, Minutes of Evidence and Appendix,* 13 June 1879.

_____, *Parliamentary Debates (Hansard),* Thursday 20 July 1989 (London: HMSO, 1989).

_____, *Parliamentary Debates (Hansard),* Monday 24 July 1989 (London: HMSO, 1989).

_____, *The Cost of Nuclear Power,* Fourth Report, Energy Committee, House of Commons, June 7, 1990 (London: HMSO, 1990).

_____, Committee of Public Accounts, *The New Electricity Trading Arrangements in England and Wales,* Second Report of Session 2003-2004, 1 December 2003.

_____, "Risk management: the nuclear liabilities of British Energy plc" (HC 354-1), 11 February 2004, uncorrected transcript of evidence. From web page of Gerry Steinberg, MP.

_____, Business and Enterprise Committee, *Energy prices, fuel poverty and Ofgem,* Eleventh Report of Session 2007-2008, Vol.1, HC 293-1, 28 July 2008.

_____, Energy and Climate Change Committee, *Energy Prices, Profits and Poverty,* Fifth Report of Session 2013-14, HC 108, 29 July 2013.

_____, Energy and Climate Change Committee, *Implementation of Electricity Market Reform*, Eighth Report of Session 2014-15, HC 664, 4 March 2015.

_____, Energy and Climate Change Committee, *Smart meters: progress or delay?,* Ninth Report of Session 2014-15, HC 665, 3 March 2015.

House of Lords, *Parliamentary Debates (Hansard*, Monday 24 July 1989 (London: HMSO, 1989).

Industry Department for Scotland, *Privatisation of the Scottish Electricity Industry*, Presented to Parliament by the Secretary of State for Scotland by command of Her Majesty (Edinburgh: HMSO, March 1988).

Inland Revenue, press release, 2 July 1997.

Monopolies and Mergers Commission, *Scottish Hydro-Electric Plc: A report on a reference under section 12 of the Electricity Act 1989* (London: HMSO, 1995).

May, Theresa, "Speech to Conservative Policy Conference," Birmingham, UK, 5 Oct. 2016.

National Audit Office, Department of Trade and Industry, "The Sale of British Energy," National Audit Press Release, 8 May 1998.

_____, Report by the Controller and Auditor General, *The Sale of British Energy*, HC694 Session 1997-98, 8 May 1998.

_____, Report by the Controller and Auditor General, *Giving Domestic Customers a Choice of Electricity Supplier,* HC 85, Session 2000-2001, 5 January 2001.

_____, Report by the Controller and Auditor General, *Pipes and Wires*, HC 723 Session 2001-2002, 10 April 2002.

_____, Report by the Controller and Auditor General, *The New Electricity Trading Arrangement*, HC 624, Session 2002-2003, 9 May 2003.

_____, Report by the Controller and Auditor General, *Risk Management: The Nuclear Liabilities of British Energy, plc*, HC 264 Session 2003-2004, 6 February 2004.

_____, Press Release, "The Department of Trade and Industry: The restructuring of British Energy, 17 March 2006.

_____, Report by the Controller and Auditor General, *The Restructuring of British Energy*, HC 943 2005-2006, March 2006.

_____, Report by the Controller and Auditor General, The Department of Energy and Climate Change, *Nuclear Power in the UK*, HC 511, Session 2016-2017, 13 July 2016.

OFFER, "Report on Distribution and Transmission System Performance 1997/1998," November 1998.

_____, *Review of Electricity Trading Arrangements, Background Paper 1*, February 1998.

_____, *Review of Energy Sources for Power Stations, Submission by the Director General of Electricity Supply*, April 1998.

_____, *Reviews of Public Electricity Suppliers 1998 to 2000, Price Controls and Competition Consultation Paper,* July 1998.

_____, *Review of Energy Sources for Power Stations Consultation Document, Response by the Director General of Electricity Supply*, August 1998.

OFGEM, Annual Report 2001-2002, (London: The Stationery Office, July 2002).

_____, *Proposed merger of National Grid Group plc and Lattice Group plc to create National Grid Transco plc, A consultation paper,* May 2002.

_____, *Domestic gas and electric supply competition. recent developments,* June 2003.

_____, *Energy Supply Probe,* 6th October 2008.

_____, "Regulator sets tough investment-led price controls on regional electricity networks," RNS Number 6533D, 7 December 2009.

_____, "Direction issued to National Grid Electricity Transmission ("NGET") by the Gas and Electricity Markets Authority pursuant to paragraph B1(b) of Part B of Schedule A to Special Condition AA5A (Balancing Services Activity Revenue Restriction) of NGET's electricity transmission licence," 21 January 2010.

_____, "Action needed to ensure Britain's energy supplies remain secure," press release, 3 February 2010.

_____, *National Grid Electricity Transmission System Operator Incentives from 1 April 2011*, 10 June 2011.

_____, *Electricity and Gas Supply Market Report*, 22 February 2010, 19 December 2011.

_____, *Financial Information Reporting: 2010 Results*, 31 May 2012.

_____, "Making the profits of the six largest electricity suppliers clear," Factsheet 123, 25 November 2013.

_____, Retail Energy Markets in 2016, 3 August 2016.

_____, "'RIIO': A New Way to Regulate Energy Networks in GB," presentation by Alistair Buchanan, August 2010.

_____, "Ofgem's Part in Britain's Energy Market Reform Package," presentation by Alistair Buchanan, Autumn 2011.

_____, *State of the Market Assessment*, 27th March, 2014.

_____, "Ofgem refers the energy market for a full competition investigation," press release, ofgem.gov.uk/press-releases, June 26, 2014.

_____, "Ofgem refers energy market for full competition investigation," news release, ofgem.gov.uk/news, June 26, 2014.

Secretary of State for Energy, *Privatising Electricity*, Presented to Parliament by the Secretary of State for Energy by Command of Her Majesty, February 1988 (London: HMSO, 1988).

UK Green Investment Bank, "UK Green Investment Bank Supports Drax's Biomass Conversion Plans," press release, 20 December 2012, and "supporting document," EIR 13/0334 – Gov.uk,, February 19, 2013.

UK Government, "Press release Long-term partnership to help UK compete in £1 trillion global nuclear industry," 26 March 2013, gov.uk.

U.K. Parliament, Select Committee on Trade and Industry, *Sixth Special Report*, 25 March 1998, Appendix 1, Annex A, P/98/240.

_____, "Report by the Joint Committee of the House of Lords and the House of Commons on Gas Prices," 1937.

U.K. Parliamentary Office of Science and Technology, "UK Electricity Networks," *postnote*, October 2001.

U.S. Department of Energy, Office of Economic, Electricity and Natural Gas Analysis, "Horizontal Market Power in Restructured Electricity Markets," March 2000.

U.S. Supreme Court, *Federal Power Commission v Hope Natural Gas Co.*, 320 US 591 (US 1944).

Water Services Association and Water Companies Association, *The cost of capital in the water industry, vol.2, Main report., vol.3, Appendices, A response by the Water Companies Association and the Water Companies Association to the OFWAT consultation Paper,* (Hartshead: WSA Publications, November 1991.

Corporate, Financial and Other Publications and Analyses

Atherton, Peter, Peter Bisztyga, Anthony White and Guy Farmer, "National Grid Group," Schroeder Salomon Smith Barney, 2 November 2012.

Baur, Chris, "Miller's power," *Scottish Business Insider*, May 1988, p. 12.

British Energy, "2002/2003 Preliminary Results, – Part 1."

_____, "Results for the six months to 30 September 2003."

_____, "Summary of Results," September 2008.

_____, "British Energy plc Proposed Disposal of Interest in Bruce Power Limited Partnership and Huron Wind Limited Partnership," press release, December 23, 2002.

Brito, Marcelo, Alexandre Kogake and Kaique Vasconcellos, "CPFL (Company Update)," Citi Research, 19 December 2012.

Brough, Martin and James Brand, "UK utilities," Deutsche Bank, 22 November 2013.

Buchanan, Alistair, "RECs: The U.K.'s Regional Electric Utilities Continue to Power Ahead," *BuySide*, Feb/Mar 1995, p. 93.

_____, "Texas Utilities," Salomon Smith Barney, May 29, 1998.

_____, "British Energy," Donaldson, Lufkin & Jenrette, July 14, 1999.

_____, "Viridian," Donaldson, Lufkin & Jenrette, July 14, 1989.

_____, "Independent Energy," Donaldson, Lufkin & Jenrette, February 22, 2000.

_____, "Independent Energy," Donaldson, Lufkin & Jenrette, May 18, 2000.

_____, "British Energy," Donaldson, Lufkin & Jenrette, July 14, 2000,

_____, "UK Electricity Sector," Donaldson, Lufkin & Jenrette, July 14, 1999.

_____, "National Power," Donaldson, Lufkin & Jenrette, July 14, 1999.

_____, "National Grid," Donaldson, Lufkin & Jenrette, April 11, 2000.

_____, "ScottishPower," Donaldson, Lufkin & Jenrette, May 10, 2000.

Burton, Nigel and Mark Loveland, "Electricity: The New Beginning," S.G. Warburg Securities, 14 May 1990.

Byrne, Alistair, "How have we done?," *Professional Investor*, May 2002.

Central Electricity Generating Board, *CEGB Statistical Yearbook, 1986/87.' 1987/88.*

_____, *Annual Report and Accounts 1987/88.*

_____, *The CEGB and Nuclear Power,* September 1986.

_____, *Drax Power Station Proposed Flue Gas Desulphurisation Plant,* January 1988.

Centrica, *Annual Report and Accounts 2012.*

Chada, Bobby, Nicholas J. Ashworth and Arsalan Obaidullah, "National Grid plc," Morgan Stanley, February 12, 2010.

Chada, Bobby, Emmanual Turpin, Igor Kuzmin, Anna Marria Scoglia, Carolina Dores, Anne N. Azzola and Alexandra A. Economides,, "Utilities: The ripple effect," Morgan Stanley, February 6, 2013.

_____, "Utilities: Rising power volatility – green shoots of recovery?," Morgan Stanley, July 29, 2013.

Credit Suisse, "European Power Breakfast, UK Regulated Utilities: CKI announces bid for EDF's UK Networks," 30 July 2010.

Deutsche Bank, "European Utilities- Idea of the week: UK utilities: going…going… gone!!," April 2, 2012.

Drax, "Biomass Financing Secured," Press Release, December 20, 2012.

Duff & Phelps, "UK Electricity Industry," Duff & Phelps Credit Rating Co., October 1998.

_____, "European Utility Industry," Duff & Phelps Credit Rating Co., October 1999.

Dumoulin-Smith, Julien and Paul Zimbardo, "PPL Corporation," UBS, 17 February 2014.

Ebadan-Bola, Gracie and Richard Hunter, "Green Light, Environmental Risk in the Power Sector," Fitch Ratings, July 2002.

East Midlands Electricity, *Operating and Development Plans 1988-1993.*

Equities International, "Unelectrifying: Britain's power privatisation," August 18, 1989.

ESB and Viridian, press release, "Joint Statement from ESB and Viridian," July 12, 2010.

Farman, Ahmed and Oliver Salvesen, "UK Utilities: Increasingly Tough Out There," Jefferies, 22 September 2016.

Fetter, Stephen M., Ellen Lapson and Kimberly A. Slawek, "U.K. Evolution Continues," Fitch Investor Services, August 1, 1997.

Financial Weekly, "Government's Shy Coal Face," October 22, 1987, p. 34.

Fitch, "Power Projects in a Less Regulated World," Fitch Investor Services, September 12, 1997.

Fitch, "British Energy plc," Credit Update, Fitch Ratings, January 2003.

Flowers, Simon, "Scottish Power," Merrill Lynch, 2 October 1998.

_____ and Ian Graham, "UK Utility Regulation," Merrill Lynch, 22 March 1998.

_____, "BG PLC," Merrill Lynch, 26 June 2000.

445

_____, "Scottish Power," Merrill Lynch, 6 August 1999.

_____, "Centrica," Merrill Lynch, 20 March 2000.

Flowers, Simon and Philip Green, "Centrica," Merrill Lynch, 8 October 2002.

_____, "Centrica," Merrill Lynch, 14 March 2003.

Flowers, Simon, Jonathan Wright and Ian Graham, "BG PLC," Merrill Lynch, 17 October 2000.

Fraulo, Francesco and Isaac Xenitides, "Avon Partners Holding, Midlands Electricity and Aquila Power Networks," Fitch Ratings, January 2003.

Freshney, Mark, Guy MacKenzie and Vincent Gilles, "UK Utilities," Credit Suisse, 25 February 2013.

_____, "National Grid," Credit Suisse, 20 March 2013.

_____, "SSE," Credit Suisse, 19 September 2014.

Goulding, A.J., and Julia Frayer, "X Marks the spot: how performance-based ratemaking (PBR) affected returns to wirescos in the UK," London Economics International, June 2001.

Graham, Ian and Adam Forsyth, "British Energy," NatWest Securities, 24 June 1996.

Graham, Ian and Simon Flowers, "National Power," Merrill Lynch, 2 April 1998.

_____, "UK Electricity," Merrill Lynch, 9 June 1998.

_____, "British Energy," Merrill Lynch, 16 March 1999.

_____, "National Power," 4 December 1998.

_____, "National Power," 14 May 1999.

_____, "Electricity Generation," Merrill Lunch, 27 January 2000.

_____, "National Grid," Merrill Lynch, Merrill Lynch, 28 February 2000.

_____, "National Power," Merrill Lynch, 16 March 2000.

_____, "National Power"; Merrill Lynch, 22 September 2000.

_____, "PowerGen," Merrill Lynch, 18 January 1999.

_____, "PowerGen," Merrill Lynch, 25 March 1999.

_____, "PowerGen," Merrill Lynch, 9 February 2001.

_____, "Innogy," Merrill Lynch, 3 April 2001.

_____, "Scottish & Southern Energy," 23 February 1999.

_____, Scottish & Southern Energy, Merrill Lynch, 14 January 2000.

Graham, Sue, "British Gas Plc," Merrill Lynch, 13 February 1991.

Green, Philip and Simon Flowers, "National Grid Transco," Merrill Lynch, 4 November 2002.

_____, "Active Union," Merrill Lynch, 26 March 2003.

Honoré, John, Thierry Bros, Andy Gboka and Didier Laurens, "UK Utilities: A £140 bn market, >55% regulated," Societe Generale, March 2010.

Hunter, Richard, Ellen Lapson and Thomas Saul, "Europe's new Powerbahn," Fitch IBCA, June 1999.

Hyman, Leonard S., "National Power," Merrill Lynch, December 20, 1991.

Liu, Ernest S., Liz Christie, James P. McFadden and Ashar Khan, "Electricity Privatisation in England and Wales: An International Perspective," Goldman Sachs, September 1990.

_____, Elizabeth A. Parrella, CFA, Debra E. Bromberg, Neil C. Choi, and W. Michael Weinstein, "Public Utility Survey: Nonregulated Business Strategies and International Investments," Goldman Sachs, September 1999.

Malik, Hasnain, Simon Taylor and Jane Hayes, "Scottish & Southern Energy," Salomon Smith Barney, 7 April 1999.

_____, "Scottish Power," Salomon Smith Barney, 4 November 1999.

Manley, John, "New Plan Tips Power Balance to Generators," *Financial Weekly,* August 24, 1989, p. 18.

Martin, Daniel, Simon Taylor and Piers Coombs, "British Energy": The Final Analysis," Barclays de Zoete Wedd, 22 May 1996.

Major UK Companies Handbook, various dates.

Martiniussen, Erik, "British nuclear losses continue to rise," Bellona, Bellona.no, June 11, 2004.

Miller-Bakewell, Robert, "ScottishPower," Merrill Lynch, 5 April 2002.

_____ and Simon Flowers, "ScottishPower," Merrill Lynch, 12 March 2002.

Monnier, Laurence and Richard Hunter, "Merchant Power Projects: Lessons from 18 Months of NETA," Fitch Ratings, 30th August 2002.

Midlands Electricity plc, *Investor Information*, "Electricity Regulation in England and Wales," Fall 1994.

Moody's Investors Services, *Moody's Electric Utility Sourcebook*, October 1993, October 1995, October 1996, October 1997.

_____, *Moody's International Manual*, various dates.

_____, *Moody's Power Company Sourcebook*, October 1999, October 2000.

_____, *Moody's Project Finance Sourcebook*, October 1997, October 2000.

_____, *Moody's Sourcebook Power and Energy Company,* October 2002, October 2003, October 2005, October 2006.

"The mutual energy company," energyireland.ie/the -mutual -energy -company, July 25, 2012.

National Grid, *Annual Review 2001/02.*

_____, "Keyspan is now part of National Grid, "*National Grid Fact Sheets,* August 2007.

_____, *GB Seven Year Statement 2009,* May 2009.

_____, "National Grid," presentation, EEI Annual Finance Meeting, 23 June 2010.

_____, "National Grid Today" *August 2011.*

National Power, *This is National Power*, August 1989.

_____, *National Power News*, February-March 1994.

_____, *National Power News*, Results Special, 1994.

_____, *Annual Review 1999.*

_____, *Report and Accounts 1999.*

_____, *The first five years*, 1985.

Open Energi, "Demand response is 'win-win' for UK economy, metering.com, September 16, 2104.

Paribas, Ltd, "The UK Electricity Supply Industry," May 1, 1989.

Pink, Nick, "British Energy," SBC Warburg, July 1996.

Placement Memorandum Dated December 11, 1990, Eastern Electricity plc, East Midlands Electricity plc, Manweb plc, Midlands Electricity plc, Northern Electric plc, NORWEB plc, SEEBOARD plc, Southern Electric plc, South Wales Electricity plc, South Western Electricity plc, Yorkshire Electricity plc, 12,306,000 ADS Package Units, Rule 144A Offering of American Depositary Shares in ADS Package Units only, each ADS Package Unit being equivalent to 1/100th of a U.K. Ordinary Share Package Unit.

PowerGen, Presentation by John Rennocks, Executive Finance Director, 17 February 1992.

_____, *Annual Report and Accounts 1989/90.*

_____, *Report and Accounts 1998.*

_____, *Report nine months ended December 1998.*

_____, *Review nine months ended December 1998.*

_____, *Annual Report and Review 2000.*

PPL, press release, "PPL to Expand Regulated Business Portfolio by Acquiring Second-Largest U.K. Electric Distribution Business," PRNewswire, March 1, 2011, energycentral.com.

Preliminary Placement Memorandum Subject to Completion Dated February 1, 1991, Rule 144A Package Offering of American Depositary Shares of National Power PLC and Powergen plc.

Prospectus, The two Scottish electricity companies share offers, 31 May 1991.

Redpoint Energy, "Redpoint Energy reports on policy options for DECC as UK Government considers shake up of national energy market," press release, December 15, 2010, energybiz.com.

Rowland, Chris and Alex Milne, "Electricity: Issues for Investors," Barclays de Zoete Wedd, September 1990.

Russell-Walling, Edward, "Power to the Distributors," *Financial Weekly*, February 9, 1989, p. 14.

_____, "A Private Duopoly," *Financial Weekly,* August 24, 1989,p. 19.

Sayers, Michael, "The Electricity Distribution Companies in England and Wales," Salomon Brothers, 1990.

_____, "The Privatisation of the Electricity Distribution Companies in England and Wales – A Summary," Salomon Brothers, September 17, 1990.

Sayers, Michael and James Hutton-Mills, "British Energy: A Unique Investment Opportunity," Morgan Stanley, 20 April 1996.

Southern Co., *Annual Report* (various dates), quarterly reports and presentations.

Standard & Poor's, *S&P Global Utilities Credit Review*, "Global Utilities," November 1996, October 1997.

"TREC Privatisation," no date or attribution.

_____, *U.S. Utilities and Power Commentary*, November 2006, November 2007.

UK Business Park, "British Energy," 2009.

Taylor, Simon, Hasnain Malik and Jane Hayes, "National Power," Salomon Smith Barney, 13 May 1999.

Thomas, Ashley, "Utilities," Societe Generale, 27 May 2015.

UK Shareholders' Association, "UK Stock Market – Background Information and Statistics, UK Stock Market Statistics, Last revised July 2007."

Venkateswaran, Deepa and Nicholas J. Green, Cosma Panzacchi, Neil Beveridge, Bob Brackett, Jean Ann Salisbury and Oswald Clint, "Competitive Auctions – What Are the Learnings and Implications for Renewable Operators?," Sanford C. Bernstein & Co., 1 August 2016.

Venkateswaran, Deepa and Gavin Kennedy, "UK Utilities: How will Budget proposals on tax deductibility of interest impact potential M&A in regulated networks?," Sanford C. Bernstein & Co., April 21, 2016.

Watton, Mark and Mark C, Lewis, "Easton Group's Bonds Rated," in *Standard & Poor's Utilities and Perspectives*, Standard & Poor's, October 20, 1997.

Wells, Alan D., CFA, *et. al.*, "European Utilities: UK Renewable Energy," Morgan Stanley, April 4, 2011.

White, Anthony, "Reshaping the Electricity Supply Industry in England and Wales," James Capel & Co. Ltd, February 1990.

_____, David Gray and Derek Lygo, "The New Electricity Companies," James Capel & Co., May 1990.

Williams, Dr. Simon, "The Electricity Company Notebook," Kleinwort Benson, September 1990.

Wilson, Dr. John, "The Electricity Industry," UBS Phillips & Drew, 28 August 1990.

Wright, Andrew and Kevin Lapgood, "UK Utility Countdown," Merrill Lynch, 24 October 1996.

_____, "British Energy plc," Merrill Lynch, 30 May 1996.

Xenitides, Isaac and Richard Hunter, "TXU Europe Ltd," Fitch Ratings, November 19, 2002.

Public Policy and Political Documents

Banks, Ferdinand E., "Energy Deregulation Unplugged," energybiz Leadership Forum, August 10, 2011, energybiz.com.

Chesshire, John, "The Privatisation of the UK Electricity Supply Industry – Introducing Competition," Electricity Consumers' Council's "Privatising Electricity – A Chance for Change?" Conference, September 1987.

Edwards, M.J., "British Electricity and British Coal," Electricity Consumer's Council's "Privatising Electricity – A Chance for Change?" Conference, September 1987.

Electricity Consumers' Council, "Electricity Privatisation: An EEC Perspective," Paper 1, July 1987.

_____, "Customer Audit and Review 1986/87," Paper 2.

_____, "Coal and the Interest of the Electricity Consumer," Paper 3, September 1987.

_____, "The National Grid and the Merit Order," Paper 4, October 1987.

"Electricity Privatisation and the Area Boards: The Case for 12 – A Report Commissioned by the 12 Area Boards of England and Wales, November 1987.

Evans, Nigel, "Electricity Privatisation – The Nuclear Connection," Electricity Consumer's Council "Privatising Electricity – A Chance for Change?" Conference, September 1987.

Gordon, Myron J. and John F. Wilson, "Don't sign the Bruce lease!," *Behind the numbers*, Canadian Centre for Policy Alternatives, April 25, 2001.

Helm, Dieter, "Regulating the Electricity Supply Industry," Electricity Consumers' Council's Conference on Regulation and Consumer Protection, 24 March 1988.

Henney, Alex, *Privatise Power: restructuring the electricity supply industry* (London: Centre for Policy Studies, 1987).

Holmes, Andrew," Privatisation in the European Context, "Electricity Consumers' Council's "Privatising Electricity – A Chance for Change?" Conference, September 1987.

Friends of the Earth, "A Briefing on Electricity Privatisation, Privatisation and Nuclear Power,," September 1988.

_____, "Energy Campaign Briefing, Electricity Privatisation and the Environmental Implications," April 1989.

_____, "'Green the Bill': Environmental Groups Challenge to MPs," press release, 31 March 1989.

_____, "Unprecedented Advertising Campaign Challenges Parkinson to Go Green on Electricity Privatisation," press release, 3 July 1989.

_____, "Friends of the Earth Labels Parkinson's Energy Efficiency Amendment 'A Devastating Let Down' ", press release, 7 July 189.

_____, "Government Encourages Nuclear Industry Incompetence," press release, 10 July 1989.

_____, Poll Shows Nuclear Power in Electricity Sell-Off Cuts Prospective Investors by Half, press release, 18 July 1989.

_____, "Government to Keep Magnox Nuclear Reactors," press release, 24 July, 1989.

Kay, John, "Cutting costs so often leads to cutting corners," *Financial Times*, June 23, 2010.

Labour election manifesto 1997, "Labour Because Britain Deserves Better, " psr.keele.ac.uk/man.

LabourParty.org.uk, "1945 Labour Party Election Manifesto."

Labour Party Campaign and Communications Directorate, "Labour Steps Up Green Campaign Tony Blair sets Out New Three Point Plan Against Pollution Government Under Fire Over Electricity Privatisation," press release, 1 July 1989.

"Living with monopoly," *Nature*, vol. 356, 19 March 1992, p. 180.

Lyons, John, "Why the CEGB Should Not Be Broken Up," Electricity Consumers' Council's "Privatising Electricity – A Chance for Change?" Conference, September 1987.

National Consumer Council, "Electricity Privatisation – NCC Policy Paper no. 2: Regulation," June 1988.

Political and Economic Planning (PEP), *Report on the Gas Industry in Great Britain* (London: PEP, March 1939).

_____, *The British Fuel and Power Industries* (London: PEP, October 1947).

Rufford, George, "Privatising Electricity: Regulating for Fair Competition," Electricity Consumers' Council's "Privatising Electricity – A Chance for Change?" Conference, September 1987.

Sykes, Allen and Colin Robinson, *Current Choices: good ways and bad to privatise electricity* (London: Centre for Policy Studies, 1987).

Toi, Richard, "Time for a celebrity regulator as Britain faces winter power cuts," *The Conversation*, November 4, 2014.

Unison, "Report Exposes Golden Hole in Energy Firm's Accounts – Inquiry Demanded into Missing Billions," press release, January 22, 2008.

_____, "Windfall Tax (United Kingdom)."

Willetts, David, "Privatising the Electricity Supply Industry," Electricity Consumers' Council's "Privatising Electricity – A Chance for Change?" Conference, September 1987.

Yarrow, George, "Notes on the Regulation of a Privatised Electricity Supply," Electricity Consumers' Council's "Privatising Electricity – A Chance for Change?" Conference, September 1987.

Newspapers, Internet News Services and General Periodicals (listed by year)

1911

"Lord Haldane on the Germans," *Grey River Argus*, 3 Whiringa- -nuku, 1911.

1970

"Second political strike planned for January 12," "Mr. Jack Jones says Chancellor's speech was provocative," *Times*, December 7, 1970, p. 1.

Routledge, Paul, "500,000 in stoppage," *Times*, December 7, 1970, p. 1.

"Power cuts affect a third of Britain as industry gets warning of shutdowns," "Prospects bleak for today," *Times*, December 9, 1970, p. 1.

Noyes, Hugh, "Hour of fury in the Commons," *Times*, December 9, 1970, p. 1.

Routledge, Paul, "Widespread stoppages had most effect on docks, newspapers and cars," *Times*, December 9, 1970, p. 1.

"Leaders of power unions reject Mr.. Carr's plea to take claim to arbitration," *Times*, December 10, 1970, p. 1.

Routledge, Paul, "Solutions nearer, ETU chief says,'" *Times*, December 10, 1970, p. 1.

Noyes, Hugh, "Government hints at court of inquiry on work-to-rule," *Times*, December 10, 1970, p. 1.

Our Political Staff, "Cabinet doubts on 'no surrender'," *Times*, December 10, 1970, p. 1.

Our Parliamentary Correspondent, "Candles and storm lamps in Commons," *Times*, December 10, 1970, p. 1.

Routledge, Paul, "Hopes of normal power supplies by weekend as work-to-rules finishes," *Times*, December 5, 1970, p. 1.

"Party leaders in angry exchange," "Widespread effects of power cuts," *Times*, December 16, 1970, p. 1.

Thomas, Michael, "Refusal to operate manning agreements is main cause of electricity cuts," *Times*, December 16,, 1970.

Our Social Correspondent, "Reassurance for kidney machine patients," *Times*, December 16, 1970.

1977

MacIntyre, David, "Blackouts as power men work to rule," *Times*, October 26, 1977, p. 1.

Labour Staff, "Industrial action leads to power reductions," *Times*, October 27, 1977, p. 5.

Our Labour Staff, "Power men black out No 10," *Times*, October 28, 1977, p. 1.

"Power men threaten even more cuts if today's talks fail," *Times*, November 3, 1977, p. 1.

Seton, Craig, "Surgeon operates by torchlight," *Times*, November 3, 1977.

Clark, George, "Benn role in power dispute criticized," *Times*, November 12, 1977, p. 1.

Routledge, Paul, "Power men's stewards vote to end dispute," *Times*, November 12, 1977, p. 1.

"Candlelight questions in the commons," *Times*, December 10. 1977, p. 1.

"Hospitals told they can call in heart and kidney patients as power cuts increase," *Times*, December 10, 1977, p. 2.

1980

Noyes, Hugh, "Government releases hold on three nationalized industries," *Times*, July 22, 1980, p. 1.

1981

Fishlock, David, "Passionate advocate takes over at Atomic Energy Authority," *Financial Times*, February 24, 1981, p. 7.
"Britain's nuclear foes get ammunition for their fight against atomic energy," *World Business Weekly*, June 8, 1981, p. 15.

1982

Riddell, Peter, "Lawson plans to end state monopoly of electricity generators," *Financial Times*, March 29, 1982, p. 1.
Cameron, Sue, "Row over CEGB chief's removal," *Financial Times*, May 6, 1982, p. 6.
Fishlock, David, "A crusader for the nuclear future," *Financial Times*, May 28, 1982, p. 10.
Dafter, Ray, "N-industry chief moves to CEGB," *Financial Times*, May 28, 1982, The Back Page.
_____, "Why Lawson made an obvious choice," *Financial Times*, October 20, 1982, p. 8.
_____, "Electricity Council's new chief," *Financial Times*, October 20, 1982, The Back Page.

1984

Garnett, Nick, "Drax Power Station," February 1, 1984, *Financial Times*, pp. 13-15.
Fishlock, David, "Channel trenches prepare way for electricity link," *Financial Times*, June 4, 1984, p. 5.

1985

"Pit strike will end tomorrow," *Financial Times*, March 4, 1985, p. 1.
Hargreaves, Ian, "Hard days ahead for coal," *Financial Times*, March 4, 1985.
Lloyd, John, "It ground them down and out," *Financial Times*, March 4, 1985, p. 29.
_____, "Industry struggles to get on its feet," *Financial Times*, March 4, 1985, p. 19
_____, "The break with past practice," *Financial Times*, March 4, 1985, p. 21.
Barnett, Philip, "How leaders turned into scabs," *Financial Times*, March 4, 1985, p. 19.
_____, "It's even harder times ahead," *Financial Times*, March 4, 1985, p. 21.
_____, "Delegates in sombre mood," *Financial Times*, March 4, 1985, p. 1.
Riddell, Peter, "Thatcher purges the traumas of the past," *Financial Times*, March 4, 1985, p. 21.
_____, "Politicians give muted welcome," *Financial Times*, March 4, 1985, p. 1.
Brindle, David, "Fierce passions on both sides," *Financial Times*, March 4, 1985, p. 21.
Wilkinson, Max, "The cost might have been higher," *Financial Times*, March 4, 1985, p. 2.

1987

Wilkinson, Max, "CEGB winning the battle to remain a single body," *Financial Times*, January 4, 1987, p. 1.
Wilkinson, Max, "Politics at odds with principle," *Financial Times*, September 14, 1987.
Samuelson, Maurice, "Power poised to return to an older generation's ways," *Financial Times*, September 17, 1987.
Kellaway, Lucy, "Encouraging the vital spark of competition," *Financial Times*, September 21, 1987.
Wilkinson, Max, "Sparks fly in the nuclear family," *Financial Times*, October 5, 1987.
Robinson, Colin, "This time, break the monopoly," *Financial Times*, October 7, 1987.

Wilkinson, Max, "CEGB attacks calls for split-up on privatisation," *Financial Times*, November 2, 1987.

Buxton, James, "Electricity boards differ over sell-off," *Financial Times*, November 10, 1987.

"Competition in electricity," *Financial Times*, November 19, 1987.

Samuelson, Maurice, "Parkinson 'cool it' plea on electricity," *Financial Times*, November 23, 1987.

McCloskey, Gerard and Miri Zlatnar, "CEGB to import supplies at half domestic price," *Financial Times*, December 14, 1987.

1988

Rudd, Roland, "ECC call for 'balanced' energy control," *Times*, January 6, 1988.

"The Electricity Industry," Financial Times Survey, *Financial Times*, January 25, 1988, pp. 13-18.

Samuelson, Maurice, "Changing the power game, p. 13.

_____, "Proud past harnessed to defend the future," p. 15.

_____, "More freedom in prospect," p. 15.

_____, "A different power debate," p. 16.

_____, "Europe's remarkably diverse power pool," p. 17.

_____, "New brain for national grid," p. 17.

_____, "West Germany shows the way," p. 18.

Wilkinson, Max, "The political posers for Mr. Parkinson," p. 14.

Robinson, Colin & Allen Sykes, "Safeguard of the consumer," p. 14.

Lyons, John, "'Intolerable contradictions,,' " p. 14.

Garnett, Nick, "Big hunger for new orders," p. 16.

Kellaway, Lucy, "UK schemes start to move off the drawing board," p. 16.

Newham, Mark, "Fairer breeze for renewables," p. 18.

Wilkinson, Max, "Electricity boards seek to pass on extra costs after sell-off," *Financial Times*, June 3, 1988.

_____, "Easier ride for nuclear power plans predicted," *Financial Times*, June 6, 1988.

_____, "Battle for commanding heights of electricity," *Financial Times*, June 6, 1988.

Samuelson, Maurice, "Power industry 'may diversify into coal mining'," *Financial Times*, June 9, 1988.

Textline, "Power privatisation generates problems" (1st September 1988), "The SSEB expresses surprise at fears over spent nuclear fuel route" (7th September 1988), "£220 m plan to build nuclear fuel store"(16th September 1988), "Scots Electricity Boards in Red' (30th September 1988), "Cecil Parkinson approves plan to pipe North Sea gas to Scottish power station" (6th October 1988), "SSEB has set aside £250m to pay its share of the decommissioning costs at Sellafield" (7th October 1988, "British Coal: chances of breaking even this year lessening" (10th October 1988), "Government facing huge bill at Sellafield" (11th November 1988), "Nuclear power will cause major difficulties for the privatisation of the electricity industry" (14th October 1988), "Scottish electricity set for a 1991 sell-off "(19th November 1998), "Scotland to export its electricity" (2nd December 1988), "The big switch on" (4th December 1988), "Taking the heat out of nuclear stations," (4th December 1988), "Nuclear issues threaten float says power chief" (12th December 1988), "Reactor at Torness Nuclear station starts up" (31st December 1988).

1989

Economist, "Electricity privatisation Sr Vac," February 25, 1989, p. 26.

Wilkinson, Max, "Spot the difference with privatised power," *Financial Times*, March 11, 1989.

Walker, Dr. Jim, "More liberal energy market to benefit all," *Glasgow Herald*, 15 March 1989.

Harrison, Michael, "Confidential report says electricity sale in trouble," *Independent*, 11 April 1989.

"Lords defeat for government on power efficiency." *Times*, May 16, 1989, p. 12.

Samuelson, Maurice, "Electricity plans will 'blaze trail' in Europe," *Financial Times*, June 14, 1989, p. 12.

Wilkinson, Max, "Power sell-off runs into problems over costs and profits," *Financial Times*, July 3, 1989, p. 8.

_____, "High costs may push down the value of nuclear power plant," *Financial Times*, July 10, 1989, p. 6.

Green, David, "Hinkley N-inspector to visit Chernobyl," *Financial Times*, July 10, 1989, p. 6.

Wilkinson, Max, "Mr. Parkinson's dilemma," *Financial Times*, July 11, 1989, p. 22.

Rutledge, Ian, "Electricity privatisation and the US experience, *Financial Times*, July 17, 1989, p. 21.

Wilkinson, Max, "Battle to put price on power," *Financial Times*, July 17, 1989, p. 10.

Samuelson, Maurice, "Government yields on N-plant sales," *Financial Times*, July 25, 1989, p. 7.

Fishlock, David, "Retiring the Magnox workhorses," *Financial Times*, July 25, 1989, p. 8.

Wilkinson, Max, "Old reactors hinder sell-off plans," *Financial Times*, July 25, 1989, p. 8.

Mason, John, "Calls for delay rejected despite Magnox move," *Financial Times*, July 26, 1989, p. 8.

Green, Davis, "Magnox ruling 'not caused by investors,'" *Financial Times*, July 27, 1989, p. 9.

Green, David, "Government seeks EC agreement on nuclear industry," *Financial Times*, July 31, 1989, p. 4.

Fishlock, David, "Reprocessing the Magnox image," *Financial Times*, July 31, 1989, p. 4.

"We're not just one of Britain's biggest businesses," advertisement, *Financial Times*, August 4, 1989, p. 21.

"Nuclear risk and reward," *Financial Times*, August 7, 1989, p. 12.

"A decade of privatisation," *Financial Times*, August 14, 1989, p. 10.

Wilkinson, Max, "Disputes threaten date of electricity sell-off," *Financial Times*, August 14, 1989, p. 5.

_____, "The dance of the dinosaurs," *Financial Times*, August 17, 1989, p. 14.

_____, "Power industry told to scale down coal threat," *Financial Times*, August 17, 1989, p. 5.

Mason, John, "PM's electricity role denied," *Financial Times*, August 29, 1989, p. 5.

"A spot market for power," *Financial Times*, August 29, 1989, p. 16.

Fishlock, David, "High costs threaten future of N-station," *Financial Times*, August 30, 1989, p. 9.

Samuelson, Maurice, "Doubt over electricity emissions clean-up," *Financial Times*, September 4, 1989, p. 7.

_____, "Brittan examines power sell-off," *Financial Times*, September 7, 1989, p. 8.

"Monopolies in power," *Financial Times*, September 13, 1989, p. 18.

Samuelson, Maurice, "Company seeks to sell UK cheap French power," *Financial Times*, September 14, 1989, p. 9.

Cookson, Clive, "Government under pressure to cut nuclear power privatisation," *Financial Times*, September 15, 1989, p. 10.

"Whilst an individual can achieve...," REC advertisement, *Financial Times*, September 20, 1989, p. 21.

Wilkinson, Max, "A clock ticks for monopoly," *Financial Times*, September 27, 1989, p. 21.

Samuelson, Maurice, "Reaction split on terms for power sell-off," *Financial Times*, September 27, 1989, p. 7.

Mason, John, "Government determined to privatise nuclear power," *Financial Times*, September 28, 1989, p. 8.

Samuelson, Maurice and James Buxton, "Power contracts seen to win cabinet vote," *Financial Times*, September 29, 1989, p. 8.

Wilkinson, Max, "Power cut in UK electricity privatisation," *Financial Times*, October 3, 1989, p. 10.

Robinson, Colin, "A programme that is going astray," *Financial Times*, October 4, 1989, p. 17.

Gribben, Roland, "Nuclear power plant building is halted," *Daily Telegraph*, October 11, 1989, p. 1.

Lublin, Joann S., "U.K. Steams Over Water and Electricity," *Wall Street Journal*, October 12, 1989, p. A 13.

Samuelson, Maurice, "When the spark loses its glow," *Financial Times*, October 25, 1989, p. 17.

_____, "Minister signals on nuclear plant sales," *Financial Times*, October 31, 1989, p. 11.

Wilkinson, Max, "Electricity industry abandons plans for wholesale market," *Financial Times*, November 1, 1989, p. 11.

_____ and Maurice Samuelson, "Nuclear industry to be excluded from power sale," *Financial Times*, November 9, 1989, p. 9.

Harrison, Michael and Colin Hughes, "N-power stations pulled out of electricity sell-off," *Independent*, 10 November 1989, p. 1.

Goodwin, Stephen, "Nuclear power 'has five years to prove its economic viability'," *Independent*, 10 November 1989, p. 6.

Fagan, Mary, "Government 'had been warned of escalating costs'," *Independent*, 10 November 1989, p. 2.

"The nuclear power fiasco," *Independent*, 10 November 1989, p. 18.

Young, Alf, "Torness plant was 'a £2500m mistake', "*Glasgow Herald*, November 10, 1989, p. 1.

Parkhouse, Geoffrey, "Wakeham explains U-turn," Glasgow Herald, November 10, 1989, p. 1.

Harrison, Michael, "End of a dream for Mr. Nuclear," *Independent*, 10 November 1989, p. 1.

McGregor, Stephen, "Pledge on power prices after nuclear decision," *Glasgow Herald*, November 10, 1989, p. 8.

"Parkinson attacked," *Glasgow Herald*, November 13, 1989, p. 1.

Harrison, Michael, "Wakeham U-turn puts privatisation on tight timetable," *Independent*, 10 November 1989, p. 2.

Fagan, Mary, "Government 'had been warned of escalating costs,'" *Independent*, 10 November 1989, p. 2.

Hughes, Colin, "Leading players absent from the drama," *Independent*, 10 November 2010, p. 2.

Douglas Home, Mark, "Scots assets stay state-owned," *Independent*, 10 November 1989, p. 2 .

"Think again, Mr. Wakeham," *Financial Times*, November 10, 1989, p. 24.

Wilkinson, Max, "The end of a nuclear dream," *Financial Times*, November 10, 1989, p. 24.

Holmes, Andrew, "Performance of nuclear reactors has been mediocre," *Financial Times*, November 11, 1989, p. 6.

Fishlock, David, "Lord Marshall leaves his position of power," *Financial Times*, November 11, 1989, p. 6.

Buxton, James, "Scotland expected to export more power to England," *Financial Times*, November 11, 1989, p. 6.

Young, Alf, "Delicious ironies in nuclear U-trip," *Glasgow Herald*, November 13, 1989, p. 13.

Harrison, Michael, "Pressure grows for more changes in electricity plans," *Independent*, 13 November 1989.

McGregor, Stephen, "MPs keep up pressure over nuclear U-turn," *Glasgow Herald*, November 14, 1989.

Goodwin, Stephen, "Calls for re-think on power rejected," *Independent*, 14 November 1989.

Newbery, David, "The power of competition," *Independent*, 15 November 1989.

Samuelson, Maurice, "Power chief calls for price rise to promote efficiency," *Financial Times*, November 17, 1989, p. 11.

Fishlock, David, "Power chief attacks plan to drop N-programme," *Financial Times*, December 1, 1989, p. 8.

Samuelson, Maurice, "Electricity chief resigns over curb on nuclear power," *Financial Times*, December 19, 1989, p. 6.

_____ and David Fishlock, "CEGB profits cut heavily by cost of nuclear power," *Financial Times*, December 22, 1989, p. 7.

Buxton, James, "Scottish power boards report heavy losses," *Financial Times*, December 22, 1989, p. 7.

Profile Information Services, Magnox Press Çommentary, "Reprocessing the MAGNOX image" (31 Jul 89), "A transport guide for a road warrior" (30 Jul 89), "Power links up to a three-way switch" (30 Jul 89), "MAGNOX bill could be 3 pounds 15 bn" (29 Jul 89), "BNFL provokes row over bid for MAGNOX stations" (29 Jul 89), "Serious doubt" (28 Jul 89), "Nuclear no-no" (28 Jul 89), "MAGNOX figures 'not accurate'" (27 Jul 89), "MAGNOX ruling 'not caused by investors'" (27 Jul 89), "Retiring the MAGNOX workhorses" (25 Jul 89), "Old reactors hinder sell-off plans" (25 Jul 89), "Electricity" (25 July 89), "MAGNOX stations stay in public ownership" (25 Jul 89), "MAGNOX decision dates from 1976" (25 Jul 89), "MAGNOX N-power stations are withdrawn from privatisation" (25 Jul 89), "Power flotation loses MAGNOX plants" (25 Jul 89), "Unwanted, unsellable" (25 Jul 89), "Why power, water and competition don't mix" (23 Jul 89), "Back to the old faithful" (20 Jul 89), "Pounds 3.5bn bill to close MAGNOX" 20 Jul 89).

_____, Press Commentary on the Electricity Industry in the UK, "Selling-off timing goes haywire" (15 Aug. 89), "600m pound power sell-off sweetener" (15 Aug. 89), "Electricity privatisation may be delayed for at least six months" (14 Aug. 89). "Consumers reject state sell-offs" (13 Aug. 89), "Small power stations hit by slower rates reductions" (12 Aug. 89), "Whip -hand over nuclear power" (10 Aug. 89), "Labour challenges on AGRs" (9 Aug. 89), "Tide-power barrage report is completed" (9 Aug. 89), "Offshore wind power project in the balance" (9 Aug. 89), "Reactors excluded from privatisation 'are more efficient'" (9 Aug. 89), "Labour fears pits will close after electricity sell-off" (8 Aug. 89), "National Grid asks Bank of England to shoulder risk" (8 August 89), "GridCo questions" (8 Aug. 89), "Electricity cash role for Bank" (8 Aug. 89), "Economics likely to generate Hinkley verdict" (7 Aug. 89), "Nuclear risk and reward "(7 Aug. 89), "Last great debate over the power-to-be" (7 Aug. 89), "Government under pressure to bear financial risks of AGRs" 7 Aug. 89), "Privatised power" (7 Aug. 89), "Power chairmen win the generation gams" (6 Aug. 89), "BNFL in plan for Magnox" (6 Aug. 89), "Shedding light on power float" (4 Aug. 89), "electricity boards may miss deadline in power cost talks" (4 Aug. 89), "Electricity boards admit lack of experience" (4 Aug. 89), "Preparing for a switch to the private sector" (4 Aug. 89), "Electricity boards shape up for sale" (4 Aug. 89), "Charter will back wind, tide power" (4 Aug. 89), "Atomic power 'bias' criticised" (3 Aug. 89), "Atomic energy chief attacks Magnox decision as 'messy'" (3 Aug. 89), "No such thing as a free shutdown" (3 Aug. 89), "Sellafield on a problem list for watchdog" (2 Aug. 89), "Shake-up ordered at HunterstonA" (2 Aug. 89), "CEGB wind farm takes a step nearer" (2 Aug. 89), "Chief safety inspector's doubts on partial sale of nuclear industry" (2 Aug. 89), "Magnox stations to shut by start of next century" (1 Aug. 89), "UK reactors among least efficient" (1 Aug. 89).

_____, Press Commentary on the Electricity Industry, "Power firms may float together" (17 Aug. 89), "Power politics hit sell-off" (17 Aug. 89), "Curbs on coal imports for power" (17 Aug. 89), "The start of the power struggle" (17 Aug. 89), Let the power shine in" (17 Aug. 89), "The dance of the dinosaurs" (17 Aug. 89), "Complex problems hinder plans for electricity sell-off" (16 Aug. 89), "Wakeham must find answers to tough questions" (16 Aug. 89).

Articles from the Financial Times Since 30/10/89 Electricity Privatisation, "Electricity" (11 Nov. 89), "High prices put paid to 20-year nuclear dream" (11 Nov. 89), "Marshall resigns from CEGB" (11 Nov. 89), Markets: A week is a long time in the City, too" (11 Nov. 89), "Chronology of nuclear power development in Britain" (10 Nov. 89), "Government says N-plants will remain in public sector" (1`0 Nov. 89), "Parliament and Politics: Attack over power sell-off 'shambles'" (10 Nov. 89), "End of the nuclear dream" (10 Nov. 89), "Politics today: Striving for the (nearly) impossible" (10 November 89), "Government cancels plans to privatise nuclear power industry" (9 November 89), "Power price kept from Hinkley C inquiry" (8 November 89), "Spot the difference with privatised power: How the Energy Secretary plans to deal with fresh obstacles" (3 November 89), "Parliament and Politics: Coal chief claims strong hand in talks" (2 November 89), "British Gas sets industry terms," (1 November 89). "Energy costs highlighted in leaked memo,"

1990

Fishlock, David, "Lord Marshall attacks U-turn on reactors," *Financial Times*, January 29, 1990, p. 8.

Thomas, David and Maurice Samuelson, "Spark of comfort for the user," *Financial Times*, February 13, 1990, p. 11.

Thomas, David, "Gas-fired power stations blow to coal industry," *Financial Times*, February 23, 1990, p. 10.

_____, "Competition rule for power privatisation 'breached,'" *Financial Times*, March 26, 1990, p. 10.

"Electricity Privatization – A Special Report," *Times*, March 26, 1990, pp. 31-39.

Young, David, "It's all systems go," p. 31.

_____, "A giant in the power game," p. 32.

_____, "Custom built for success in a tough market-place," p. 32.

_____, "The big shares switch on," p. 33.

_____, "Battle for the National Grid," p. 33.

Wilson, John, "Will electricity be a good investment?," p. 33.

Hobson, Rodney, "The wheels of fortune," p. 35.

Hatfield, Michael, "Public role for private sector," p. 35.

_____, "Why coal is still king,"p. 36.

Hobson, Rodney, "Links to the new power base," p. 36.

_____ and Nick Nuttall, "Trying to catch the wind," p. 37.

Hatfield, Michael, "Position of power with new responsibilities," p. 37.

Advertisements by Thames Power, p. 31, Scottish utilities, p. 33, PowerGen, p. 34, British Coal, p. 35, Nuclear Electric, p. 36, National Grid, p. 38, RECs, p. 39.

Thomas, David and Maurice Samuelson, "Power to some of the people," *Financial Times*, March 28, 1990, p. 17.

Kellaway, Lucy, "EC allows UK to pay power subsidies," *Financial Times*, March 29, 1990, p. 20.

Samuelson, Maurice, "French sign deal to supply power across Channel," *Financial Times*, March 29, 1990, p. 10.

"Electricity Industry – Financial Times Survey," *Financial Times*, March 29, 1990, pp. I-XVI.

Thomas, David "Battered by green issues," p. I.

_____, "A privatised Goliath emerges," p. II.

Garnett, Nick, "Confusion for equipment makes," p. II.

Cragg, Chris, "Trying to turn ambitious plans into reality," p. III.

Thomas, David, "Facing up to a larger rival," p. III.

Plaskett, Lucy, "On the road to the market," p. IV.

Thomas, David, "Unique among exotic creatures," p. IV.

Buxton, James, "A light at the end of the queue," p. IV.

Fishlock, David, "Considering options," p. V.

Gapper, John, "Pay machinery in doubt," p. V.

Wilkinson, Max, "Free market fears," p. XII.

McCloskey, Gerard, "Slow start for brave new world," p. XIV.

Pearson, Clare, "Water sale sets fine precedent," p. XIV.

Thomas, David, "Throwing light on dimly lit areas," p. XVI.

_____, "Academic in the hall of the private sector," p. XVI.

Advertisements for National Grid, p. I, British Coal, p. III, British Nuclear, p. IV, Nuclear Electric, p. VI, British Gas, p. VII, RECs, p. X, PowerGen, p. XIII, Thames Power, p. XIV, Scottish utilities, p. XVI.

Samuelson, Maurice, "Enron of US to build £700m power station," *Financial Times*, March 30, 1990, p. 8.

Thomas, David, "Marketing of electricity will remain centralised," *Financial Times*, April 2, 1990, p. 8.

_____, "Power body reviews limits on competition," *Financial Times*, April 17, 1990, p. 12.

Garnett, Nick, "Power struggle," *Financial Times*, April 20, 1990, p. 21.

Thomas, David, "Volatile profits predicted for new power companies," *Financial Times*, May 3, 1990, p. 11.

Samuelson, Maurice, "Pollution and costs prompt gas-burner plan for coal station," *Financial Times*, May 9, 1990, p. 12.

Robinson, Colin, "Cleaning up the nuclear debate," *Financial Times*, May 9, 1990, p. 21.

"Prototype N-station to close," "Honda to buy local power," *Financial Times*, May 11, 1990, p. 10.

Samuelson, Maurice, "Cheap natural gas threat to 'green' power station," *Financial Times*, May 12, 1990, p. 4.

Thomas, David, "Electric supply industry facing market challenge," *Financial Times*, May 21, 1990, p. 7.

_____ and Maurice Samuelson, "Power competition limits reduced," *Financial Times*, May 22, 1990, p. 9.

Gapper, John, "Nuclear industry warned of 2,500 job cuts," *Financial Times*, May 22, 1990, p. 9.

Samuelson, Maurice, "Pitfalls threaten the 'ultimate' privatisation," *Financial Times*, May 23, 1990, p. 11.

Thomas, David, "Electricity generators told to hold £1 bn coal stocks," *Financial Times*, May 30. 1990, p. 11.

_____, "Finnish and Irish companies seek contract to run UK power station," *Financial Times*, May 30, 1990, p. 11.

Green, Richard, "Competition in power generation," *Financial Times*, May 31, 1990, p. 19.

Samuelson, Maurice, "ICI awarded right to buy its own electricity supplies," *Financial Times*, June 1, 1990, p. 11.

"Marshall warning on Sizewell cost," *Financial Times*, June 8, 1990, p. 8.

Samuelson, Maurice, "French unveil cross-Channel cable project," *Financial Times*, June 14, 1990, p. 6.

Thomas, David, "Sizewell costs top £2bn, leaked document shows," "Regulator investigates electricity market," *Financial Times*, June 25, 1990, p. 9.

"Scrutiny of nuclear power," *Financial Times*, June 27, 1990, p. 12.

Thomas, David, "BNFL to view nuclear schemes," "Ministers face attack over N-power privatisation," *Financial Times*, June 28, 1990, p. 6.

_____, "Texaco plans £450m Welsh power station," *Financial Times*, July 13, 1990, p. 22.

Thomas, David, "Government fights power share plan," *Financial Times*, July 16, 1990, p. 5.

Samuelson, Maurice, "British Gas enters power market with French group," *Financial Times*, July 17, 1990, p. 7.

Thomas, David, "Power companies fail to meet targets ahead of sell-off," *Financial Times*, July 20, 1990, p. 8.

Thomas, David, "Plugging in to a double-edged prospect," *Financial Times*, July 23, 1990, p. 1.

_____ and David Owen, "UK electricity privatisation facing increasing disarray," *Financial Times*, July 23, 1990, p. 1.

"Hanson's desire for power," *Financial Times*, July 24, 1990, p. 24.

"Late switch for power," *Financial Times*, July 24, 1990, p. 22.

Samuelson, Maurice, "PowerGen attracts power players," *Financial Times*, July 24, 1990, p. 10.

Thomas, David, "Electricity companies' debt put at nearly $2 billion [sic],"*Financial Times*, July 24, 1990, p. 10.

_____, "Hanson may bid for UK power generating group," *Financial Times*, July 24, 1990, p. 1.

_____, "British energy group plans to close two old power stations," *Financial Times*, July 25, 1990, p. 14.

_____, "Breathing space for Power-Gen purchaser," *Financial Times*, July 25, 1990, p. 6.

_____ and Maurice Samuelson, "British electricity workers urge buyout of generating company," *Financial Times*, July 31, 1990, p. 22.

Thomas, David and Andrew Hill, "Conditions revealed for PowerGen buyers," *Financial Times*, August 1, 1990, p. 12.

Buxton, James, "How a switch in focus is generating a profit," *Financial Times*, August 1, 1990, p. 8.

"Power sell-off in Scotland," "PowerGen to carry debt," *Financial Times*, August 9, 1990, p. 8.

Lascelles, David, "PowerGen buy-out scheme moves closer," *Financial Times*, August 13, 1990, p. 14.

Atkins, Ralph, "Government attacked over energy policy," *Financial Times*, August 14, 1990, p. 8.

Goodling, Kenneth and Andrew Hill, "UK energy group's buy-out study cleared," *Financial Times*, August 14, 1990, p. 16.

Goodling, Kenneth, "Scheme to cut long-term costs of N-power," *Financial Times*, August 15, 1990, p. 8.

"How not to sell public assets," *Nature*, Vol. 346, 16 August 1990, p. 593.

Urry, Maggie, "Hanson decision draws privatisation saga to a close, *Financial Times*, August 24, 1990, p. 8.

Pearson, Clare, "Government salvages 'favour' for taxpayers," *Financial Times*, August 24, 1990, p. 8.

Hunt, John, "Windpower receives new boost," *Financial Times*, August 24, 1990, p. 8.

"Power groups lose supply," "Energy savings of £1.5 bn urged," *Financial Times*, August 31, 1990, p. 8.

McCartney, Scott, "Lincoln Savings Memo: 'The Weak, Meek and Ignorant Are Good Targets'," *AP News Archive (Beta)*, September 9, 1990.

Pearson, Clare, "Frankenstein brings a spark to electricity privatisation," *Financial Times*, September 11, 1990, p. 10.

"One cheer for electricity," *Financial Times*, September 12, 1990, p. 1.

Thomas, David, "ICI, Enron set for big power plant," *Financial Times*, September 12, 1990, p. 18.

_____, "Investors concerned over lack of power sell-off information," *Financial Times*, September 13, 1990, p. 11.

_____, "A monster campaign to sell electricity," *Financial Times,* September 13, 1990, p. 12.

"Now you could buy into what you plug into," advertisement for REC share offering, *Financial Times*, September 13, 1990, p. 5.

_____, "Power regulator rules on test case," Financial Times, September 14, 1990, p. 11.

_____, "The hows and watts of the electricity privatisation," "A safety first promotion," *Financial Times*, September 15, 1990, p. Weekend FT III.

_____, "Labour power plan threatens sector profits, broker warns," *Financial Times*, September 17, 1990, p. 8.

_____, "Principal player in the electricity game," *Financial Times*, September 17, 1990, p. 19.

_____, "Prospects differ for companies in electricity sale," *Financial Times,* October 1, 1990, p. 11.

_____, "Tenfold rise sought for renewable energy," *Financial Times*, October 1, 1990, p. 12.

_____, "State electric group plans revival of nuclear industry," *Financial Times*, October 3, 1990, p. 9.

Thomas, David and Clare Pearson, "Investors given incentives to buy electricity shares," *Financial Times,* October 4, 1990, p. 9.

Pearson, Clare, "Timetable of power sell-off gives scope for profit," *Financial Times*, October 4, 1990, p. 9.

Thomas, David, "A renewed source of scepticism," *Financial Times*, October 10, 1989, p. 20.

Hunt, John and Ivo Dawney, "Labour Party pledges to phase out nuclear power," *Financial Times*, October 16, 1990, p. 10.

Hargreaves, Deborah, "Foreign investors get ready to be switched on," *Financial Times*, October 17, 1990, p. 33.

Waller, Martin, "Advisers oppose partial power float," *Times*, October 19, 1990, p. 23.

Sychrava, Juliet, "Small generators threatened by privatisation says report," *Financial Times*, October 22, 1990, p. 7.

Thomas, David, "British Gas plans power generation," *Financial Times*, November 1, 1990, p. 10.

Pearson, Clare, "Share price in electricity sell-off expected to be 240p at flotation," *Financial Times*, November 19.1990, p. 8.

Riley, Barry, "The wrong way to privatise," *Financial Times,* November 30, 1990, p. 19.

"Few sparks," Financial times, November 22, 1990, p. 18.

Thomas, David and Juliet Sychrava, "Electricity companies switch on to sell-off," *Financial Times*, November 22, 1990, p. 10.

Hunt, John, "Greenpeace criticises cost of reprocessing nuclear fuel," *Financial Times*, December 3, 1990, p. 10.

McLain, Lynton, "A new generation of power station," *Financial Times*, December 6, 1990, p. 11.

Sychrava, Juliet, "Light at the end of the electricity tunnel," *Financial Times*, December 6, 1990, p. 11.

_____, "Grid system enters commercial world," *Financial Times*, December 10, 1990, p. 9.

"So farewell then, privatisation," *Financial Times*, December 10, 1990, p. 16.

"Too much power for comfort," *Financial Times,* December 11, 1990, p. 18.

"The shocking price of electricity," *Financial Times*, December 12, 1990, p. 16.

Buxton, James, "Two key companies impress analysts," *Financial Times*, December 14, 1990, p. VI.

McCloskey, Gerald, "Battle to stem coal import flow," *Financial Times,* December 17, 1990, p. VI.

Green, David, "Getting to the heart of the nuclear question," *Financial Times*, December 17, 1990, p. 6.

1991

Dawkins, William, "Giant straining at the leash," *Financial Times*, January 8, 1991, p. 15.

Sychrava, Juliet, "Power to the people," *Financial Times*, January 10, 1991, p. 8.

"The power offer is now on," advertisement, generator offering, *Financial Times*, January 11, 1991, p. 5.

"Regenerating interest in electricity," *Times*, January 11, 1991, p. 21.

Thomas, David and Clare Pearson, "Profit curbs announced on generator sell-off," *Financial Times*, January 11, 1991, p. 8.

Pearson, Clare and Juliet Sychrava, "Electricity generating sell-off aims to avoid 'give-away' jibe," *Financial Times*, January 11, 1991, p. 8.

Sychrava, Juliet, "Power generators seek bigger market share," *Financial Times*, January 14, 1991, p. 6.

Thomas, David, "Companies face 20% electricity price rise," *Financial Times*, January 14, 1991, p. 8.

London, Simon, "Trickle of paper likely from RECs," *Financial Times*, January 17, 1991, p. 22.

Sychrava, Juliet, "Electricity watchdog's national standards," *Financial Times*, January 21, 1991, p. 7.

_____, "Sparks fly as electricity price rises are finalised," *Financial Times*, February 21, 1991, p. 10.

Thomas, David, "N-power operator fails to hive off high cost of shutdowns," *Financial Times*, January 22, 1991, p. 10.

"Nuclear energy plan urged," "Official fuel policy attacked," *Financial Times*, February 12, 1991, p. 8.

"When you get the power will you also get the experience?," East Midlands Electricity advertisement, *Financial Times*, February 12, 1991, p. 7.

"Generating companies' share price set at 175p," *Financial Times*, February 22, 1991.

Pearson, Clare, "Investors offered share perks in electricity generators' flotation," *Financial Times*, January 23, 1991, p. 9.

"PowerGen to sign sales deal," *Financial Times*, January 23, 1991, p. 23.

Thomas, David, "Flotation code omits coal station flues," *Financial Times*, January 24, 1991, p. 10.

_____, "A powerful reckoning," *Financial Times*, January 28, 1991, p. 11.

Sychrava, Juliet, "Why price will be the ultimate determinant," *Financial Times*, January 30, 1991, p. 10.

Robinson, Colin, "Two cheers for power privatisation," *Financial Times*, January 30, 1991, p. 13.

Thomas, David, "Concession over state power sell-off," *Financial Times*, February 4, 1991, p. 6.

Owen, Geoffrey, "Twin-track talent at the top," *Financial Times*, February 4, 1991, p. 34.

"If you want Pool Price electricity, we won't throw you in at the deep end," Midlands advertisement, *Financial Times*, February 6, 1991, p. 7.

Pearson, Clare, "Generating a source of interest," *Financial Times*, February 15, 1991, p. 27.

Sychrava, Juliet and David Thomas, "British Coal and the big black veil," *Financial Times*, March 4, 1991, p. 13.

Fishlock, David, "Nuclear power's difficult rebirth," *Financial Times*, March 5, 1991, p. 13.

Heathfield, P. E., "Coal cannot be run solely according to market forces," *Financial Times*, March 11, 1991, p. 11.

"A mysterious surge in power," *Financial Times*, March 13, 1991, p. 13.

Leadbeater, Charles and David Thomas, "Unequal struggle over power," *Financial Times*, March 13, 1991, p. 14.

Robinson, Colin, "Benefits of privatising British Coal," *Financial Times*, March 13, 1991, p. 15.

"UK-French venture wins power deal," "Share deal for Scots workers,"*Financial Times*, March 31, 1991, p. 8.

Pearson, Clare and David Thomas, "Big profits made in power sell-off," *Financial Times*, March 13, 1991, p. 7.

"Longer life for N-stations," *Financial Times*, March 14, 1991, p. 8.

Sychrava, Juliet, "Foreign investors move for power shares," *Financial Times*, March 14, 1991, p. 28.

Pearson, Clare, "Safety features to encourage honesty," *Financial Times*, March 14, 1991, p. 28.

Our City Staff, "National Power buys gas abroad," *Times*, April 12, 1991, p. 21.

Thomas, David, "Facing a deadline of megawatt proportions," *Financial Times*, April 15, 1991, p. 10.

Buxton, James, "Scale back urged if Scottish electricity oversubscribed," *Financial Times*, April 24, 1991, p. 23.

"Electricity Industry – Financial Times Survey," *Financial Times*, April 25, 1991, pp. 13-16.

Thomas, David, "New market pressures," p. 13.

Sychrava, Juliet, "Greens urge sales reduction," p. 13.

_____, "A small but fierce competitor," p. 14.

Pearson, Clare, "Flotation not without headaches," p. 14.

Thomas, David, "Moulding a management culture," p. 14.

Plaskett, Lucy, "Investment will cut the bills," p. 16.

Advertisements for Eastern Electricity, p. 13, East Midlands, p. 14, Manweb, p. 15, National Grid, p. 16, Yorkshire Electricity, p. 16.

Buxton, James and Clare Pearson, "Government expected to sell all shares in Scots electricity companies," *Financial Times*, May 4/May 5, 1991, p. 4.

Sychrava, Juliet, "Generators hope to cut orders for UK coal," "A fight for power in the generation game," *Financial Times*, May 11/May 12, 1991, p. 5.

Hunt, John, "Power failure in energy efficiency," *Financial Times*, May 16, 1991, p. 14.

Sychrava, Juliet, "National Power to axe 2,000 jobs," *Financial Times*, May 22, 1991, p. 19.

"Tight price for the Scots," *Financial Times*, May 31, 1991, p. 14.

Sychrava, Juliet, "A canny deal on offer for Scottish investors," *Financial Times*, June 1/June 2, 1991, p. WEEKEND FT III.

"East Midlands," *Financial Times*, June 20, 1991, p. 18.

Gapper, John, "A survivor with shrinking assets," *Financial Times*, June 21, 1991, p. 13.

Olins, Rufus, "Power firms in race," *Times*, June 23, 1991, section 4, p. 2.

"No problem for National Power in hitting target, *Times*, June 26, 1991, p. 23.

Waller, Martin, "National Power generates £479m profit," *Times*, June 26, 1991, p. 22.

Sychrava, Juliet, "Midlands Electricity to compete with British Gas," *Financial Times*, August 2, 1991, p. 7.

_____, "Nuclear companies to challenge government on research," *Financial Times*, August 14, 1991, p. 7.

Freeman, Andrew, "An experiment in the sale of power," *Financial Times*, September 26, 1991, p. BUSINESS BOOKS 7.

Hargreaves, Deborah, "Regulator threatens inquiry over wholesale power price," *Financial Times*, October 4, 1991, p. 13.

"Nuclear waste can be contained," advertisement for British Nuclear Fuels, *Financial Times*, October 4, 1991, p. 5.

Green, David, "Rules that bar industry from power generation," *Financial Times*, October 9, 1991, p. 13.

Sychrava, Juliet, "Generators say industry under-charged for power," *Financial Times*, November 13, 1991, p. 8.

"The nuclear waste cover-up," advertisement for British Nuclear Fuels, *Financial Times*, November 13, 1991, p. 11.

Baker, John, "Spiky pattern for pool prices," *Financial Times*, December 2, 1991, p. 15.

Sychrava, Juliet, "Report accuses power generators of inflating prices," *Financial Times*, December 21/December 22, 1991, p. 22.

_____, "Lit-up and glowing revellers sobered by regulator," *Financial Times*, December 21/December 22, 1991, p. 8.

1992

Owen, David and Deborah Hargreaves, "Labour plans radical scheme to regain control of National Grid," *Financial Times*, January 4/January 5, 1992, p. 22.

Hargreaves, Deborah, "Gas power stations 'may be barred'," *Financial Times*, January 9/January 10, 1992, p. 6.

Newbery, David, "More power needed in electricity competition," *Financial Times*, January 9/January 10, 1992, p. 13.

"Competition in electricity," *Financial Times*, January 20, 1992, p. 10.

Samuelson, Maurice, "Youthful new look for pylons," *Financial Times*, January 30, 1992, p. 10.

Madden, Dave, "Light at the end of the tunnel," *Financial Times*, January 30, 1992, p. 10.

Wilson, Richard, "Communications in power struggle," *Financial Times*, January 30, 1992, p. 10.

Prokesh, Steven, "Mere Plug at the Wire's End: In England, That's Progress," *N Y Times*, January 31, 1992.

Hargreaves, Deborah, "Big electricity users to fight nuclear subsidy," *Financial Times*, February 4, 1992, p. 6.

_____, "Power users face legal action over bills threat," *Financial Times*, February 5, 1992, p. 8.

"Electricity," Financial times, February 7, 1992, p. 17.

Lascelles, David, "Curbs on direct power sales by generators are eased," *Financial Times*, February 8/ February 9, 1992, p. 4.

Sychrava, Juliet, "Power generators under attack," *Financial Times*, March 6, 1992, p. 27.

Dawney, Ivo, "Labour defends power plans," *Financial Times*, March 20, 1992, p. 8.

Lascelles, David, "Approaches to the future from opposite poles," *Financial Times*, April 3, 1992, p. 6.

Sychrava, Juliet, "Outcome crucial for British Coal," *Financial Times*, April 3, 1992, p. 6.

Cookson, Clive, "A shock to the system," *Financial Times*, April 9, 1992.

Sychrava, Juliet, "Electricity surplus of 60% seen by 1997," *Financial Times*, April 24, 1992, p. 9. Porter, David, "Diversity benefits electricity supply, despite surplus," *Financial Times*, April 28, 1992, p. 15.

Sychrava, Juliet, "Power chief sees threat to coal sell-off," *Financial Times*, May 5, 1992, p. 7.

Waller, Martin, "Surging power profits expected to prompt outcry," *Times*, May 11, 1992, p. 17.

Sychrava, Juliet, "Coal deal provides spark for results," *Financial Times*, June 9, 1992, p. 23.

Buckley, Neil, "National Grid meets forecasts with 32% rise," *Financial Times*, June 12, 1992, p. 27.

"National Power," *Financial Times*, June 17, 1992, p. 20.

Sychrava, Juliet, "National Power advances 18%," "Norweb almost doubles to £137.9m," *Financial Times*, June 17, 1992, p. 28.

_____, "UK power group profits rise 96%," *Financial Times*, June 17, 1992, p. 21.

_____, "National Grid charges may be cut," "Northern Electric surges to £98.2," *Financial Times*, July 3, 1992, p. 22.

"National Grid," *Financial Times*, July 8, 1992, p. 14.

"Power prices," *Financial Times*, July 10, 1992, p. 14.

"Nuclear fall-out," *Financial Times*, July 30, 1992, p. 12.

"Old N-reactor wins reprieve," *Financial Times*, August 11, 1992, p. 6.

Owen, David, "Privatisation policy could net £8bn," "Family silver prepared for final auction," *Financial Times*, August 17, 1992, p. 4.

Lascelles, David, "Industry receives ammunition for battle on power prices," *Financial Times*, September 3, 1992, p. 8.

"Nukes and taxes," *Financial Times*, September 7, 1`992, p. 12.

Burt, Tim, "Safer plug for electric appliances," *Financial Times*, September 8, 1992, p. 15.

"UK electricity," *Financial Times*, October 9, 1992, p. 16.

"UK electricity," *Financial Times*, October 14, 1992, p. 22.

Lascelles, David, "Generators defend fuel switch," *Financial Times*, October 15, 1992, p. 10.

Tomkins, Richard, Lisa Wood and Ivo Dawney, "Pit closures threaten railway jobs," *Financial Times*, October 15, 1992, p. 10.

Donkin, Richard, "Miners see grim future at end of tunnel," *Financial Times*, October 15, 1992, p. 10.

Atkins, Ralph and Philip Stephens, "Cabinet to meet today on coal industry crisis," *Financial Times*, October 19, 1992, p. 1.

Lascelles, David, "Search for a balance of power," *Financial Times*, October 22, 1992, p. 14.

"In search of an energy policy," *Financial Times*, October 29, 1992, p. 14.

Lascelles, David, "Review sparks nuclear reaction," *Financial Times*, October 31/November 1, 1992, p. 7.

_____, "Coal on his Christmas list," *Financial Times*, November 30, 1992, p. 10.

Smith, Michael, "Generators face music on results," *Financial Times*, November 17, 1992, p. 26.

Jackson, Tony, "The pendulum swings," *Financial Times*, November 20, 1992, p. 14.

Lascelles, David, "Gas generators cleared over price charge," *Financial Times*, December 10, 1992, p. 8.

Smith, Michael, "Report finds 11 pits on hit-list 'profitable'," *Financial Times*, December 10, 1992, p. 8.

Buckley, Neil, "Hurdles in the path of the dash for gas," *Financial Times*, December 10, 1992, p. 8.

Goodhart, David, "Pit jobs protest marked by fresh cuts," *Financial Times*, December 10, 1992, p. 8.

Lascelles, David, "Watchdog accuses generators of inflating electricity prices," *Financial Times*, December 19/December 20, 1992, p. 22.

1993

Lascelles, David and Paul Abrahams, "Big industrial power users seek electricity price cuts," *Financial Times*, January 15, 1993, p. 6.

Smith, Michael, "Subsidies could undermine coal imports," *Financial Times*, January 29, 1993, p. 8.

"Power sector setback," *Financial Times*, January 30/January 31, 1993, p. 13.

"The next small step for National Power," advertisement for National Power, *Financial Times*, February 24, 1993, p. 5.

"Breaking up British Gas," *Financial Times*, March 2, 1993, p. 15.

Smith, Michael, "Generators advised to sell off surplus power stations," *Financial Times*, March 10, 1993, p. 8.

Cassell, Michael, "Dark descent to a political minefield," *Financial Times*, March 25, 1993, p. 14.

Helm, Dieter, "Rewrite the rules for regulation," April 7, 1993, *Financial Times*, April 7, 1993, p. 15.

Hargreaves, Deborah, "Monopoly under a microscope," *Financial Times*, April 26, 1993, p. 14.

Rogaly, Joe, "Howard's way on energy," *Financial Times*, April 30, 1993, p. 14.

Newbery, David, "Fossil fuel levy fails efficiency test," *Financial Times*, May 6, 1993, p. 13.

Lascelles, David, "The high cost of cleaning up," *Financial Times*, May 26, 1993, p. 12.

Dixon, Hugh, "Too high, too low, or just about right," *Financial Times*, June 10, 1993.

Smith, Michael, "Critics cause a ripple in the pool," *Financial Times*, July 30, 1993, p. 17.

_____, "Generators may face controls," *Financial Times*, July 31. August 1,1993, p. 7.

_____, "Electricity regulator calls for reform of power market," *Financial Times*, July 31/August 1, 1993, p. 24.

Maddox, Bronwen, "Thorp reprocessing plant will earn UK £950m," *Financial Times*, August 5, 1993, p. 5.

Lascelles, David, "A balancing act," *Financial Times*, August 11, 1993, p. 6.

Smith, Michael, "Competition hits Nat Power, *Financial Times*, August 20, 1993, p. 18.

_____, "Pressure to switch from appliances," *Financial Times*, August 31, 1993, p. 16.

Maddox, Bronwen, "Dilution measures for acid rain," *Financial Times*, August 31, 1993, p. 12.

Fisher, Andrew, "Power to the people," *Financial Times*, November 2, 1993, p. 14.

Fells, Ian and Nigel Lucas, "Nuclear answers need right questions," *Financial Times*, November 2, 1993, p. 19.

Helm, Dieter, "Nuclear review conclusions not obvious," *Financial Times*, November 4, 1993, p. 12.

Smith, Michael, "The chance to strike it rich," *Financial Times*, November 8, 1993, p. 11.

_____, "Finer balance of power," *Financial Times*, November 9, 1993, p. 16.

_____, "PowerGen move to cut prices," *Financial Times*, November 25, 1993, p. 18.

Buxton, James, "Scots are developing the power to invade England," *Financial Times*, November 25, 1993, p. 22.

Lascelles, David, "The heat is on for UK power," *Financial Times*, December 6, 1993, p. 13.

Davison, MJ, "Euro-plug plan would be dangerous," *Financial Times*, December 7, 1993, p. 14.

Baxter, Andrew, "A shock to the system," *Financial Times*, December 7, 1993, p. 11.

"Breaking the power duopoly," *Financial Times*, December 7, 1993, p. 15.

"Euro-plugs," *Financial Times*, December 10, 1993, p. 13.

"Watchdog boost for generators," "First industrial wave generator," *Financial Times*, December 16, 1993, p. 9.

Maddox, Bronwen and Roland Rudd, "N-plant gets go-ahead for reprocessing," *Financial Times*, December 16, 1993, p. 8.

Maddox, Bronwen, "BNF sees £500m profit for new plant," *Financial Times*, December 16, 1993, p. 8.

Terazono, Emiko, Quentin Peel and Bronwen Maddox, "Germany and Japan welcome go-ahead," *Financial Times*, December 16, 1993, p. 8.

Smith, Mike, "An unexpected sector of bright sparks," *Financial Times*, December 22, 1993, p. 18.

1994

Smith, Michael, "Suppliers offer discounts on electricity business," *Financial Times*, January 12, 1994, p. 7.

_____, "How to distribute the cash mountain," *Financial Times*, January 19, 1994, p. 21.

_____, "Cost-saving option could cause big power cut," *Financial Times*, January 31, 1994, p. 6.

_____, "Generators in deal to sell off plants," *Financial Times*, February 12/February 13, 1994, p. 1.

"Power play," *Financial Times*, February 12/February 13, 1994, p. 24.

Stevenson, Richard W., "The Pain of British Privatizations Has Yielded a String of Successes," *New York Times*, February 22, 1994, p. 1.

Smith, Michael, "Power deal disappoints competitors," *Financial Times*, February 12/February 13, 1994, p. 6.

_____, "Generators in deal to sell off plants," *Financial Times*, February 12/February 13, 1994, p. 1.

Lascelles, David, "Industrial users to gain most from 7% reduction," *Financial Times*, February 12/February 13, 1994, p. 6.

Littlechild, Stephen, ' 'Customers set to benefit by up to £500m," *Financial Times*, February 12/February 13, 1994, p. 6.

Smith, Michael, "Power market may see more competition," *Financial Times*, March 4, 1994, p. 7.

Rudd, Roland and Michael Smith, "Treasury urges prompt sale of nuclear power industry," *Financial Times*, March 12/March 13, 1994, p. 1.

Lascelles, David, "Wind farmers warned noise may count against them," *Financial Times*, March 12/March 13, 1994, p. 22.

Smith, Michael, "Power prices forced down by competition," *Financial Times*, March 14, 1994, p. 7.

_____, "The power to choose," *Financial Times*, March 16, 1994.

Lascelles, David, "Counting the cost," *Financial Times*, March 18, 1994, p. 11.

Smith, Michael, "British Coal pits for sale in five groups," "Turning foreign eyes to the court of King Coal," *Financial Times*, April 14, 1994, p. 8.

"Your Ticket to the Privatization Party," *Business Week*, April 18, 1994.

Smith, Michael, "All eyes on efficiency," *Financial Times*, April 27, 1994, p. 10.

Smith, Michael, "Early nuclear power sell-off 'impractical'," *Financial Times*, May 20, 1994, p. 8.

Maddox Bronwen, "N-waste disposal review welcomed by green groups," *Financial Times*, May 20, 1994, p. 8.

Green, David and Jenny Luesby, "One bump bad, three domes good say aesthetes," *Financial Times*, May 20, 1994, p. 8.

Smith, Michael, "Nuclear Electric seeks support for Sizewell," *Financial Times*, May 31, 1994, p. 7.

_____, "Electricity's poor relation makes good," *Financial Times*, June 2, 1994, p. 24.

_____, "Boost for National Grid flotation," *Financial Times*, June 8, 1994, p. 22.

Adonis, Andrew, "Diverging views on UK utilities' identities," *Financial Times*, June 17, 1994, p. 19.

"Littlechild underpowered," *Financial Times*, August 12, 1994, p. 11.

Smith, Michael, "Power price controls lift shares," *Financial Times*, August 12, 1994, p. 13
_____, "Renationalisation of coal ruled out," *Financial Times*, November 9, 1994, p. 10.
"High wire act," *Financial Times*, December 15, 1994, p. 21.

1995

Lascelles, David, "UK unveils power sales details," *Financial Times*, January 11, 1995, p. 16.
"Northern Electric Seeks to reject a $1.9 Billion Bid," *NY Times* (Archives), January 24, 1995.
Peggy Hollinger and Kevin Brown, "Electricity grid sale becomes a current problem," *Financial Times*, January 27, 1995, p. 9.
Hollinger, Peggy, "Different pulls of power," *Financial Times*, January 30, 1995, p. 18.
Wighton, David, "Recs add spark to acquisition speculation," *Financial Times*, February 2, 1995, p. 20.
Luesby, Jenny, "Large buyer of electricity says prices are too high," *Financial Times*, February 8, 1995, p. 9.
"Trafalgar House Drops Hostile Utility Bid," *NY Times* (Archives), March 11, 1995.
Lascelles, David, "More than one way to go," *Financial Times*, March 15, 1995, p. 13.
"Power," *Financial Times*, March 20, 1995, p. 16.
Simonian, Haig, "Power politics," *Financial Times*, March 20, 1995, p. 15.
Wallis, Edmund, "Three misconceptions about PowerGen," *Financial Times*, March 21, 1995, p. 15.
Smith, Michael, "Nuclear Electric braced for break-up," *Financial Times*, March 30, 1995, p. 10.
Lascelles, David, "Nuclear power station wins extension," *Financial Times*, April 4, 1995, p. 9.
Peston, Robert, "UK moves closer to sale of nuclear power industry," *Financial Times*, April 20, 1995, p. 16.
_____ and David Lascelles, "Fossil fuel subsidy for nuclear power to be scrapped," *Financial Times*, April 26, 1995, p. 8.
Lascelles, David, "PM frets about selling nuclear power stations," *Financial Times*, April 27, 1995, p. 9.
Smith, Michael, "Back to basics for the utilities?," *Financial Times*, April 28, 1995, p. 20.
"Trafalgar House Ends Northern Electric Bid," *NY Times* (Archive), August 6, 1995.
Smith, Michael and David Wighton, "UK power giant in talks on takeover of regional group," *Financial Times*, September 18, 1995.
Lascelles, David, "Delight greets death knell of nuclear expansion," *Financial Times*, December 12, 1995.
"Shares in National Grid Set to Be Sold," *NY Times* ((Archives), December 12, 1995.

1996

Baker, John, "Obituary: Lord Marshall of Goring," Independent, 26 February 1996, independent.co.uk.
Thomas, Emory, Jr. and Kimberley A. Strasset, "U.S. Utility Weighs a Major U.K. Bid," *Wall Street Journal*, April 18, 1996, p. A 15.
Harverson, Patrick, "Bent on vertical integration," *Financial Times*, April 18, 1996, p. 22.
"National Power Rejects Talks With Southern," *NY Times* (Archives), April 19, 1996.
Mittelstaedt, Martin, "Nuclear shutdown stirs questions," *Globe and Mail*, April 23, 1996, p. A6.
"British Utility Parrying a Bid by Southern," *NY Times*, April 23, 1996, p D16.
Strom, Stephanie, "British Reject 2 Power-Industry Takeovers," *NY Times*, April 25, 1996, p. D7.
"Britain Says It Will Not Allow Takeovers of 2 Big Electric Utilities," *NY Times*, May 3, 1996, p. D5.
Calian, Sara, "Britain Makes a Move to Prevent Takeovers of Big Power Producers," *Wall Street Journal*, May 3, 1996.
Wighton, David, Patrick Harverson and Simon Holberton, "UK to block bids with 'golden share'," *Financial Times*, May 3, 1996.
Harverson, Patrick, "US utilities line up for Midlands Electric," *Financial Times*, May 3, 1996, p. 21.
"Two British Utilities May Pay New Dividends," *NY Times* (Archives), May 4, 1996.
Salpukas, Agis, "A $2.6 Billion U.S. Offer for British Utility," *NY Times*, May 8, 1996, p. D1.

Adonis, Andrew, "Diverging views on UK utilities identities," *Financial Times*, June 17, 1996.

"Low Price in British Energy Privatization," *New York Times*, July 15, 1996, p. D4.

"U.K. Nuclear Sell-Off Disappoints," *Wall Street Journal*, July 15, 1996.

"Regulator to Review Powergen-Midlands Deal," *NY Times* (Archives), September 23, 1996.

1997

Kranhold, Kathryn, "U.S. Utilities In U.K. Take Big Tax Hit," *Wall Street Journal*, October 29, 1997, p. A3.

1999

Searjeant, Graham, "Byers heralds euthanasia for utilities," *Times*, July 15, 1999, p. 35.

Harrison, Michael, "National Power sells Drax plant to US energy company AES for pounds 1.9 bn," *The Independent,*, independent.co.uk., 19 August 1999.

Corzine, Robert and Anna Minton, "National Power hopes split will improve its current standing," *Financial Times*, November 19, 1999, p. 26.

Baker, David, "Light up with a clear conscience," *Financial Times*, November 20/November 21, 1999, p. 7.

2000

Harrison, Michael, "British Energy halves dividend and warns of possible slide into losses," independent. co.uk, 11 May 2000.

Hughes, David, *Daily Telegraph News Blog*, August 25, 2000.

2001

Ainger, Will, "U.K. Power Trading System Crashes on Test Run," *power finance & risk*, February 26, 2001, p. 1.

Chipman, Andrea, "U.K. Power Firms See Continental Predators," *Wall Street Journal*, July 24, 2001, p. B4B.

2002

Fagan, Mary, "Innogy bows to £3 bn RWE bid," telegraph.co.uk, 17 March 2002.

Barker, Sophie, "Grid and Lattice form utility supergroup," telegraph.co.uk, 23 April 2002.

Kapner, Suzanne with Andrew Ross Sorkin, "A British Diversion in a British Utility's Strategy," *New York Times*, April 23, 2002, p. W1.

Platt, Gordon, "UK's National Grid Gases Up for US Bid'," *Global Finance*, June 2002, p. 18.

Macalister, Terry, "National Grid backs Urwin," guardian.co,uk, 29 July 2002.

N-Base Briefings, "The crisis starts," "Lightning move, "Wilson supports rates demands," (17th August 2002); "Quick collapse" (1st September 2002); "British Energy desperately looks for help" (7th September 2002); British Energy – the crisis deepens"(15th September 2002); "British Energy's fight for survival, "Reactor shut," "Let BE go says SERA," "Legal challenge," "Coal's cry for help" (21st September 2002); "Loan extended and more money," "Public ownership call," "BNFL stake?," "EC investigation?" (28th September 2002); "A quiet week" (5th October 2002); "Tax avoidance less likely" (12th October 2002).

Nissé, Jason, "Why British Energy is running out of steam," independent.co.uk, 25 August 2002.

Ainger, Will, "British Energy Management Takes Rap for Financial Crisis," *power finance & risk*, September 9, 2002, p. 1.

Macalister, Terry, "British Energy crisis deepens," *The Guardian*, Guardian Unlimited, September 11, 2002.

Moore, Malcolm, "Brit Energy seeks to sell its 80pc of Bruce Power," telegraph.co.uk, 16 November 2002.

"Retailing Could Be Next U.K. Crisis Point," *Power Finance & Risk*, November 25, 2002.

Nissé, Jason, "British Energy sells Bruce in cut-price deal," independent.co.uk, 15 December 2002.

Mackintosh, James, "British Energy close to agreeing Canadian sale," *Financial Times*, December 16, 2002, p. 16.

2003

Nicholson, Mark and Andrew Taylor, "Brit Energy kills 'bail-out' hopes," *Financial Times*, January 15, 2003, p. 21.

Searjeant, Graham, ""Whitehall kills another British industry," *Times*, February 21, 2003, p. 34.

Taylor, Andrew, "British Energy to write down nuclear stations," *Financial Times*, June 2, 2003, p. 19.

Timmons, Heather, "British Energy Takes Big Write-Down on Nuclear Plants," *New York Times*, June 4, 2003, p. W1.

Kemeny, Lucinda, "Centrica chief woos investors after burning his fingers, "*The Sunday Time*s, TimesOnline, June 15, 2003.

Shelley, Toby, "Electricity prices boost British Energy," *Financial Times*, September 29, 2003, p. 23.

"British Energy In Agreement On Bailout," Bloomberg News, *New York Times*, October 2, 2003, p. W1.

Taylor, Andrew, "Midlands buy is 'final piece of the jigsaw' for Powergen," *Financial Times*, October 22, 2003, p. 22.

Timmons, Heather, "British Plan Major 'Wind Farm' To Generate Power Along Coasts," *New York Times*, *December 19, 2003.*

2004

Sovich, Nina, "Ofgem Hits EDF and E.On In Price Review," *Dow Jones Newswires*, November 29, 2004.

2005

Eaglesham, Jean, "UK to consider more reliance on nuclear power," *Financial Times*, May 14/15, 2005, p. 3.

_____, "Britain to decide on new nuclear power stations next year," *Financial Times*, September 29, 2005, p. 3.

Bream, Rebecca, "Britain presses for equal access to rivals' markets," *Financial Times*, October 4, 2005, p. 3.

Catan, Thomas, "UK faces obstacles in move to revive its nuclear age," *Financial Times*, October 10, 2005, p. 4.

Bream, Rebecca and Thomas Catan, "Centrica seeks freer markets," *Financial Times*, November 7, 2005, p. 22.

Adams, Christopher and Jean Eaglesham, "Blair review to look at private investment in nuclear plants," *Financial Times*, November 30, 2005, p. 4.

Kay, John, "Nuclear power for Britain, yes; ineptitude and lies, no," *Financial Times*, December 6, 2005, p. 19.

Lea, Robert, "Drax powers ahead with £2.3 billion floatation," *Evening Standard*, This is Money (website), 15 December 2005.

2006

Blitz, James, "Nuclear renewal in Britain is essential, says Blair," *Financial Times*, May 17, 2006, p. 2.

Adams, Christopher and Rebecca Bream, "No more splitting the difference: why Blair foresees a commercial nuclear age," *Financial Times*, July 11, 2006, p. 12.

John Kay, "A nuclear energy plan that would truly benefit Britain," *Financial Times*, July, 18, 2006, p. 11.

2007

Watt, Nicholas, Oliver Morgan and Robin McKie, "Brown's vision for a nuclear Britain," *Observer*, guardian.uk.com, 20 May 2007.

Griffiths, Katherine, ""White paper sets out nuclear strategy," telegraph.co.uk, 24 May 2007.

Potter, Mark, "Government cuts stake in british Energy," *Reuters*, May 30, 2007.

Crooks, Ed, "BE output setbacks disappoint investors," *Financial Times*, August 17, 2007, p. 15.

Northedge, Richard, "How Britain's nuclear chief Bill Coley left the US under a cloud," telegraph.co.uk, 9 September 2007.

2008

Harvey, Fiona, and Jim Pickard, "Britain a step closer to a new generation of nuclear reactors," *Financial Times*, January 9, 2008, p. 4.

Pickard, Jim and Ed Crooks, "UK gives new plants green light," *Financial Times*, January 11, 2008, p. 5.

"The nuclear button," *Financial Times*, January 11, 2008, p. 8.

Herman, Michael, "British Energy wins fight to keep power station," TIMESONLINE,, February 7, 2008.

Jameson, Angela, "Centrica joins with EDF in British Energy deal," Times Online, May 9, 2008.

Werdigier, Julia, "British Energy Is Said to Be Near Sale to French rival," *New York Times*, July 26, 2008, p. 3.

Bream, Rebecca, "Centrica eyes British Energy," *Financial Times*, August 4, 2008.

Crooks, Ed, Kate Burgess and Matthew Green, "UK minister says EDF remains top option in British Energy sale," *Financial Times*, August 25, 2008, p. 13.

Crooks, Ed, "EDF wins battle for control of British Energy, *Financial Times*, September 24, 2008, p. 22.

BBC News, "EDF agrees to buy British Energy," 24 September 2008.

Pagnamenta, Robin, Lewis Smith and Adam Sage, "Anger as France becomes a nuclear power – in Britain," TIMESONLINE, September 25, 2008.

Crooks, Ed, Kate Burgess and Peggy Hollinger, "Nuclear fusion," *Financial Times*, September 26, 2008, p. 13.

2009

Sen, Neil, "Centrica weighs bid for Venture Petroleum," *The Daily Deal*, March 18, 2009, p. 9.

Crooks, Ed, "Centrica looks at radical steps to spark growth," *Financial Times*, March 23, 2009, p. 17.

Hollinger, Peggy, "EDF weighs UK distribution unit sale," *Financial Times*, May 4, 2009, p. 15.

Whitfield, Paul, "In France, EdF and Areva may unload assets," *The Daily Deal*, May 5, 2009, p. 7.

_____, "Investors cheer Centrica acquisition,"*The Daily Deal*, May 12, 200

_____, "Centrica takes British Energy stake," *The Daily Deal*, May 11, 2009, p. 4.

Hollinger, Peggy, "EDF to sell 20% British Energy stake to Centrica," *Financial Times*, May 11, 2009, p. 15.

_____ and Ed Crooks, "EDF sells 20% stake in British Energy," *Financial Times*, May 12, 2009, p. 19.

Crooks, Ed, "Both sides claim success in a rare 'win-win' deal," *Financial Times*, May 12, 2009, p. 19.

Hollinger, Peggy, "Acquisitions have led to balance sheet pressure,"*Financial Times*, May 12, 2009, p. 19.

"NIE's owner is in 'weak position' ," BBC News, 11 June 2009, bbc.com.

"Energy firms 'fail to pass billions in savings to customers',"telegraph.co.uk, 25 June 2009.

Swaine, Jon, "Ofgem chief Alistair Buchanan claimed £50 for 'fuel poverty' drinks," telegraph.co.uk, 25 August 2009.

"Alternating current," *Financial Times*, October 12, 2009, p. 8.

"Giant Centrica Deal Gives Hope to Future Funding of Green Energy," *Evening Standard*, October 28, 2009, *Energy Central*.

"New nuclear power could supply every home in the UK," M2 Presswire, *Energy Central*, October 28,

"UK plans to build 16 GW of new nuclear energy capacity," Datamonitor, *Energy Central*, November 26, 2009.

"Regulator questions nuclear reactor plans," UPI, November 27, 2009, *Energy Central*.

2010

"Scottish power: Crossed wires," *Economist*, January 14th, 2010.

Pagnamenta, Robin, "Labour prepares to tear up 12 years of energy policy," TIMESONLINE, February 1, 2010.

"Centrica: unfair criticism for record profits," Datamonitor, February 25, 2010, *Energy Central*.

Crooks, Ed, "Energy suppliers challenge Ofgem figures," *Financial Times*, February 22, 2010, ft.com.

"UK Government Departments Publish Plans to Tackle Climate Challenges," M2 Newswire, March 31, 2010, *Energy Central*.

"British Gas: driving UK residential smart metering," Datamonitor, April 1, 2010, *Energy Central*.

"UK: over-legislation is likely to affect energy market integrity," Datamonitor, April 28, 2010, *Energy Central*.

"BG Group: backing out of US and UK generation," Datamonitor, April 30, 2010, *Energy Central*.

Brooks, David, "National Grid may exit New Hampshire over rates," *Telegraph* (Nashua, NH), May 25, 2010, *Energy Central*.

Sanderson, Bill, "Power Co. To Fight Refunds," *New York Post*, May 29, 2010, *Energy Central*.

UK energy: power from renewables falls, along with emission target hopes," Datamonitor, July 7, 2010, *Energy Central*.

"UK energy bills: consumers need to take advantage of switching," Datamonitor, July 11, 2010, *Energy Central*.

Lum, Rosy, "National Grid snuffing out rumors about sale of US operations," *SNL*, July 20, 2010.

Lodge, Tony, "Set out your priorities, Mr. Huhne, and give Britain the energy policy it needs," *Yorkshire Post*, yorkshirepost.co.uk,, 22 July 2010.

Volkery, Carsten, "Britain's Nuclear Renaissance in Doubt under New Government," *Spiegel Online International*, 7/21/2010.

Simpson, John, "Company Snapshot: Mutual Energy Ltd," *Belfast Telegraph*, belfasttelegraph.co.uk, 24 August 2010.

"UK energy suppliers have been warned over unpublicized price rises," *energybiz*, September 24, 2010.

Mason, Rowena, "Britain's new power chiefs reveal nuclear blueprints," "Government blinks first in UK's nuclear stand-off," telegraph.co.uk, 13 November 2010.

Blair, David, "Scepticism greets retail energy prices review," *Financial Times*, November 27/28, 2010, p. 3.

Pfeifer, Sylvia, "Customers see red as bills are set to rise amid global gas glut," *Financial Times*, November 27/28, 2010, p. 3.

Mason, Rowena, "Utilities fear extra costs from Coalition 'Green Deal' plan," "EDF, Centrica inject £435m into nuclear consortium," 5 December 2010, telegraph.co.uk.

_____, "UK taxpayers face unlimited nuclear waste bills if costs spiral," telegraph.co.uk, 8 December 2010.

de Rivaz, Vincent and Sam Laidlaw, "Far from being a drain on the public purse, new power plants will be a major boost to UK economy," telegraph.co.uk,, 14 December 2010.

Huhne, Chris, "The biggest energy market shake-up in 25 years," telegraph.co.uk, 16 December 2010.

Mason, Rowena, "UK government agrees to subsidise nuclear power companies' prices," 16 December 2010, telegraph.co.uk.

"UK energy policy," *Financial Times*, December 17, 2010, p. 10.

"A Climate Change Levy exemption for nuclear in the UK is both problematic and unlikely," Datamonitor, December 17, 2010, *Energy Central.*

2011

Groom, Brian, "Flamboyant libertarian who helped Thatcher defeat the miners," *Financial Times*, January 8/January 9, 2011, p. 11.

Stacey, Kiran, "Nuclear industry windfall feared," *Financial Times*,, February 14, 2011, ft.co.

Prince, Rosa, "Nick Clegg: Britain's proposed nuclear plants may not be built," 29 March 2011, telegraph. co.uk.

"A lighter shade of green," *Economist*, May 7th, 2011, p. 61.

Hennessy, Mark, "Sellafield included among eight new nuclear plants," *Irish Times*, June 24, 2011, energybiz. com.

Jones, Adam, "Tank drivers may think twice at UK plc's accounting revolution," *Financial Times*, June 28, 2011, p. 16.

"Poles Apart," *The Economist*, July 14, 2011, p. 14.

Blair, David, "EDF makes UK nuclear pledge," *Financial Times*, July 20, 2011, ft.com.

Pfeifer, Sylvia and David Blair, "Future dims for UK energy revamp," *Financial Times*, August 13/14, 2011, p. 10.

Wright, Robert, "Union Pacific back on the rails," *Financial Times*, August 23, 2011, p. 13.

"UK energy suppliers call bluff on referral to Competition Commission," energycentral.com, September 21, 2011.

"Ofgem: new powers are unlikely to stabilize energy prices," energycentral.com, September 28, 2011.

Delingpole, James, "'Let's commit suicide more slowly', suggests Osborne," blogs.telegrapoh.co.uk, October 4th, 2011.

Murray, James, "Chris Huhne: no change in direction on carbon budgets, *Guardian*," theguardian.co.uk, 10 October 2011.

"Market power," *Financial Times*, October 17, 2011, p. 12.

"Power for recovery," *Financial Times*, October 19, 2011, p. 10.

"Trouble turning up the heat," *The Economist*, October 22, 2011, p. 69.

Airlie, Catherine and Matthew Carr, "U.K.'s Cheapest Way to 2050 Carbon Goal Would Triple Nuclear," *Bloomberg News*, December 1, 2011, 12:07 PM EST.

2012

McGhie, Tom, "'Leftovers for Britain' in nuclear power deal," *Financial Mail on Sunday*, energybiz.com, February 20, 2012.

"Drax is poised for biomass push if subsidies are boosted," *Evening Standard*, energycentral.com, February 21, 2012.

Kerr, Simeon and Camilla Hall, "Bahrain's Arcapita faces debt challenge," *Financial Times*, February 27, 2012, p. 19.

Pfeifer, Sylvia, "German setback for Britain's nuclear plans," *Financial Times*, March 30, 2012, p. 6.

Boxell, James and Anousha Sakoui, "GDF offer for rest of UK power group," *Financial Times*, March 30, 2012, p. 16.

Reilly, Bill, "UK launches CCS funding competition, long-term CCS plan," *SNL*, April 3, 2012, snl.com.

"Volt from the blue," *Economist*, May 26th, 2012, p. 57.

"UK energy reform," *Financial Times*, May 26/27, 2012, p. 8.

"Subsidising wind," *Financial Times*, June 19, 2012, p. 10.

"UK nuclear future buoyed by possible bid for Horizon," *Datamonitor*, July 18, 2012, energycentral.com.

"UK energy reform," *Financial Times*, July 21/July 22, 2012, p. 6.

Helm, Dieter, "Energy proposals losing their spark," *Financial Times*, June 25, 2012, p. 8.

Shotter, James, "Proposed carbon floor sparks chemical outcry," *Financial Times*, June 25, 2012, p. 16.

"Something doesn't add up," *Economist*, June 30th, 2012, p. 69.

Rulison, Larry, "National Grid power failure," *Times Union*, energycentral.com, July 22, 2012.

"EDF wants taxpayers' cash to pay for nuclear power," *Daily Mail*, August 14, 2012, energycentral.com.

Pickard, Jim and Anousha Sakoui, "UK left with little option for funds to build nuclear plants," *Financial Times*, August 20, 2012, p. 3.

Sakoui, Anousha and Jim Pickard, "Call to limit Chinese stake in UK nuclear deal," *Financial Times*, August 20, 2012, p. 1.

"UK nuclear power," *Financial Times*, August 21, 2012, p. 6.

Macalister, Terry, "China could take key role in UK nuclear infrastructure through Hinkley Point," *Guardian*, 2 September 2012, guardian.co.uk.

"Power prices," *Financial Times*, November 13, 2012, p. 8.

Clark, Pilita, "Carbon capture plans choked by upfront capital costs," *Financial Times*, November 19, 2012, p. 18.

Harvey, Fiona and Patrick Wintour, "Simpler energy prices may not be cheaper, say experts," *Guardian*, 20 November 2012, guardian.co.uk.

2013

"Centrica writes off £200m to quit nuclear power project," *Independent*, 8 January 2013, independent.co.uk.

Jowit, Juliette and Fiona Harvey, "EDF confirms it wants 40-year contracts to build nuclear plants," *Guardian*, 19 February 2013, guardian.co.uk.

"Electric avenues," *Financial Times*, February 20, 2013, p. 12.

Campbell, Peter, "On the brink of a nuclear breakdown," Daily Mail, March 8, 2013, energycentral.com.

"Where the wind blows," *Economist*, March 9th 2013, p. 59.

Ahmed, Kamal, "Want growth, Chancellor? Then back UK nuclear," *Telegraph*, 16 March 2013.

Chan, Szu Ping, "UK nuclear power station given green light," *Telegraph*, 19 March 2013.

Gosden, Emily, "Government extends new nuclear power station timetable by five years, confirms first plant will cost up to £14 bn, " *Telegraph*, 27 March 2013.

"Fears grow over mounting cost of nuclear deal with EDF," *Daily Mail*, April 1, 2013, energycentral.com.

"Big energy users pin hopes of green levy cut on new minister," *Daily Mail*, April 1, 2013, energycentral. com.

BBC News, "Smart meter project is delayed," 10 May 2013, bbc.com.

Pfeifer, Sylvia, "Clean-up quandary," *Financial Times*, July 2, 2013, p. 7.

Sakoui, Anousha, Guy Chazan and Jim Pickard, "UK's flagging nuclear sector set for NuGen stake sale boost," *Financial Times*, July 13/July 14, 2013, p. 8.

Kay, John, "A real energy revolution needs us to look beyond sound bites," *Financial Times*, July 24, 2013, p. 9.

"UK energy policy, *Financial Times*, July 30, 2013, p. 6.

Carnegy, Hugh, "EDF to quit US nuclear market," *Financial Times*, July 31, 2013, p. 12.

"Fracking protests," *Financial Times*, August 21, 2013, p. 6.

"UK energy security," *Financial Times*, September 4, 2013, p. 6.

Chazan, Guy, "Russia in push to build UK reactors," *Financial Times*, September 6, 2013, p. 14.

"UK power politics," *Financial Times*, September 25, 2013, p. 10.

"Energy bill threat weighs on Centrica and SSE," *Financial Times*, September 26, 2013, p. 23.

"Tilting at windmills," *Economist*, September 28th, 2013, p. 53.

Simons, Martin E., "Centrica is no exception to UK under-investment," *Financial Times*, October 3, 2013, p. 8.

Luff, Nick, "More to Centrica's performance than return on capital," *Financial Times,* October 7, 2013, p. 14.

"UK carbon flaw," *Financial Times,* October 8, 2013, p. 10.

Clark, Pilita, "UK wind farm jobs blow over to Europe," *Financial Times,* October 13/October 14, 2012, p. 11.

Sanghani, Radhika, "Hinkley Point nuclear plant: 10 things to know," telegraph.co.uk, 21 October 2013.

Gosden, Emily, "EDF Hinkley Point nuclear deal: an overview," telegraph.co.uk, 21 October 2013.

Silverstein, Ken, "Nuclear Heavyweights Enter Next Round as Great Britain Adds Nuclear Energy," *energybiz,* October 21, 2013, energycentral.com.

_____, "Great Britain Sparks Hopes by Infusing its Nuclear Program," *energybiz,* October 22, 2013, energycentral.com.

"Going nuclear over rising power costs," *Financial Times,* October 22, 2013, p. 10.

"UK energy market needs perestroika," *Financial Times,* October 28, 2013, p. 10.

Kay, John, "Britain's 'great leap forward' was the start of a nuclear power failure," *Financial Times,* October 30, 2013, p. 9.

Evans, Richard, "New power station deal 'threatens higher prices for decades to come," telegraph.co.uk, 30 October 2013.

Mason, Rowena, "David Cameron at centre of 'get rid of all the green crap' storm," theguardian.com, 21 November 2013.

Macalister, Terry, "Electricity networks told to reduce costs by regulator," theguardian.com, 22 November 2013.

Huhne, Chris, "'Greenest government ever' or 'green crap:' which way will David Cameron jump?," *Guardian,* 24 November 2013, theguardian.com.

Rankin, Jennifer, "Ofgem's acting chief executive pulls out of race for top job," theguardian.com,, 29 November 2013.

"A big blow," *Economist,* November 30th, 2013, p. 8.

Chazan, Guy, "Forest fuels'" *Financial Times,* December 10, 2013, p. 8.

Gosden, Emily, "Brussels 'state aid' inquiry to hit Hinkley Point new nuclear plans," telegraph.co.uk, 11 December 2013.

"Huge energy shake-up to hike bills," *Daily Mail,* December 16, 2013, energycentral.com.

Gosden, Emily, "Subsidies for UK nuclear plant could reach £17bn and 'may be unnecessary'," telegraph. co.uk, 18 December 2013.

Thompson, Dorothy, "Critical role of biomass in UK's energy needs," *Financial Times,* December 16, 2013, p. 10.

2014

"Rueing the waves," *Economist,* January 4th, 2014, p. 41.

Trotman, Andrew, "UK nuclear project to be up and running in 2024," telegraph.co.uk, 14 January 2014.

Chazan, Guy, "Power down," *Financial Times,* February 20, 2014, p. 7.

"EDF Energy rakes in £1.7bn from UK unit as nuclear output grows," *City A.M. (UK),* energycentral.com.

"SSE freezes energy prices until 2016," BBC News, 26 March 2014, bbc.com.

"UK energy market needs a rethink," *Financial Times,* March 27, 2014, p. 8.

Kent, Sarah and Tapan Panchal, "Ofgem Proposes U.K. Energy Market Competition Investigation," *Wall Street Journal,* March 27, 2014, wsj.com.

Luce, Edward, "Zealot who saw virtue in a thrift and left savers with zero," *Financial Times,* April 5/April 6, 2014, p. 5.

"Energy Drax follows Centrica in issuing profits warning," Guardian, May 9, 2014, energycentral.com.

"Chairman says SSE not part of problem as profits hit pound(s) 1.6 bn," *Herald* (Scotland), May 22, 2014, energycentral.com.

"UK electricity," *Financial Times,* June 4, 2014, p. 12.

"Sun, wind and drain," *Economist*, July 26th, 2014, p. 63.

"Drax gains after EU clears renewable subsidies," *Financial Times*, July 24, 2014, p. 21.

Gosden, Emily, "Ofgem defends profit estimates as it pressures firms to cut prices," telegraph.co.uk, 30 July 2014.

"Britain's nuclear plans given boost by EU's green light for Hinkley C," *Guardian*, September 23, 2014, energycentral.com.

Kay, John, "Governing by announcement leaves all out in the cold," *Financial Times*, October 1, 2014, p. 9.

"Europe backs Hinkley nuclear plant," *BBC News*, 8 October 2014, bbc.com.

Reed, Stanley, E.U. Approves British Plan for Nuclear Power Plant," *NY Times*, October 9, 2014, p. B5.

"Cash all gone," *Economist*, October 11, 2014, p. 73.

Palmer, Kate, "Latest predictions: how to beat energy bill rises," telegraph.co.uk, 20 August 2014.

"Scotland's opportunity to build shale industry," *Financial Times*, November 21, 2014, p. 8.

Straw, Will, "If he wants to cut green taxes, Cameron should axe the flawed Carbon Price Floor," telegraph.co.uk, 24 October 2013.

Rankin, Jennifer, "Ofgem's acting chief executive pulls out of race for top job," theguardian.com, 29 November 2014.

Guthrie, Jonathan, "It's time for Britain's national champions to raise their game," *Financial Times*, December 8, 2014, p. 8.

Adams, Chris, "Chinese nuclear group CGHN to buy UK wind farms," *Financial Times*, December 15, 2014, p. 15.

Macalister, Terry, "Consumers face £750m subsidy scheme bill for generators to keep lights on," *Guardian*, 18 December 2014, theguardian.com.

2015

Johnson, Mark, "Keeping the lights on," *Economist*, The World in 2015, p. 108.

De Clercq, Geert, "UPDATE 3-EDF says UK Hinkley Point investment decision will take time," *Reuters*, February 12, 2015.

Stothard, Michael, "Areva issues further profit warnings as nuclear group expects 4.9bn loss," *Financial Times*, 24 February 2015, p. 13.

Hollinger, Peggy, "Paris looks to EDF to power up flagging Areva," *Financial Times*, 5 March 2015, p. 16.

'EDF and Areva: nuclear family," *Financial Times*, 10 March 2015, p. 12.

Burke, Kevin, "British Households could save £200 a year by switching energy supplier," *energybiz, March 19, 2015*, energybiz.com.

2016

Carrington, Damian, "Hinkley Point C nuclear deal contains £22 bn 'poison pill' for taxpayer," theguardian.com, 18 March 2016.

"Centrica: green party," *Financial Times*, 6 May 2016, p. 10.

Stothard, Michael, "EDF warns Hinkley cost to rise £2.7 bn," *Financial Times*, 13 May 2016, p. 12.

_____, "France attempts top calm post-Brexit fears over Hinkley Point power project," *Financial Times*, 29 June 2016, p. 19.

_____, "Trade unions urge EDF to delay Hinkley Point decision," ft.com, 1 July 2016.

"Hinkley Point presents cost and security issues," *Financial Times*, 30 July/13 July 2016, p. 8.

Stacey, Kiran, Kate Allen and Lucy Hornby, "China surprise at UK nuclear delay," *Financial Times*, 30 July/31 July 2016, p. 1.

Stacey, Kiran, George Parker, Anne-Sylvaine Chassany and Lucy Hornby, "EDF left in dark over British warning to France," *Financial Times*, 30 July /31 July 2016, p. 3.

Butler, Nick, "Britain can find opportunities beyond Hinkley," *Financial Times*, 1 August 2016, p. 9.

Erlanger, Steven, "British Nuclear Plant Reversal Angers China and France," *New York Times*, August 1, 2016, p. A6.

Grubb, Michael, "May is right to review Hinkley Point contract," *Financial Times*, 4 August 2016, p. 8.

"UK energy consumers deserve transparency," *Financial Times*, 4 August 2016, p. 8.

Liu Xiaoming, "Hinkley Point is a test of mutual trust between China and Britain," *Financial Times*, 9 August 2016, p. 9.

"China flexes its muscles over Hinkley Point deal," *Financial Times*, 10 August 2016, p. 8.

Lowry, David, "Chinese ambassador's nuclear safety claims are open to challenge," *Financial Times*, 11 August 2016, p. 8.

"It's not easy being green," *Economist*, August 13th, 2016, p. 37.

Stacey, Kiran, "Generating criticism," *Financial Times*, 19 August 2016, p. 7.

Stacey, Kiran and Robert Williams, "France's EDF split as £18 bn project for UK nuclear plant approved," *Financial Times*, 29 August 2016, p. 3.

Martin, RCF, "Consider CCS technology as alternative to Hinkley," *Financial Times*, 13 September 2016, p. 10.

Castle, Stephen, "Britain Gives O.K. to Build Nuclear Plant Tied to China," *New York Times*, September 16, 2016, p. A 10.

Ward, Andrew, Jim Pickard and Michael Stothard, *Financial Times*, 16 September 2016, p. 1.

Stothard, Michael, "Power plant development will make or break EDF," *Financial Times*, 16 September 2016, p. 4.

Ward, Andrew, "Decision follows years of fierce debate," *Financial Times*, 16 September 2016, p. 4.

Parker, George and Jim Pickard, "May praised over nuclear compromiser," *Financial Times*, 16 September 2016, p. 4.

Pickard, Jim, George Parker and Gill Plimmer, "Downing St signals tougher approach to big deals with focus on 'critical infrastructure,'" *Financial Times*, 16 September 2016, p. 4.

"An amber light for China in UK nuclear power," *Financial Times*, 16 September 2016, p. 10.

Wolf, Martin, "Big energy decisions are best taken by government, not the market," *Financial Times*, 16 September 2016, p. 11.

"Hinkley hangover?," *Economist*, September 24th 2016, p. 56.

"Investors cold shoulder Centrica and SSE as utilities feel May heat," *Financial Times*, 6 October 2016, p. 21.

"May's revolutionary conservatism," *Economist*, October 8th 2016, p. 54.

Gross, Robert, "Legacy of UK's existing power stations is keeping energy prices down," *Financial Times*, 17 October 2016, p. 8.

Vaughn, Adam, "Green subsidies to push UK energy bills higher than expected," theguardian.com, 18 October 2016.

Liebreich, Martin, "Green industrial strategy is not about subsidies," *Financial Times*, 2 November 2016, p. 8.

Ward, Andrew, "Small energy groups lose power to big six," *Financial Times*, 6 December 2016, p. 16.

Fedor, Lauren, Don Weinland and Yuan Yang, "National Grid's China-Australia deal tests UK stance on infrastructure sales," *Financial Times*, 9 December 2016, p. 1.

2017

"EDF: glowing," *Financial Times*, 5 January 2017, p. 10.

Ward, Andrew and Jim Pickard, "Toshiba under pressure over finance for UK nuclear project," *Financial Times*, 23 January 2017, p. 1.

"The case for public funding of nuclear power," *Financial Times*, 14 February 2017, p. 8.

Index